Elementary Quantitative Chemistry

A SERIES OF BOOKS IN CHEMISTRY

EDITOR **Linus Pauling**

Elementary Quantitative Chemistry

ESMARCH S. GILREATH

Washington and Lee University

W. H. FREEMAN AND COMPANY

San Francisco

Printed in the United States of America.
Library of Congress Catalog Card Number: 78-78923
Standard Book Number: 7167 0146-4

9 8 7 6 5 4 3 2

Contents

Laboratory Exercises

Preface

This manual is designed as the basis for a one-semester course in elementary quantitative analysis on either a freshman or sophomore level, depending upon the chemistry background of the students in the class.

The materials included constitute complete classroom and laboratory programs. The theoretical section is presented on an elementary level, since no knowledge beyond general chemistry is assumed. The experimental section contains methods that have been developed or chosen from time-tested procedures, which illustrate the analytical principles contained in the theoretical portion.

The rapid advancements of recent years in chemical instrumentation are having a profound effect upon the methods of chemical analysis. However, the number of instruments and instrumental procedures used in an elementary course should be dictated by a realistic appraisal of the objectives of the course in relation to the cost of the equipment. Satisfactory experiments for which a pH meter and a simple spectrophotometer must be used are included, but there is an adequate number of experiments for which the instrumental procedures are optional. On the other hand, no attempt has been made to provide a comprehensive coverage of quantitative experiments to meet every need of every type of institution. Many teachers prefer to develop some of their own experiments.

I wish to express my appreciation to the many students who have undergone the painful ordeal of using a textbook in its development stages. I am grateful to Professors W. J. Watt and T. C. Imeson for helpful criticisms and suggestions, and particularly to Professor J. B. Goehring, who wrote parts of the original drafts of Chapters 12, 14, and 15. Finally, I am deeply indebted to Mrs. Betty Lou Duff, who typed and proofread the entire manuscript, and to my wife, Sally Gilreath, for her encouragement and assistance throughout the preparation of the text.

E. S. Gilreath

THEORETICAL ASPECTS

Introduction

Analytical chemistry consists of two major divisions, qualitative analysis and quantitative analysis. The laboratory procedures of *qualitative analysis* are designed for identifying and determining the *approximate* amounts of the constituents present in a substance, whereas the procedures of *quantitative analysis* are concerned with determining the *exact* amounts of constituents present. A complete analysis of a sample must include both qualitative and quantitative determinations. However, the qualitative analysis must precede the quantitative, since the former serves as a basis for selecting methods for the latter.

The Methods of Quantitative Chemistry

Quantitative methods may be conveniently subdivided into four general divisions: gravimetric, titrimetric, photometric, and electrical methods. However, such a classification is more arbitrary than accurate since, frequently, an analytical procedure may employ more than one method in arriving at the percentage composition of a substance. It seems pertinent to summarize the nature of these quantitative measurements.

Gravimetric Methods. A gravimetric analysis is one based entirely upon weight. In such a procedure the original substance is weighed, and from it the constituent to be determined is isolated and also weighed. From the two weights the amount of the desired constituent is calculated.

Titrimetric Methods. A titrimetric determination is made by measuring the volume of a standard reagent reacting with the desired constituent in a definite chemical reaction. The amount of the constituent may be obtained from the weight of the original sample and the milliequivalents of the standard solution used in the titration.

Photometric Methods. The substance to be determined is converted to a compound which imparts a distinctive color to its solution. The intensity of the color of an unknown solution is compared with standard solutions containing known amounts of the colored compound. Such a comparison may provide an estimation of percentage composition.

Electrical Methods. Among these many methods can be listed the measurements of certain basic electrical properties such as potential, conductance, quantity, and capacitance. A correlation of these physical measurements with concentrations, for knowns and unknowns, provides a means for quantitative evaluations.

The Objectives of Quantitative Chemistry

The laboratory procedures in quantitative chemistry are of disciplinal, educational, and practical value to a student.

Disciplinal Values.   In the analysis of a substance a student strives to obtain results that are close to percentages that are known only to the instructor. In seeking a high degree of accuracy certain personal qualities are essential:

(1) The student must have a sense of objective honesty in the evaluation of laboratory measurements.

(2) He must be able to follow directions carefully and intelligently, and to make accurate observations and readings.

(3) He must acquire neat and precise working habits, and always record laboratory data following a systematic form in a suitable notebook.

Educational Values. Of the students who complete a course in quantitative chemistry, very few will ever enter a profession that requires routine analytical determinations. However, the educational values within the study are fully as important as any other values which may be attributed to the course. Among the educational advantages are the following:

(1) The student is given an opportunity to use certain aspects of applied mathematics.

(2) He supplements and expands his knowledge of some basic principles of analytical chemistry.

(3) He becomes familiar with the literature of analytical chemistry.

Practical Values.   In most chemistry curriculums, quantitative chemistry is the first course in which the attainment of precise and accurate results is a primary objective.

(1) To attain accurate results a student must acquire manipulative ability and reasonable speed in the handling of chemical equipment.

(2) He must appreciate the limitations of the methods and equipment he uses, and the magnitude of possible errors that may be involved.

(3) He must be able to make rapid calculations from analytical observations to a precision warranted by the data included.

Finally, a course in quantitative chemistry will provide the student with confidence in his laboratory skills. This confidence facilitates the determination of the proper approach to almost any type of laboratory problem; thus it is essential in advanced courses, in research, and in the problems that may be faced in professional and industrial laboratories.

CHAPTER 1

The Evaluation of Analytical Data

There are few, if any, absolute values in science. Most values refer to a standard that has been arbitrarily chosen for purposes of reference. For example, molecular and atomic weights are based on the mass of an isotope of carbon, which has been assigned the value 12.0000. The number 12.0000 was chosen arbitrarily, and has no absolute significance. In quantitative chemistry many results are expressed in percentages that are obtained by scientific observations. However, such observations involve many possible errors which can be properly evaluated only by a statistical approach. The evaluation of scientific measurements, and of the attendant errors, is the subject of this chapter. In an elementary treatment of this topic, some oversimplification is unavoidable, but a dedicated student will find a more profound treatment of this subject in many advanced textbooks in analytical chemistry.

SCIENTIFIC MEASUREMENTS

1·1 Precision versus Accuracy

In analytical chemistry the term *precision* is used to describe the reproducibility of results. It can be defined as the degree of agreement among several identical measurements, observations, or other types of quantitative results. When a student is able to reproduce two or more measurements with only slight differences in the results, his work can be described as precise. For example, if one measures the length of a room with a meter stick several times to obtain values that are extremely close, this is an example of good precision. Yet, a high degree of precision gives no assurance that the results are accurate. In taking the room measurement, there is always the possibility that the meter rule is inaccurate in its construction. In fact, all instruments are inaccurate to some extent.

Accuracy has been described as the degree of agreement between a measured value and the probable true value. Unfortunately, there are no true values in science since no measurement is completely accurate. We must content ourselves by stating that true values can be approached but never attained. Thus, accurate results are assumed to be those measured values which have been obtained by skilled scientists using instruments of the best quality. To approach a true value it is necessary to have a number of observers and several instruments, both observers and instruments being interchanged for the final *average* measurement.

In quantitative chemistry a student's average analysis is frequently checked against a known value which has been established by several reputable analysts. However, both the student and the instructor should realize that the known value is not absolute. For this reason, the instructor will probably establish a grading scale that allows for small inaccuracies in a known value.

1·2 Determination of Precision

In elementary quantitative analysis it is customary to perform analyses in triplicate, using three portions of the same sample. There is no satisfactory method by which a student can establish the accuracy of his results; however, he can express the precision of his analytical observations. If the average of the three results shows close agreement, more reliance can be placed on the average than on any one or two results. For example, a 50 ml buret is calibrated such that an approximation must be made in the last decimal place, and this approximation entails a built-in error of about 2 parts per thousand (e.g., 20.42 ml could easily be read as either 20.43 ml or 20.41 ml). The student should attempt to keep his margin of error within this range. In an elementary course, a precision that is less than 2 parts per thousand may be the result of a coincidence. On the other hand, a precision that is more than 5 parts in a thousand (involving the analysis of a homogeneous substance) indicates unreliable laboratory observations.

The steps involved in the computation of precision of analytical results are given in detail in the experimental section of this text. To illustrate this computation briefly, consider the three results obtained in the analysis of a soda ash sample.

Portion	Percentage found	Deviation
1	31.83	0.07
2	31.81	0.05
3	31.66	0.10
Average	31.76	0.07

To express these results as precision in parts per thousand, the average deviation is multiplied by 1000 and the result divided by average percentage.

$$\frac{0.07 \times 1000}{31.76} = 2 \text{ parts per thousand}$$

The foregoing method yields a rough approximation of the probable accuracy. Statistical methods for the evaluation of analytical data are more accurate, and are now customarily employed for the analyses of analytical measurements. A simplified explanation of these methods is given in Sections 1·5 through 1·7.

ERRORS

Errors in any set of measurements can be divided into two broad categories, determinate and indeterminate errors. The distinction between the two is not always tangible. Usually a determinate error is one of a definite value that can be assessed and controlled by corrective measures or compensating factors. If, however, the determinate error cannot be controlled, it may become an indeterminate error. For example, if the knife edge of a balance has become worn, a determinate error may be caused, but if the defective part produces erratic results the error can be classified as indeterminate.

1·3 Determinate Errors

Determinate errors are those which occur with definite regularity owing to faulty methods of technique or defective measuring instruments. Determinate errors produce results that are fairly constant in value, and that may be evaluated and corrected by proper compensations. Determinate errors fall into four major types:

Methodic Errors. These are errors inherent in a quantitative measurement. For example, in gravimetric analysis it is recognized that all precipitates are soluble to some extent. In the analysis of calcium by precipitation with the oxalate ion, the precipitate is washed, and because of the solubility of CaC_2O_4, a definite error is introduced. Such an error can be determined and corrected by a compensating factor.

Observational Errors. Personal errors belong in this classification. These are largely the results of improper techniques and carelessness. Some observers appear incapable of making careful observations, or do not realize the need for them. Errors in reading a buret, or failure to wash and ignite a precipitate properly, belong in this category.

Instrumental Errors. All instruments are imperfect to some extent, and when the imperfection is significant, a gross error may be introduced. Also, instruments are affected by temperature changes, especially if the change is rapid. Such errors can be recognized and partially controlled.

Reagent Errors. An analysis may be inaccurate if impure reagents are used. If pure reagents are not available, the degree of impurity must be measured in order that a suitable correction may be made. This is often accomplished by running a *blank*. Such a procedure involves going through all of the analysis, using the same solvent and reagents in the same quantities, but omitting the unknown component. Reagent impurities usually cause determinate errors. However, if the material for analysis is not homogeneous in composition, the error, or errors, caused by the lack of homogeneity constitutes an indeterminate error.

The two classes of indeterminate errors may be denoted as random errors, which are usually accidental in nature. If a large number of observations is made, the occurrence and magnitude of the error can be predicted by the laws of probability. A statistical analysis of random errors reveals these three facts: (1) random errors of frequent occurrence are those of small magnitude; (2) large errors rarely occur; (3) most random errors, positive and negative deviations occur in about the same frequence. A plot of random errors produces what is known as a normal distribution curve; such a curve is illustrated in Figure 1·1.

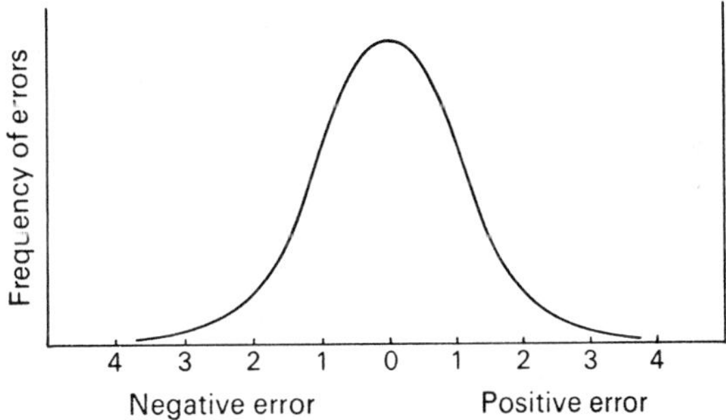

FIGURE 1·1 Normal distribution curve for random errors

1·4 Indeterminate Errors

These errors are difficult to define, but usually it is possible to divide them into two classes:

Variations within Determinate Errors. As previously stated, a worn knife edge in a balance can produce a determinate error, but when the error varies, owing to load (or other causes), it becomes an indeterminate error. The variations humidity and temperature in a balance room produce determinate errors, but if they cannot be controlled the errors become indeterminate.

Erratic Errors. These errors are difficult to pinpoint, and usually the observer is unaware of their presence. Vibration in a balance room can cause erratic errors in weighing. Accidental losses of materials during an analysis produce erratic errors, of which the analyst is unaware.

STATISTICAL TREATMENT OF ERRORS

1·5 Average Deviation and Standard Deviation

Various methods are used to indicate the precision of measurements. In the past a quantity denoted as the average deviation was widely used in specifying analytical results. More recently the application of statistics to deviations among measurements has increased the significance of a measure known as standard deviation.

Average Deviation. An average deviation for a set of quantitative results is obtained by adding the absolute values of deviations (deviations from the mean value), and dividing by the number of determinations. For example, the

TABLE 1·1   Analysis of a sample of soda ash

Determination	% Na_2CO_3	Deviation from mean (Δ)	$(\Delta)^2$
1	59.46	0.20	0.0400
2	59.97	0.31	0.0961
3	59.62	0.04	0.0016
4	59.71	0.05	0.0025
5	59.56	0.10	0.0100
	59.66	0.14	0.1502
	(mean value)	(average deviation)	($\Sigma(\Delta)^2$)

analysis of a sample of soda ash gave values shown in Table 1·1. From the mean value and average deviation, the percentage of Na_2CO_3 could be expressed as 59.66 ± 0.14.

Standard Deviation.   A more precise representation of the foregoing percentage for the analysis of soda ash is obtained through the use of the standard deviation, s. This quantity is obtained from the following expression:

$$s = \sqrt{\frac{\Sigma(\Delta)^2}{N}}$$

where $\Sigma(\Delta)^2$ is the sum of the squares of the individual deviations and N is the number of determinations. When the number of determinations is small, it is appropriate to modify the above equation to

$$s = \sqrt{\frac{\Sigma(\Delta)^2}{N-1}}$$

The divisor, $N - 1$, is the number of *independent* deviations from the mean which arises from N determinations.*

* There is a subtle distinction between the two equations for the standard deviation,

$$s = \sqrt{\frac{\Sigma(\Delta)^2}{N}} \quad \text{and} \quad s = \sqrt{\frac{\Sigma(\Delta)^2}{N-1}}$$

The former equation defines the *population standard deviation,* which refers to the statistical behavior of a very large population of measurements — essentially of an *infinite* number of measurements, exhausting all of the material under analysis. In this equation, Δ is the deviation from the mean of the entire population — the mean of an infinite number of separate measurements, comprising the entire material.

From the results given in Table 1·1, the standard deviation for these data is

$$s = \sqrt{\frac{0.1502}{4}} = \sqrt{0.0375} = 0.19$$

Thus, making use of the standard deviation, the percentage analysis of soda ash can be expressed as 59.66 ± 0.19.

From a statistical viewpoint, the standard deviation is more reliable than the average deviation, because, by squaring the deviations, more weight is given to larger deviations.

1·6 Confidence Limits

The object of a chemical analysis is to determine the most probable true percentage of a constituent. True values are not known, and if they were, it is unlikely that a student's average determination would be the same value. However, through the use of statistics it is possible to arrive at a prediction on whether a series of

continuation of footnote

The latter equation defines the *sample standard deviation,* which refers to the statistical behavior within a *finite* number of measurements on a finite sample, a portion of the entire material. In this equation, Δ is the deviation from the mean of the finite number of measurements. $N - 1$ is the number of *independent deviations* from the mean which arises out of N determinations on the finite sample.

In quantitative analysis, the analyst obtains a few measurements on a sample of a large quantity of material. On the basis of his measurements on the (finite) sample, he draws conclusions about the composition of the entire material (the infinite population). So, the latter equation applies when he analyzes his data statistically. How the measured mean for a finite sample relates to the probable mean of the entire population, is the subject of the discussion in Section 1·6.

TABLE 1·2   Values for *t* for calculating confidence interval

Number of determinations	Factor for confidence interval of		
	90%	95%	99%
2	6.314	12.706	63.657
3	2.920	4.303	9.925
4	2.353	3.182	5.841
5	2.132	2.766	4.604
6	2.015	2.571	4.032
7	1.943	2.447	3.707
8	1.895	2.365	3.449
9	1.860	2.306	3.355
10	1.833	2.262	3.250
11	1.812	2.228	3.169
12	1.796	2.201	3.106
13	1.782	2.179	3.055
14	1.771	2.160	3.012
15	1.761	2.145	2.977

determinations (without determinate errors) is substantially correct. Statistical treatment will not produce a prediction of 100% probability, but it will give risk or probability levels below 100% (see Table 1·2).

Although it is impossible to determine a true value it is possible to assign an interval, around the measured mean, that will contain within a certain probability the true value. This interval, called the *confidence interval (C.I.)*, may also be defined as the range on either side of the measured mean in which the probable true value may be expected to fall.

The limits of the confidence interval are called *confidence limits.*

The percentage probability that the true value will fall within the confidence limits is denoted as the *confidence level.*

Thus, the probable true value, T_p, is expressed in a manner such as the following:

$$T_p (90\%) = 4.272 \pm 0.009$$

or

$$T_p (95\%) = 4.272 \pm 0.016, \text{etc.}$$

In the latter equation, for example, the 95% confidence level indicates that there is a 95% probability that the true value lies between the stated limits. The number 4.272 expresses the measured mean, that is, the actual mean value of the measurements obtained in analysis of the sample. The confidence limits are 4.272 + 0.016 and 4.272 − 0.016. The confidence interval is the range between these two limits.

Confidence intervals for a series of measurements are calculated by use of a table of *t* values (see Table 1·2). The confidence interval (*C.I.*) is calculated from the expression

$$C.I. = \pm \frac{ts}{\sqrt{N}}$$

where *s* is the standard deviation, *N* is the number of determinations, and *t* is the appropriate statistical constant for the % confidence interval chosen in Table 1·2. The *t* values are merely proportionality constants,* which relate the confidence interval to the number of measurements, the standard deviation, and the degree of probability for the percentage of correctness desired.

If the average (mean) value is denoted as X, and T_p is the probable true value for a given confidence interval, the foregoing expression may be expanded to

$$T_p = X \pm \frac{ts}{\sqrt{N}}$$

---

* The *t* values were originally determined in a classic study by W. S. Gossett (under the pseudonym of Student) in 1908, in which he accounted for the manner in which a measured mean of a sample might deviate from the probable true mean of the entire material, and how a sample standard deviation might differ from a population standard deviation.

TABLE 1·3   Critical values* for rejection quotient Q

Number of observations	Q (90% confidence)
3	0.94
4	0.76
5	0.64
6	0.56
7	0.51
8	0.47
9	0.44
10	0.41

* Selected values given by Dean and Dixon, *Anal. Chem.*, **23**, 636, (1951).

The procedure for the evaluation of the probable value within a confidence interval is illustrated by the following problem:

The analysis of impure potassium acid phthalate gave a series of percentage values of 48.34, 48.43, 48.33, and 48.47. From these results calculate the probable true value, T_p, within a 95% confidence interval.

Observation	Deviation(Δ)	Δ^2
48.34	0.05	0.0025
48.43	0.04	0.0016
48.33	0.06	0.0036
48.47	0.08	0.0064
Mean 48.39	Mean deviation 0.06	$\Sigma(\Delta)^2$ 0.0141

From these results, the standard deviation is

$$s = \sqrt{\frac{\Sigma(\Delta)^2}{N-1}}$$

$$s = \sqrt{\frac{0.0141}{3}} = 0.068$$

The probable true value, T_p, is

$$T_p = X \pm \frac{ts}{\sqrt{N}}$$

and

$$T_p(95\%) = 48.39 \pm 3.182 \times \frac{0.068}{\sqrt{4}}$$

$$= 48.39 \pm 0.22$$

The probable true value at confidence intervals of 90% and 99% could be calculated similarly:

$$T_p(90\%) = 48.39 \pm 2.353 \times \frac{0.068}{\sqrt{4}}$$

$$= 48.39 \pm 0.16$$

$$T_p(99\%) = 48.39 \pm 5.841 \times \frac{0.068}{\sqrt{4}}$$

$$= 48.39 \pm 0.40$$

1·7 Rejection of Results

Practically all students will find, at some time or other, that in a series of replicate determinations one or more values appear to be out of line with the others. The question arises as to the justification of discarding a questionable value.

If a student is aware of a gross error occurring in one of a series of replicate determinations, he is justified in discarding this particular determination. In the absence of the knowledge of a definite error it is desirable to follow arbitrary rules for the rejection of a measurement. Two such rules are widely used by analytical chemists.

The Statistical Q Test for Retention or Rejection of Extreme Values. The symbol Q denotes a quotient which is obtained by the following method: Compute the difference separating the highest and lowest observation in a series. Divide this difference into the spread between the doubtful value and its nearest neighbor, to obtain a quotient denoted as Q. Compare the value of Q with the appropriate figure in Table 1·3. The doubtful determination may be rejected with 90% confidence if Q is greater than the expected value given in the table. The Q Test is not valid unless three or more measurements are made.

Consider the statistical Q Test for a series of results obtained in the analysis of iron ore:

33.78%
33.84%
33.60%
33.15%

The difference between the highest and lowest values (33.84 − 33.15) is 0.69%. The spread between the most doubtful value and its nearest neighbor (33.60 − 33.15) is 0.45%. Therefore, Q is equal to 0.45/0.69 = 0.65. In Table 1·3 the number tabulated for Q from 4 results in 0.76; consequently, the value 33.15% may not be discarded.

The 4 d Rule for the Rejection of One Value in a Series. The application of this rule requires at least four results, and only one value in a series may be rejected. The criterion for the rejection of a suspected value is as follows: omit the suspected value temporarily and determine the average value and average deviations of the other members of the series; if the deviation of the suspected value from the average value is 4 times the average deviation, or greater, the uncertain number should be discarded.

If the 4 d rule is applied to the results obtained for the analysis of iron ore as given in the Q Test, an analysis reveals the following:

Results	Deviations
33.78%	0.04
33.84%	0.10
33.60%	0.14
[33.15]%	[0.59]
33.74	0.09
(average of 3 best results)	(average deviation of 3 best results)

Since the deviation of the suspected result [0.59] is more than four times the average deviation, 0.09, the value of 33.15% should be discarded. (Compare the Q Test.)

The 4 d rule has been criticized as being too severe, and likely to exclude measurements that should be retained. On the other hand, this rule has the advantage of simplicity as compared with the Q Test, inasmuch as a student may apply the rule without reference to a table derived for statistical evaluations.

SIGNIFICANT FIGURES

The significant figures in a number are all the digits whose values are known with reasonable certainty plus one additional digit (but only one) whose value is uncertain. For example, the numbers 0.456, 4.56, 45.6, and 456 all contain three significant figures. But, if these values have been obtained in a computation involving multiplication or division of measured quantities, the values 4 and 5 are reasonably certain, whereas the 6 is an uncertain value.

A zero is not a significant figure when used to locate a decimal; however, terminal zeros are always significant. For example, in the number 0.02500, the zeros to the left of the digit 2 are not significant since they serve only to locate the decimal point. On the other hand, the zeros beyond the digit 5 are significant; thus the number 0.02500 contains four significant figures.

1·8 Rounding Numbers

A few simple rules are usually applied in the computation of analytical data:

(1) In expressing an experimental measurement, never retain more than one doubtful digit.

(2) The rules for rejecting superfluous digits are these: (a) When the last digit dropped is greater than 5, increase by one the last digit retained. Thus, if the last digit in the number 456.7 is rejected, the new value becomes 457. (b) When the last digit dropped is less than 5, the last digit retained remains unchanged. For example, if the last figure in 6.543 is rejected, the retained value remains 6.54. (c) If the last rejected digit is 5, increase by one the last digit retained if it is an *odd* number. Thus the value 875.5 would be rounded to 876. But, if the last digit is an *even* number, it remains unchanged.

For example, the number 876.5 would be rounded to 876.

1·9 Addition and Subtraction

One rule, which is applied to any series of additions or subtractions, states that the precision of the answer is determined by the least precise of the numbers involved in the computation. In other words, the sum or difference of the quantities must be rounded to the same number of digits to the right of the decimal as is contained in the quantity with the fewest such digits. However, premature rounding of all components to this extent is not recommended. A safe procedure is to retain one doubtful figure in each component (this would be one digit more than the component with the least number of decimal places) and add. The result is then rounded to contain the same number of decimals as the most uncertain component. The procedure is illustrated as follows:

	Not	
Problem	recommended	Recommended
+743.2	743.2	743:2
+4.14	4.1	4.14
+0.231	0.2	0.23

Sum $\longrightarrow$ 747.57 $\longrightarrow$ rounded
to $\longrightarrow$ 747.6

In the final sum, it is meaningless to retain digits beyond the first decimal point since the value for the first term is unknown beyond this digit.

1·10 Multiplication and Division

In multiplication or division, the precision (in parts per thousand) of the product or quotient cannot be greater than the precision of the least precise number entering into the computation. However, it may not be possible to state the result with the same number of digits as contained in the least precise number.

Consider the multiplication of $22.8 \times 0.094 \times 1.234$. In computing the result it is assumed that each number has a deviation of 1 in the last decimal place. The deviation of each number, in parts per thousand is then

$$\frac{1}{228} \times 1000 = 4.4 \text{ parts per thousand}$$

$$\frac{1}{94} \times 1000 = 10.6 \text{ parts per thousand}$$

$$\frac{1}{1234} \times 1000 = 0.8 \text{ parts per thousand}$$

The final answer should indicate a deviation within a magnitude of approximately 10 parts per thousand, which is the precision of the most uncertain number, 94. In evaluating the result, we should reemphasize the definition of significant figure: *the significant figures* in a number from multiplication and division should include any *digits whose values are known with reasonable certainty plus one digit whose value is uncertain.*

The answer to the problem is 2.64. In this result the digits 2 and 6 are of reasonable certainty and the 4 is uncertain. The answer 2.64 implies a deviation of $1/264 \times 1000 = 3.8$ parts per thousand.

If the result had been rounded to 2.6 the implied deviation would have been $1/26 \times 1000 = 38$ parts per thousand, which is much greater than the 10 parts implied in the figure 94. If the result contained an additional digit and were stated as 2.645, the deviation implied would have been $1/2645 \times 1000 = 0.3$ parts per thousand. Thus a certain amount of sound judgment is involved in accepting the proper number of digits to be retained.

TYPES OF EXERCISES

Type 1. The Determination of Precision

Problem: What is the precision (in parts per thousand) for the series of figures 46.19, 46.11, 46.47, and 45.94?

Solution:

	Measurement	Deviation
	46.19	0.01
	46.11	0.07
	46.47	0.29
	45.94	0.24
Average	46.18	0.15

$$\frac{0.15 \times 1000}{46.18} = 3 \text{ parts per thousand precision}$$

Type 2. Standard Deviation

Problem: Calculate the standard deviation s for the values 31.18, 31.45, 31.53, and 31.58.

Solution:

$$s = \sqrt{\frac{\Sigma(\Delta)^2}{N-1}}$$

	Values	Δ	$(\Delta)^2$
	31.18	0.26	0.0676
	31.45	0.01	0.0001
	31.53	0.09	0.0081
	31.58	0.14	0.0196
Average	31.74		$\Sigma(\Delta)^2 = 0.0954$

$$s = \sqrt{\frac{0.0954}{3}} = \sqrt{0.0318} = 0.18$$

Type 3. Confidence Interval

Problem: What is the confidence interval (90%) for the values 38.48, 38.59, 38.49, and 38.21?

Solution:

$$C.I. = \pm \frac{ts}{\sqrt{N}}$$

s is determined as in Type 2 and the result is 0.16
t from Table 1·2 is 2.353, and $\sqrt{N} = ?$

$$C.I. = \pm \frac{2.353 \times 0.16}{2} = \pm 0.19$$

Type 4. Confidence Limits

Problem: What are the confidence limits of a 95% level for the values 27.94, 27.80, 27.70, and 27.62?

Solution:

$$\text{Confidence limits} = X \pm \frac{ts}{\sqrt{N}}$$

where X is the mean (average value), s is the standard deviation, N is the number of values, and t is obtained from Table 1·2.

The confidence interval, $C.I. = \pm(ts/\sqrt{N})$, is obtained as in Type 3, and is found to be ± 0.22. The average is 27.76. Therefore the confidence limits are 27.54 and 27.98.

Type 5. Probable True Value

There is no difference between the terms *confidence limits* and *probable true value* except the manner in which the result is stated. Confidence limits indicates two numbers between which the true value will fall. The probable true value, T_p, indicates one number with a $\pm$ factor. In other words, it is the average value, X, plus or minus the confidence interval.

Problem: What is the probable true value for the figures 15.33, 14.98, 15.25, and 15.23 for a 90% level?

$$T_p = X \pm \frac{ts}{\sqrt{N}}$$

The confidence interval is calculated as given in Type 3, and has a value of ± 0.18. The mean, X, is computed as 15.20. Therefore,

$$T_p = 15.10 \pm 0.18.$$

Type 6. Rejection of a Result by the *Q* Test

Problem: The following percentages were obtained in the analysis of an iron ore sample:

$$44.56\%$$
$$44.34\%$$
$$43.91\%$$
$$44.17\%$$

Apply the *Q* Test to determine if the rejection of the outlying result is justified.

Solution: The difference between the highest and lowest values $(44.56 - 43.91)$ is 0.65%. The spread between the most doubtful number and its nearest neighbor $(44.17 - 43.91)$ is 0.26%. Therefore, Q is equal to $0.26/0.65$ and the quotient is 0.40 (see Section 1·7). Upon comparing this quotient with the Q value for 4 values, given in Table 1·3, it is found that 0.40 is less than the Q value 0.76 (in Table 1·3), and, consequently, the rejection of the value 43.91% is not justified.

Type 7. Rejection of a Result by the 4 *d* Test

Problem: A student's analyses of a sample of soluble antimony yielded the following percentages: 12.48, 12.36, 12.30, and 12.08. Apply the 4 *d* test to determine if any result should be discarded (see the discussion on the 4 *d* rule in Section 1·7).

Solution: The suspected value is 12.08, and is omitted in determining the average value and the average deviation for the other three values.

	Result	Deviation
	12.48%	0.10
	12.36%	0.02
	12.30%	0.08
	[12.08]%	[0.30]
Average	12.38	0.07

The average deviation of the three best results is 0.07. The deviation of the suspected value from the mean is 0.30. Since the deviation of 0.30 is more than 4 times the average deviation of 0.07, the suspected value (12.08) should be discarded.

Type 8. Significant Figures in Addition and Subtraction

Problem: Perform the following computation and express the result with the proper number of significant figures:

$$89.89 + 8.989 - 0.8989$$

Solution: Round the values as described in Section 1·9. The computation then becomes:

$$89.89 + 8.989 - 0.899$$

And the result is 97.98.

Type 9. Significant Figures in Multiplication and Division

Problem: Perform the following computation and express the result with the proper number of significant figures:

$$\frac{564.5 \times 0.08340}{0.032}$$

Solution: Round all values to the number of digits of greatest certainty, retaining one doubtful value if necessary to prevent a decrease in accuracy. The computation is thereby reduced to the following:

$$\frac{5.6 \times 10^2 \times 8.3 \times 10^{-2}}{3.2 \times 10^{-2}}$$

and the answer is 146. In this result the figures 1 and 4 are of reasonable certainty, whereas the 6 is doubtful. The result could be expressed as 1.46×10^2.

EXERCISES

1. In the analysis of stibnite (an ore containing antimony) a student obtained percentage values of antimony as follows: 15.35, 15.28, 15.29, and 15.15. Evaluate the precision of these results in parts per thousand.
2. The following results were obtained in the analysis of iron ore for iron: 34.79%, 34.86%, 34.80%, 34.53%, and 34.76%. From these values calculate the following:
    (a) The average deviation in parts per thousand;
    (b) The standard deviation for the analysis.
3. The gravimetric analysis of impure sodium chloride for the chloride content gave percentages values 33.76, 33.70, 33.69, and 33.52. Determine the 95% confidence interval for these values.
4. Calculate the 95% confidence limits for the following measurements: 5.061, 5.039, 5.146, 5.027, and 5.152.
5. In the analysis of impure copper oxide for copper the following percentage results were obtained: 4.70, 4.79, 4.92, and 4.59. By means of the Q Test, determine if any result may be rejected.
6. The percentage analysis of soda ash for Na_2CO_3 yielded the values 45.99, 46.90, 46.60, and 46.46. Using the $4\,d$ test, determine if any value should be discarded.
7. Express the result of the following calculations with the proper number of significant digits:
    (a) $44.12 + 2.32 + 22.0$
    (b) $777.1 + 4.10598 + 47.12$
    (c) $172.4 - 42.345$
    (d) $0.0051 \div 0.421$
    (e) 4.92×1.2345

8. Express the result of the following calculations with the proper number of significant figures:

(a) $\dfrac{9.42 + 0.410}{1.0001}$

(b) $\dfrac{724.3 - 4.72}{2.0}$

(c) $\dfrac{88.412 + 2.40 - 67.1}{0.0345}$

(d) $\dfrac{21.724 \times 0.340 \times 0.012}{2.0 \times 10^{-3}}$

(e) $\dfrac{42.7 \times 14 \times 0.72}{5 \times 10^{2}}$

9. Indicate how many parts per thousand the following results deviate from each other: (a) 37.46 and 37.59; (b) 5195 and 5187.

10. In the analysis of impure calcium carbonate, a student obtained percentage values for calcium as follows: 39.27, 40.19, 39.89, and 39.78. Evaluate the precision of these results in parts per thousand.

11. In the standardization of NaOH against five weighed samples of potassium acid phthalate, the following normalities were obtained: 0.1058, 0.1044, 0.1055, 0.1067, and 0.1036. Calculate the mean, the average deviation, and the standard deviation of these results.

12. Using the normalities given in problem 11, (a) determine by the $4\,d$ test if any value may be discarded; (b) determine by the Q Test if it is possible to reject any value.

13. The analysis of three samples of impure calcium carbonate with permanganate gave the following percentage values: 12.10, 12.27, and 12.34. Determine the 95% confidence interval for these values.

14. Four determinations of iron ore yielded percentages of 40.46, 40.41, 40.40, and 40.38. Establish the 90% confidence interval for the four values.

15. What is the probable true value for the figures, 24.73, 24.32, 24.53, 24.20, and 24.60?

16. Rewrite the following additions and subtractions in a vertical form and express the answers with the proper number of significant figures:
    (a) $1.23 + 12.34 + 123.4 + 1234$
    (b) $9.876 + 98.76 + 987.6 + 9876$
    (c) $875.4 - 87.54 - 8.754$
    (d) $999.9 - 99.99 - 9.999$

17. Express the results of the following calculations with the proper number of significant figures:

(a) $\dfrac{123.4 \times 56.7}{7.87}$ (b) $\dfrac{89.76 \times 5.43}{23.4}$ (c) $\dfrac{89.7 - 65.4}{0.320}$ (d) $\dfrac{12.34 - 5.67}{2.89}$

CHAPTER 2

Acid–Base Concepts, Concentration Units and Stoichiometry

The nature of acids and bases has been a matter of speculation for a period of more than three hundred years; and the problem of explaining the fundamental actions of these substances has been approached in many different ways. There are, at present, many interpretations of acid–base systems; and hopefully, better and broader ones will be developed in the future. This is not to say, however, that the existing interpretations of acid–base systems are conflicting or controversial, but simply that all of them are somewhat limited in their applications. Each theory possesses certain distinctive merits that appear most applicable to the system for which it was conceived.

CONCEPTS

Of the many acid–base systems, some have important applications, and others possess intriguing implications. These are described in almost all advanced inorganic textbooks, and a student in quantitative chemistry should find them of interest. In this manual only two concepts will be discussed: one because of its historical impact, and the other on account of its wide acceptance. Both concepts will be used, as needed, throughout this text.

2·1 The Arrhenius Concept

In 1887 a young Swedish chemist, Svante Arrhenius, as a result of studying the conduct-

ance of acid solutions, advanced the theory of ionization to account for the properties of aqueous solutions of electrolytes. This theory was the first progressive approach to an explanation of acid–base behavior. According to the Arrhenius theory, *an acid is defined as a hydrogen compound that in water solution gives hydrogen ions, and a base is a hydroxide compound that in aqueous solution yields hydroxide ions.* The process of neutralization results from the union of hydrogen ions and hydroxide ions to form water molecules.

The Arrhenius interpretation of the nature of acids and bases is subject to a number of objections. The theory defines acids and bases in terms of aqueous solutions, and not in terms of the substances involved. It places emphasis upon hydrogen ions in aqueous solution, and provides no explanation for the acid properties of certain nonaqueous solutions in which no hydrogen ions are present. The theory is inadequate to explain properly either the basicity of ammonia solution or the acidity of solutions containing salts such as aluminum chloride. The idea of the bare proton (hydrogen ion) existing in solution is untenable in view of the large amount of energy released when hydrogen ions enter into solution.

In spite of the many objections to the Arrhenius theory of acids and bases, it is unrivaled in its simplicity, and is widely used in a modified form. Other acid–base theories can also be criticized, but usually for different reasons. The importance of the Arrhenius concept is the fact it represents one of the great breakthroughs in chemical theory. The theory of ionization won for Arrhenius the 1903 Nobel prize in chemistry.

2·2 The Brønsted–Lowry Definitions

In an attempt to overcome some of the limitations of the Arrhenius concept of acid–base behavior, J. M. Brønsted in Denmark and T. M. Lowry in England proposed, almost simultaneously, a more general definition for acids and bases in 1923. According to this definition *an acid is a substance having a tendency to lose one or more protons, and a base is a substance having a tendency to add protons.* Since Brønsted has done much more than Lowry to develop this approach, it is generally designated as the Brønsted theory.

The Brønsted–Lowry theory for acid–base phenomena postulates that the reaction of an acid with a base produces another acid and base. The fundamental relationship between an acid and a base is indicated by the following three equations:

$$acid_1 \rightleftharpoons H^+ \text{ (proton)} + base_1$$

$$base_2 + H^+ \text{ (proton)} \rightleftharpoons acid_2$$

and the sum of these equations is

$$acid_1 + base_2 \rightleftharpoons acid_2 + base_1$$

Therefore, any acid–base reaction involves two acids and two bases, and these acids and bases are called *conjugate pairs.* In the preceding equation, the conjugate acid and base are designated by the same subscript; consequently, $acid_1$ and $base_1$ form a conjugate pair, and $base_2$ and $acid_2$ are also a conjugate pair. To illustrate with a specific example, the compound hydrogen chloride in the liquid state is a poor conductor of electricity and may be considered as a covalent compound, but in water solution it produces a strong electrolyte. Hydrogen chloride is an acid since it produces protons in water solution, and the water molecule can be considered to be a base inasmuch as water molecules accept protons to form hydrated protons (frequently designated as hydronium ions). The Cl^- ion is

the conjugate base of the acid HCl; and the H_3O^+ is the conjugate of the base, the water molecule. The reaction is

$$acid_1 + base_2 \longrightarrow acid_2 + base_1$$

$$HCl + H_2O \longrightarrow H_3O^+ + Cl^-$$

Ions may also act as acids; for example, the ammonium ion is an acid in aqueous solution. The equation for this reaction is

$$acid_1 + base_2 \rightleftharpoons acid_2 + base_1$$

$$NH_4^+ + H_2O \rightleftharpoons H_3O^+ + NH_3$$

The ammonium ion is a weak acid, since comparatively few hydronium ions are produced in this reaction; consequently, the equation is indicated as being reversible.

In the foregoing illustrative equations the base that accepted protons was the molecule water; however, according to the original definition by Brønsted and Lowry, any ion or molecule that accepts protons is a base, as is illustrated by the following equations:

$$acid_1 + base_2 \rightleftharpoons acid_2 + base_1$$

$$HCl + NH_3 \rightleftharpoons NH_4^+ + Cl^-$$

$$H_2O + NH_3 \rightleftharpoons NH_4^+ + OH^-$$

$$H_3O^+ + OH^- \rightleftharpoons H_2O + H_2O$$

$$H_3O^+ + CO_3^= \rightleftharpoons HCO_3^- + H_2O$$

According to the Brønsted definitions all negative ions, anions, are classified as bases, since they combine with protons. Those anions which form slightly dissociated substances with protons are considered strong bases, and those which form weak linkages with protons are weak bases. Thus, in aqueous solution, the hydroxide ion is a very strong base, and the Cl^- ion is one of the weakest bases. Strength of a base is measured by its ability to capture protons in competition with other bases.

In Table 2·1 are a number of acids and their conjugate bases. The acids are arranged in the order of decreasing strength, whereas the conjugate bases are in the order of increasing strength. In this table water appears listed as an acid and also as a conjugate base. When water functions as an acid, its conjugate base is the hydroxide ion,

$$acid_1 \rightleftharpoons base_1 + proton$$

$$HOH \rightleftharpoons OH^- + H^+$$

When water functions as a base, its conjugate acid is the hydronium ion,

$$base_1 + proton \rightleftharpoons acid_1$$

$$HOH + H^+ \rightleftharpoons H_3O^+$$

A few ions also can function, like water, either as an acid or as a base. One example is the $H_2PO_4^-$ ion. When the ion acts as an acid, typical of ions containing hydrogen atoms, the conjugate base is the HPO_4^{--} ion.

$$acid_1 \rightleftharpoons base_1 + proton$$

$$H_2PO_4^- \rightleftharpoons HPO_4^{--} + H^+$$

When the ion acts as a base, typical of negatively charged ions, the conjugate acid is molecular H_3PO_4.

$$base_1 + proton \rightleftharpoons acid_1$$

$$H_2PO_4^- + H^+ \rightleftharpoons H_3PO_4$$

The Brønsted–Lowry interpretation of acid–base phenomena is a much broader concept than that presented by the Arrhenius theory. It defines acids and bases in terms of the substances themselves, and not in terms of their ionization in aqueous solution. Moreover, the concept of hydrolysis is not necessary to explain the acid behavior of the NH_4^+ ion or the basic character of the CO_3^{--} ion. In terms of Brønsted definitions the basic nature of NH_3 is indicated by the ability of this molecule to accept protons in forming the acid NH_4^+. This is a more comprehensive viewpoint than that of the Arrhenius theory, in which a base is defined as a substance that produces hydroxide ions. Basic properties are indicated by the ability of a substance to capture protons. Furthermore, the Brønsted approach recognizes that acid–base behavior is neither restricted to any particular solvent nor dependent upon it.

As indicated in Table 2·1, the Brønsted concept serves as a method for comparing the strengths of various acids and bases in water solution. It is also useful when comparing the behavior, as well as the strength, of substances in several solvents.

The original Brønsted theory did not explain the acidity of various hydrated cations. However, acid ions such as the hydrated aluminum and ferric ions do exist in water solution, and a few of these are included in Table 2·1. The acidity results from the loss of a proton by a water molecule hydrating the metal cation. Hydrated cations with high positive charge on the metal atom usually are stronger acids than those with low metallic charge, because of the stronger electrostatic repulsion between the highly charged metal atom and the proton of the water molecule.

The Brønsted concept leaves the impression that the hydronium ion can contain only one water molecule of hydration. Inasmuch as experimental evidence indicates that higher hydrates of the proton are often present, it appears feasible to use the symbol H^+ for the solvated ion without designating the number of attached water molecules. In this text, the terms hydrogen ions, hydronium ions, and hydrated protons, as well as the symbols H^+ and H_3O^+, will be used interchangeably as seems best suited to a particular application.

CONCENTRATION UNITS

In an aqueous solution, the weight of the solute in a given quantity (weight or volume) of water is known as the *concentration* of the solution. The concentration of an aqueous solution is expressed by two general methods, namely, the weight of the solute in a given volume of solution, and the weight of the solute in a given weight of water or solution.

2·3 Weight of Solute in a Given Volume of Solution

Three systems are widely employed for designating the weight of solute in a given volume of

TABLE 2·1   The relative strengths of some acids in water solutions

Acid	Conjugate base	Ionization constant	pK
HCl	Cl^-		
HNO_3	NO_3^-		
H_2SO_4	HSO_4^-		
H_3O^+	H_2O	55.3	-1.74
HIO_3	IO_3^-	1.69×10^{1}	0.77
$H_2C_2O_4$	$HC_2O_4^-$	5.97×10^{2}	1.23
SO_2 (aq)*	HSO_3^-	1.72×10^{2}	1.76
H_3PO_3	$H_2PO_3^-$	1.6×10^{2}	1.80
HSO_4^-	SO_4^{--}	1.2×10^{2}	1.92
H_3PO_4	$H_2PO_4^-$	7.5×10^{3}	2.12
$Fe(H_2O)_6^{3+}$	$Fe(H_2O)_5(OH)^{++}$	6.0×10^{3}	2.22
H_3AsO_4	$H_2AsO_4^-$	5.0×10^{3}	2.30
HNO_2	NO_2^-	4.6×10^{4}	3.37
HF	F^-	3.53×10^{4}	3.45
$Cr(H_2O)_6^{3+}$	$Cr(H_2O)_5(OH)^{++}$	1.54×10^{4}	3.82
$HC_2O_4^-$	$C_2O_4^{--}$	6.40×10^{5}	4.19
$HC_2H_3O_2$	$C_2H_3O_2$	1.75×10^{5}	4.75
$Al(H_2O)_6^{3+}$	$Al(H_2O)_5(OH)^{++}$	1.5×10^{5}	4.85
CO_2 (aq)	HCO_3^-	4.3×10^{7}	6.37
HSO_3^-	SO_3^-	1.07×10^{7}	6.91
H_2S	HS^-	5.7×10^{8}	7.24
H_2PO_4	HPO_4	6.2×10^{8}	7.21
$HOCl$	OCl^-	3.5×10^{8}	7.46
H_3BO_3	H_2BO_3	5.8×10^{10}	9.24
NH_4^+	NH_3	5.34×10^{10}	9.27
HCN	CN	4.93×10^{10}	9.31
HCO_3	CO_3	5.6×10^{11}	10.25
HPO_4	PO_4^3	4.8×10^{13}	12.32
HS	S	1.2×10^{15}	14.92
HOH	OH	1.9×10^{16}	15.72
NH_3	$NH_2\cdot$		
OH	O		

* The molecular acids H_2SO_3 and H_2CO_3 are unstable and decompose into simpler molecular species, SO_2, CO_2, and H_2O.

solvent, and each has its own distinctive advantage. These systems, expressed as *formality*, *molarity*, and *normality*, are described as follows:

Formality. A *formal* solution contains 1 *gram-formula weight* of the solute dissolved in sufficient water to produce a liter of solution. The formality (F) of a solution is the number of gram-formula weights of solute used in preparing one liter of solution. The formula weight (fw) is the sum of the atomic weights of the elements appearing in the accepted chemical formula of the solute, and this weight, expressed in grams, is the gram-formula weight (gfw). For example,

the formula weight of Na_2CO_3 is $(2 \times 22.99) + (12.01) + (3 \times 16.00)$ or 105.99. The mass of one gram-formula weight of Na_2CO_3 is 105.99 grams. A 0.500 F solution of Na_2CO_3 is prepared by dissolving 53.0 grams (0.500 gfw) in sufficient water to give 1.000 liter of final solution.

This method of expressing a given concentration has the advantage that no information about the properties of the substance in solution is required. It applies equally well to ionic and molecular substances in solution. Because of its simplicity, this method of expressing concentration will be used extensively throughout this text.

TABLE 2·2   Comparison of formality and molarity for selected solutes

Solute	Grams per liter of solute	Formality in gfw	Molarity of molecular species (approximately)	"Molarity" of ionic species (approximately)
NaCl	58.44	1 F	0.0 M	1 M Na$^+$ 1 M Cl$^-$
HSO$_3 \cdot$NH$_2$ (sulfamic acid)	97.09	1 F	0.72 M	0.28 M H$^+$ 0.28 M SO$_3 \cdot$NH$_2^-$
HC$_2$H$_3$O$_2$ (acetic acid)	60.05	1 F	0.996 M	0.004 M H$^+$ 0.004 M C$_2$H$_3$O$_2$
CH$_3$OH (methanol)	32.03	1 F	1 M	0.0 M

Molarity.   A *molar* solution contains 1 *gram-molecular weight* of the solute dissolved in enough water to make a liter of solution. The molarity (M) of a molecular substance is its concentration in moles of molecules per liter of solution. The molecular weight is the sum of the atomic weights of the elements indicated in the formula of the substance, and this weight expressed in grams, is the gram-molecular weight.

By the above definition of "molarity," the molarity concept was originally restricted to substances composed of molecular species. Therefore, a 1 M solution of an acid HA and a 1 F solution of the same acid are not equivalent unless HA is an undissociated solute. For weak electrolytes the difference is slight; however, for strong, or moderately strong, electrolytes the difference between molarity and formality of the solute is usually large. The formality concept gives a precise statement of a solution in terms of its preparation; there is no implication as to the species contained in the resulting solution. The confusion from trying to equate the two different concepts is indicated by selected examples in Table 2·2; consequently, it seems desirable to avoid the term molarity on account of the ambiguous assumption in its original definition.

Normality.   A *normal* solution contains 1 *gram-equivalent weight* of the solute in a liter of solution. A gram-equivalent weight of an acid or a base is that weight (in grams) of the solute that will furnish or react with 1 gram-formula weight of hydrogen ions (hydrated protons). Since an equivalent weight is a large quantity, it is more often convenient to refer to a smaller quantity, the milliequivalent weight, which is one-thousandth of an equivalent weight.

2·4 Weight of Solute in a Given Weight of Solvent or Solution

A number of methods are used for expressing weight of solute in a given weight of solvent or solution; however, they are seldom used in quantitative chemistry. The two most common of these systems are those of molality and percentage composition.

Molality.   A molal solution contains 1 gram-molecular weight (1 mole) of the solute dissolved in 1000 g of solvent. It differs from a molar solution in that the ratio of solute to solvent is determined by weight.

Percentage Composition.   The percentage concentration of a solution is the weight in grams (or any other weight unit) of the solute contained in 100 g (or corresponding weight unit) of the solution. Thus 10 g of sodium hydroxide in 100 g of solution is a 10 percent solution.

Percentage composition is a rather loose term, since it also may indicate the volume of concentrated liquid solute in a given volume of solvent. For example, 10 ml of glacial acetic acid diluted in water to 100 ml may be designated as a 10 percent solution of acetic acid. Again, percentage composition may indicate weight of the solute in 100 ml of solution. Accordingly, it is possible to say that the 10 g of KOH dissolved in 100 ml of water is a 10 percent solution. If percentage composition methods are used, the amounts of solute and solvent should be clearly indicated, with emphasis on the dimensional units involved.

Other concentration units such as mole ratios, mole fractions, and mole percentages are employed for special purposes. A chemist must be familiar with the various methods of expressing concentrations; however, in quantitative chemistry it is rarely necessary to express concentration units by other than the molar, formal, and equivalent methods.

2·5 Formal Methods for the Expression of Concentration

A formal solution is defined as one containing one gram-formula weight (gfw) of the solute diluted with enough water to produce exactly one liter of solution. The term *formality* may be defined as the number of gram-formula weights per liter of solution and designated by the symbol F. The definition of formality may be stated in dimensional units as

$$F = \frac{\text{gfw of solute}}{\text{liter of solution}}$$

This expression can be expanded to

$$F = \frac{\text{g of solute}}{\text{fw of solute}} \times \frac{1}{\text{liter of solution}}$$

Since a liter of solution is not suitable for most titrimetric observations, it is better to reduce the dimensional unit of the foregoing expression to milliliters. This is done by dividing the formula weight by 1000, and, at the same time, multiplying 1 liter by 1000 to give

$$F = \frac{\text{g of solute}}{\text{fw/1000}} \times \frac{1}{\text{liter} \times 1000}$$

consequently

$$F = \frac{\text{g of solute}}{\text{milliformula wt}} \times \frac{1}{\text{ml}}$$

Other relations from the preceding equation are:

$$\text{grams of solute} = F \times \text{ml} \times \text{milliformula wt}$$

and

$$\text{mf} = \frac{\text{grams of solute}}{\text{mfw}} = F \times \text{ml}$$

In dilution procedures involving the same solute, the following relations exist:

$$\underset{\text{before dilution)}}{F_1 \times \text{ml}_1} = \underset{\text{(after dilution)}}{F_2 \times \text{ml}_2}$$

This relationship is true because the number of mfw remains the same before and after dilution regardless of volume or formality.

2·6 Equivalent Methods

The *normality*, designated as N, of an acid or base is defined as the number of gram-equivalent weights of the solute per liter of solution. As stated previously, the gram-equivalent weight of an acid or base is the weight in grams of the solute that will furnish, or react with, 1 gram-formula weight of hydrogen ions. Normality, as a unit of concentration, provides the basis for a direct comparison of the reactive capacity of most acid and base solutions. Equivalent weight is always equal to formula weight or to a rational fraction thereof. Therefore, the normality of a solution is equal to the formality of the solution or an integral multiple therefrom. For example, a $1\,F$ and a $1\,N$ solution of HCl are identical because only one proton is involved. On the other hand, a $1\,F$ solution of H_2SO_4 is equivalent to a $2\,N$ solution of the same solute, since the acid contains two protons. Additional comparisons between formality and normality are provided in Table 2·3.

The value of the normality of a solution, or that of the equivalent weight of a solute, depends importantly upon the reaction which the solute undergoes. For example, Table 2·3 indicates that the substance H_3PO_4 may lose three, two, or only one proton, depending upon the conditions of the reaction. The normality of a H_3PO_4 solution varies accordingly. Whenever normality or

TABLE 2·3   Comparisons of normality and formality for some common acids and bases

Substance	Proton change	Eq. wt equals	Normality equals
$NaOH$	$H^+ + OH \rightleftharpoons H_2O$	fw	formality
$Ba(OH)_2$	$2H^+ + 2OH \rightleftharpoons 2H_2O$	$\frac{1}{2}$ fw	2 × formality
Na_2CO_3	$H^+ + CO_3 \rightleftharpoons HCO_3$	fw	formality
Na_2CO_3	$2H^+ + CO_3 \rightleftharpoons CO_2 + H_2O$	$\frac{1}{2}$ fw	2 × formality
$HC_2H_3O_2$	$H^+ + OH \rightleftharpoons H_2O$	fw	formality
H_2SO_4	$2H^+ + 2OH \rightleftharpoons 2H_2O$	$\frac{1}{2}$ fw	2 × formality
NH_3	$H^+ + NH_3 \rightleftharpoons NH_4^+$	fw	formality
H_3PO_4	$H_3PO_4 + OH \rightleftharpoons H_2PO_4 + H_2O$	fw	formality
H_3PO_4	$H_3PO_4 + 2OH \rightleftharpoons HPO_4 + 2H_2O$	$\frac{1}{2}$ fw	2 × formality
H_3PO_4	$H_3PO_4 + 3OH \rightleftharpoons PO_4^3 + 3H_2O$	$\frac{1}{3}$ fw	3 × formality

equivalent weight is used, the exact nature of the intended chemical reaction must be clearly understood.

Normality can be stated in dimensional units, as in the following expressions:

$$N = \frac{\text{equivalents of solute}}{\text{liter of solution}}$$

and

$$N = \frac{\dfrac{\text{g of solute}}{\text{formula weight}}}{\text{no of protons donated or accepted per fw}} \times \frac{1}{\text{liter of solution}}$$

or

$$N = \frac{\text{g of solute}}{\text{eq. wt}} \times \frac{1}{\text{liter}}$$

The last equation may be converted to milli-equivalent (meq) wt and milliliters,

$$N = \frac{\text{g of solute}}{\text{eq. wt}/1000} \times \frac{1}{\text{liter} \times 1000}$$

to give the following:

$$N = \frac{\text{g of solute}}{\text{meq wt}} \times \frac{1}{\text{ml}}$$

Other relationships from the foregoing equations are:

$$N \times \text{ml} = \frac{\text{g of solute}}{\text{meq wt}} = \text{meq (milliequivalents)}$$

A comparison of acids and bases results in

$$\begin{array}{ccc} \text{meq}_1 & = & \text{meq}_2 \\ \text{(acid or base)} & & \text{(acid or base)} \end{array}$$

consequently,

$$\begin{array}{ccc} N_1 \times \text{ml}_1 & = & N_2 \times \text{ml}_2 \\ \text{(acid or base)} & & \text{(acid or base)} \end{array}$$

From the last two expressions, it is possible to state that *one equivalent of an acid or base always reacts with one equivalent of its opposing acid or base*, in an acid–base chemical reaction.

In the analysis of impure substances and in preparing formal-concentration solutions from concentrated reagents, the student and the analyst must make interconversions between the two general methods of expressing concentration, discussed in Sections 2·5 and 2·6. In addition, procedures for accomplishing some important concentration- and composition-calculations are described below.

Density. The handbooks of chemistry contain tables showing percent by weight of various acid and base solutions as a function of density. The term density denotes the *weight in grams of 1 ml of a substance or solution.* For example, concentrated hydrochloric acid, containing 37.2% hydrogen chloride, has a density of 1.19 at room temperature; concentrated ammonia, containing 28.7% NH_3, has a density of 0.90.

In dilution procedures involving concentrated acids and bases, it is necessary to include density and percentage factors to compute either formality or normality. Thus, if concentrated nitric acid contains 70.0% HNO_3 and has a density of 1.41, its formality can be obtained from the following expression:

$$F = \frac{\text{wt of 1 ml} \times \text{fractional purity}}{\text{vol of 1 ml} \times \text{mfw}}$$

and

$$F = \frac{1.41 \times 0.700}{1.00 \times 0.0630} = 15.7\ F$$

Similarly, if concentrated sulfuric acid contains 96.7% H_2SO_4 and has a density of 1.84, its normality may be computed from the equation,

$$N = \frac{\text{wt of 1 ml} \times \text{fractional purity}}{\text{vol of 1 ml} \times \text{meq wt}}$$

and

$$N = \frac{1.84 \times 0.976}{1.00 \times 0.049} = 36.3\ N$$

Composition. Many laboratory procedures involving acid–base analyses include determinations of purity for unknown acids and bases. Usually, but not always, such substances are solids. Among these processes may be listed the determination of the composition of impure potassium acid phthalate, and that of impure sodium carbonate, which is frequently designated as soda ash.

Samples of the above substances are dissolved in water; then they are titrated with standard acid or base and a visual indicator is used to determine the end point. The end point should be very near the equivalence point; that is, the amount of substance titrated is equivalent to the volume of the standard titrating solution added (see Section 4·1).

One of the mathematical expressions in Section 2·6 is restated as

$$\text{ml} \times N = \frac{\text{g of solute}}{\text{meq wt}}$$

If the solute is impure, a factor for purity converts the equation to

$$\text{ml} \times N = \frac{\text{g of solute} \times \text{fractional purity}}{\text{meq wt}}$$

or

$$\frac{\text{ml} \times N \times \text{meq wt}}{\text{g of solute}} = \text{fractional purity}$$

Percentage composition may be obtained by expanding the equation to

$$\frac{\text{ml} \times N \times \text{meq wt} \times 100}{\text{g of sample}} = \%\ \text{purity}$$

TYPES OF EXERCISES

Type 1. Preparation of a Formal Solution

Problem: How many grams of pure KOH are necessary to prepare 500 ml of 0.1000 F solution?

Solution: The formula weight of KOH is 56.11. Since a formal solution contains the formula weight expressed in grams diluted to a liter, the problem is calculated as:

$$F = \frac{\text{g of solute}}{\text{ml} \times \text{mfw}}$$

or

$$0.1000 = \frac{\text{g}}{500 \times 0.0561}$$

and

$$\text{g of KOH} = 0.1000 \times 500 \times 0.0561$$

$$\text{g} = 2.805$$

Type 2. Preparation of a Normal Solution

Problem: How many g of pure barium hydroxide are necessary to prepare 750 ml of 0.2000 N solution?

Solution: The formula weight of $Ba(HO)_2$ is 171.35; therefore, the milliequivalent weight is $171.35/2000 = 0.0857$ (see Table 2.3), and from the general equation,

$$N = \frac{\text{g of pure solute}}{\text{ml} \times \text{meq wt}}$$

$$g = 0.2000\ N \times 750\ \text{ml} \times 0.0857$$

$$g = 12.86$$

Type 3. Comparison of Acid and Base Solutions

Problem (1): How many g of NaOH are needed to react with 50.0 ml of 0.2000 N H_2SO_4 solution?

Solution:

$$\text{ml} \times N = \frac{\text{g of solute}}{\text{meq wt}}$$

$$50.00 \times 0.2000 = \frac{\text{g of NaOH}}{0.0400}$$

$$\text{g of NaOH} = 0.400$$

Problem (2): How many g of MgO will react with 5.00 meq of H_2SO_4?

Solution:

$$\text{meq}_1 = \text{meq}_2$$

$$\text{meq of } H_2SO_4 = \frac{\text{wt of MgO}}{\text{meq wt of MgO}}$$

$$5.00 = \frac{\text{g of MgO}}{0.02016}$$

$$\text{g of MgO} = 5.00 \times 0.02016 = 0.1080$$

Type 4. Dilution of Acids and Bases

Problem: How many ml of 12.00 N HCl are necessary to prepare 2.000 liters of 0.1000 N solution?

Solution:

$$\text{ml}_1 \times N_1 = \text{ml}_2 \times N_2$$

$$\text{ml}_1 \times 12.00 = 2000 \times 0.1000$$

$$\text{ml}_1 = 16.66 \text{ of } 12.00\ N \text{ HCl}$$

Type 5. Density of Acids and Bases

Problem: What is the N of concentrated ammonia, containing 28.7% NH_3, and having a density of 0.90?

Solution:

$$N = \frac{\text{wt of 1 ml} \times \text{fractional purity}}{\text{vol of 1 ml} \times \text{meq wt}}$$

$$N = \frac{0.90 \times 0.287}{1.00 \times 0.0170} = \frac{0.258}{0.017} = 15.2$$

Type 6. Composition of Impure Acids and Bases

Problem (1): A sample of impure potassium acid phthalate weighing 0.9945 g is dissolved in water and is titrated with 0.1039 N NaOH, 23.30 ml being required. What is the percentage of KHP in the sample?

Solution:

$$\frac{\text{ml} \times N \times \text{meq wt} \times 100}{\text{wt of sample}} = \% \text{ KHP}$$

$$\frac{23.30 \times 0.1039 \times 0.2042 \times 100}{0.9945} = 49.71\%$$

Problem (2): A sample of soda ash (impure Na_2CO_3) which weighs 0.4018 is dissolved in water and 43.65 ml of 0.1033 N HCl added. The excess acid is titrated with 2.12 ml of 0.0969 N NaOH. What is the percentage of Na_2CO_3 in the sample?

Solution:

$$\frac{[(\text{ml of acid} \times N) - (\text{ml of base} \times N)] \times \text{meq wt} \times 100}{\text{wt of sample}} = \% \text{ composition}$$

$$\frac{[(43.65 \times 0.1033) - (2.12 \times 0.0969)] \times 0.0530 \times 100}{0.4018} = 56.76\% \ Na_2CO_3$$

EXERCISES

1. What fraction $(1, \frac{1}{2}, \frac{1}{3},$ and so on) of the formula weight represents the equivalent weight of each of the following substances acting as acids and bases (assuming complete neutralization in each case):

   (a) $KHC_2O_4 \cdot H_2C_2O_4 \cdot 2H_2O$ (n) HNH_2SO_3
   (b) N_2O_5 (o) Cr_2O_3
   (c) Na_2HPO_4 (p) NH_3
   (d) ZnO (q) Al_2O_3
   (e) As_2O_5 (r) SO_2
   (f) CO_2 (s) ThO_2
   (g) Li_2O (t) P_2O_5
   (h) $CaCO_3$ (u) $KIO_3 \cdot HIO_3$
   (i) SO_3 (v) $KHCO_3$
   (j) $HC_2H_3O_2$ (w) Na_2CO_3
   (k) $H_2C_2O_4 \cdot 2H_2O$ (x) Ag_2O
   (l) $Ba(OH)_2$ (y) SeO_2
   (m) MnO_2 (z) CrO_3

2. Fill in the correct answers in the following spaces pertaining to acids and bases (assuming complete neutralization in each case):

Substance	$Al(OH)_3$	KH_2AsO_4	H_3PO_4	SO_3	NH_3
amount of solute in g (approximate)	156				
volume of solution in ml	1000	500	1000	500	1000
fw (approximate)	78	180	98	80	17
eq wt (approximate)					

Substance	$Al(OH)_3$	KH_2AsO_4	H_3PO_4	SO_3	NH_3
gfw		1.00	2.00		
equivalents				2.00	
meq					2000
formality					
normality					

3. How many grams of acid or base are required to make the following solutions in the given concentrations?

	Solution	Concentration
(a)	3000 ml HCl	0.2000 N
(b)	2500 ml H_2SO_4	0.1000 F
(c)	1000 ml NaOH	0.1250 N
(d)	750 ml Na_2CO_3	0.0500 F
(e)	500 ml $NaHCO_3$	0.1000 N
(f)	250 ml NH_3	0.1000 F

4. How many grams of substance B are required to neutralize completely 100 ml of substance A (assuming complete neutralization in all cases)?

	A	B
(a)	0.1000 N HCl	________g Na_2CO_3
(b)	0.1000 N HCl	________g Na_2O
(c)	0.1000 F H_2SO_4	________g BaO
(d)	0.1000 N $Ba(OH)_2$	________g H_2SO_4
(e)	0.1000 F H_2SO_4	________g $KHCO_3$
(f)	0.1000 F H_2SO_4	________g KOH
(g)	0.1000 N HNO_3	________g MgO
(h)	0.1000 F HNH_2SO_3	________g $Ca(OH)_2$
(i)	0.1000 N NH_3	________g $HClO_4$
(j)	0.1000 F Na_2CO_3	________g HBr

5. How much of substance B is needed to react with a given amount of A (assuming complete neutralization in all cases)?

	A	B
(a)	2.5 meq H_2SO_4	____ml of 0.1000 F $Ba(OH)_2$
(b)	0.0053 g Na_2CO_3	________mf of H_2SO_4
(c)	0.1250 mfw MgO	________ml of 0.1000 F HCl
(d)	1.000 liter of 0.1000 N NaOH	________g of HI
(e)	100 ml of 0.1000 N $NaHCO_3$	________meq of $HClO_4$
(f)	100 ml of 0.1000 F Na_2CO_3	________g of HNH_2SO_3
(g)	100 ml of 0.1500 F H_2SO_4	________meq $CaCO_3$

6. Compute the formality of each of the following acids and bases (percentage is by weight, and density is in g/ml):

(a) Ammonia: density 0.923, percentage 20.0
(b) Hydrochloric acid: density 1.040, percentate 8.16
(c) Phosphoric acid: density 1.700, percentage 85.6
(d) Potassium hydroxide: density 1.491, percentage 44.9
(e) Fuming sulfuric acid: density 1.956, percentage of SO_3 is 88.8

7. Calculate the normality of each of the following acids and bases (percentage is by weight):

   (a) Acetic acid: density 1.053, percentage 99.5
   (b) Sodium hydroxide: density 1.526, percentage 50.7
   (c) Formic acid: density 1.024, percentage 90.0
   (d) Sulfuric acid: density 1.395, percentage 50.0
   (e) Nitric acid: density 1.187, percentage 31.2

8. A 0.7952 g sample of primary standard potassium acid phthalate is dissolved in water and titrated with 34.65 ml of sodium hydroxide. Calculate the normality of the NaOH solution.

9. A 0.1800 g sample of a pure organic acid requires 36.52 ml of NaOH, 50.00 ml of the latter being equivalent to 0.5300 g of pure Na_2CO_3. What is the equivalent weight of the organic acid?

10. A 1.200 g sample of vinegar (impure acetic acid) requires 32.00 ml of NaOH; 1.000 ml of the latter is equivalent to 2.042 mg of potassium acid phthalate. Calculate the percentage of acetic acid in the vinegar.

11. A 0.3946 g sample of soda ash was over-titrated with 45.18 ml of 0.1017 N HCl and back-titrated with 1.61 ml of NaOH. The composition of Na_2CO_3 was calculated to be 59.62%. What was the normality of the NaOH used?

12. What weight of sample is required in order that 40.00 ml of 0.1000 N HCl is required to titrate soda ash of a purity of 60.00%?

13. What weight of sulfamic acid is required to prepare one liter of 0.05 F solution?

14. How many gfw of NaOH are in two liters of 0.2 F solution?

15. Calculate the formality of 30 g of benzoic acid dissolved in 750 ml of solution.

16. Determine the normality of 900 ml of solution containing 52 g of potassium tetraoxalate.

17. What weight of potassium bitartrate is equivalent to 30 ml of 0.1100 F NaOH?

18. How many ml of 0.02 N $Ca(OH)_2$ will react with 15 ml of 0.5 F H_2SO_4?

19. Determine the normality of a liter of solution containing 25 g of potassium binoxalate.

20. What is the formality of a liter of solution containing 20 grams of tartaric acid?

21. What weight of oxalic acid is equivalent to 500 ml of 0.100 F NaOH?

22. To what volume must a solution containing 20 grams of sodium carbonate be diluted so that the final concentration will be 0.4 N?

23. How many grams of potassium biiodate are equivalent to 34.45 ml of 0.1500 F KOH?

24. How many ml of concentrated ammonia are necessary to prepare one liter of 0.1200 F solution?

CHAPTER 3

Acid–Base Equilibria and pH of Solutions

CHEMICAL EQUILIBRIA

All ionic reactions are reversible to some degree, inasmuch as the products, unless completely removed from the field of action, tend to reproduce the original substances. Therefore, once any ionic reaction has been initiated, it may be considered the result of two opposing reactions: one is the decomposition into the products, and the other is the recombination of the products into the initial reactants. If the external conditions of the system undergo no change, the rates at which the two opposing reactions progress will eventually become equal. Such a condition is said to be in a state of equilibrium, but this equilibrium is dynamic, implying that the two opposing reactions are still in progress.

3·1 Equilibrium Constants

If two substances, A and B, of known formula weight concentrations react, then

$$A + B \longrightarrow \text{products}$$

and if m molecules or ions of A react with n molecules or ions of B, the initial reaction becomes

$$mA + nB \longrightarrow \text{products}$$

and the reaction velocity is mathematically defined as

$$V = k^* \times [A]^m \times [B]^n$$

If we assume that the products of the initial reaction are o molecules or ions of C and p molecules or ions of D, then the velocity of the reverse reaction may be indicated as

$$V = k_2 \times [C]^o \times [D]^p$$

The overall reaction is indicated as

$$mA + nB \rightleftharpoons oC + pD$$

and when equilibrium is attained the reaction rates are equal to each other,

$$k_1 \times [A]^m \times [B]^n = k_2 \times [C]^o \times [D]^p$$

or

$$\frac{[C]^o [D]^p}{[A]^m [B]^n} = \frac{k_1}{k_2} = K_e$$

where K_e is the equilibrium constant for the reaction. The brackets indicate formula-weight concentrations, and each concentration is raised to the power indicated by the number of reacting molecules or ions. As a matter of convention the products are always placed in the numerator.

Since the numerical value of K_e varies with temperatures of the reaction, it is necessary to

---

* The term k is a constant which contains all factors other than concentration, which may influence the speed of the chemical reaction.

specify temperature for each value of K_e. The foregoing mathematical expression may be stated in words as follows:* *For a reversible chemical reaction at a fixed temperature and in a state of equilibrium, the product of the formula weight concentrations of the substances formed in the reaction divided by the product of the formula weight concentrations of the reactants, each raised to the power indicated by the number of molecules or ions in the balanced equation, is equal to a constant.*

3·2 Application of Equilibrium Constants

The equilibrium constant is a mathematical value which refers *quantitatively* to a ratio of the products and reactants of a chemical reaction as indicated in Section 3·1. The numerical value of K_e indicates the extent to which a reaction proceeds toward completion, and under certain conditions this value may be of use in calculating the component concentrations involved in the equilibrium. A positive value above unity for K_e indicates that the equilibrium for the reaction of the written equation occurs to the right, whereas a value smaller than unity indicates that the equilibrium lies to the left. Thus, the course and extent of chemical reactions can be predicted by the values of the equilibrium constants for the reactions.

There are five types of chemical equilibria that are of importance in analytical chemistry, namely, the dissociation of water, the dissociation of weak acids and bases, the dissociation of complex ions, redox equilibria, and heterogeneous equilibria between slightly soluble substances and their ions in saturated solutions. A discussion of the first two types will be included in this chapter.

3·3 Limitations to the Equilibria Concept

In the previous discussion of equilibrium constants it was implied that the value of an

equilibrium constant changed only with temperature. Actually, the equilibrium concept can be applied exactly only to dilute solutions. Departures from this concept are observed in any aqueous solution containing appreciable concentrations of electrolytes. It has been observed that the dissociation of a weak acid, and the solubility of a fairly insoluble substance are increased by the presence of electrolytes such as sodium chloride (Section 9·9). On the other hand, the dissociation of strong electrolytes, such as HCl, is decreased as the concentration of the electrolyte is increased. Since the dissociation values are valid only in dilute solutions, the deviations produced by increased ionic concentrations may be loosely ascribed to a change in the nature of the solvent. The change is produced by the interionic attraction between closely spaced ions, which results in changing the dissociation of the substance under observation.

To compensate for the difference between an apparent concentration and an actual concentration, the apparent (effective) concentration is designated as the activity of the particular ion. The coefficient that is used to convert actual concentration to effective concentration is called the activity coefficient. If the actual concentration is C, the effective concentration is a, and the activity coefficient is f, then the mathematical relationship of the three factors is

$$a = fC$$

or

activity =
activity coefficient × actual concentration

The foregoing discussion is an oversimplification for "activity" and "activity coefficient." A student planning to major in chemistry is urged to expand this brief explanation with readings in more advanced textbooks in analytical chemistry. The use of activity instead of concentration is necessary in the accurate calculations of advanced chemistry. However, activity coefficient is not the numerical value of the constant, but it is dependent upon the concentration and nature of the solute, as well as other factors; therefore, relatively few values for this term are known. Until more exact numerical

---

* The concept of equilibrium is sometimes called the Law of Chemical Equilibrium.

TABLE 3·1   Change of K_w with temperature

Temperature, °C	K_w	Temperature, °C	K_w
0	0.11×10^{-14}	50	5.5×10^{-14}
18	0.55×10^{-14}	75	19×10^{-14}
25	1.01×10^{-14}	100	48×10^{-14}

data are available for activities, it is sufficient to use concentrations in the calculations of quantitative chemistry. It so happens that activities and concentrations are approximately the same for concentrations below $0.1\ F$ and, consequently, the equilibrium constants for substances within this concentration range are fairly accurate.

ACID–BASE EQUILIBRIA

As will be recalled from Section 3·2, it was stated that the concept of chemical equilibria may be applied to the dissociation of water, and to the dissociation of weak acids and bases. The latter substances are few in number, but the applications of their equilibria extend to fields that are broader than quantitative analysis. The equilibria of weak acids (and their salts) are of great importance in controlling life functions in living organisms. Blood, milk, digestive juices, and other fluids that are produced, or used, in living tissues are highly buffered solutions, that is, solutions of weak acids and their salts. The control of the hydrogen-ion concentration of living protoplasm is necessary for the maintenance of life, and the mechanism for this control is the same in biological fluids as in inorganic solutions.

3·4 Ionization of Water

Water may be considered as a weak electrolyte since the purest water obtainable will conduct an electric current to a measurable extent. The conductance results from the slight dissociation of water into hydronium and hydroxide ions. Under appropriate conditions, water may be regarded as either an acid or a base, since it may act as either a proton donor or a proton acceptor. This is due to the interaction of water

molecules to produce an ionization reaction (see Section 2·2), similar to that of any other acid or base

$$H_2O + H_2O \rightleftharpoons H_3O^+ + OH^-$$

The extent of this reaction to the right is small but definite in quantity, and it is dependent, as are all other ionization equilibria, upon temperature as is indicated in Table 3·1.

The equilibrium expression for the ionization of water is

$$\frac{[H_3O^+][OH^-]}{[H_2O]^2} = K_e$$

Since the weight of a liter of water is 997 g at 25°, the formula-weight concentration of water at this temperature is $997/18.02 = 55.3\ F$. This value is effectively constant, which results in an expression in which the product of the ions of water is equal to a new constant K_w,

$$[H_3O^+][OH^-] = K_e \times (55.3)^2 = K_w$$

or simply,

$$[H^+][OH^-] = K_w$$

K_w is a special ionization constant and is comparable to other equilibrium constants except that no denominator is involved; consequently, K_w is usually referred to as the ion product of water. At room temperature it has a value of 1×10^{-14}. Because most laboratory procedures in quantitative chemistry are performed at room temperature with dilute solutions, the value of 1×10^{-14} can be considered as a constant for calculations involving K_w.

The equilibrium between water and its ions is important in calculating hydrogen- or hydroxide-ion concentrations of aqueous solutions, since the ion product is practically the same for any dilute solution as that of pure water. In other words, the relationship of

$$[H^+][OH^-] = 1 \times 10^{-14}$$

holds true for any dilute aqueous solution regardless of whether it is an acid, a base, or a salt solution. Moreover, since K_w does not change for a stated temperature, it follows that the formula-weight concentrations of either the hydronium or hydroxide ion may be varied but the product will always be the same. For example, in a $0.1\ F$ solution of hydrochloric acid the hydrogen ion is $0.1\ F$, and the hydroxide-ion concentration is calculated as

$$0.1 \times [OH^-] = 1 \times 10^{-14}$$

$$[OH^-] = 1 \times 10^{-13}\ F$$

The same relationship is true for a solution of a strong base, except in this case the hydrogen-ion concentration is reduced to satisfy the constant K_w.

3·5 Ionization Constants for Weak Monoprotic Acids and Bases

A weak acid, such as acetic acid, is composed largely of undissociated molecules in water solution. However, a reaction between the molecular acid and water does proceed to a slight extent to yield hydronium ions and acetate ions. This reaction is reversible, the reversibility being due to the comparative strengths of the conjugate acids and bases of the reaction.

$$HC_2H_3O_2 + HOH \rightleftharpoons H_3O^+ + C_2H_3O_2^-$$

The reaction may be considered as a competition between two bases, water molecules and acetate ions, for protons. Since the acetate ion is a stronger base than the water molecule, the extent of the reaction to the left is greater than that to the right.

The equilibrium expression for the foregoing reaction is

$$\frac{[H_3O^+][C_2H_3O_2^-]}{[HC_2H_3O_2][H_2O]} = K_e$$

Inasmuch as the formula-weight concentration of water in a dilute solution varies only slightly from pure water, it is possible to substitute $55.3\ F$ as a constant for $[H_2O]$ and the equation becomes,

$$\frac{[H_3O^+][C_2H_3O_2^-]}{H_2C_2H_3O_2} = K_e \times 55.3 = K_i$$

The new constant K_i is the ionization constant of the weak acid, acetic acid, in dilute solutions. It is obvious that this equation cannot be true for more concentrated solutions in which the formality of water varies appreciably from $55.3\ F$.

A weak base such as ammonia reacts slightly with water molecules to produce hydroxide ions according to the equation:

$$NH_3 + HOH \rightleftharpoons NH_4^+ + OH^-$$

The chemical equilibrium for the reaction is

$$\frac{[NH_4^+][OH^-]}{[NH_3][H_2O]} = K_e$$

which may be reduced to

$$\frac{[NH_4^+][OH^-]}{[NH_3]} = K_e \times 55.3 = K_i$$

The term K_i denotes ionization constant and may be applied to either a weak acid or a weak base. To differentiate between the two, it is customary to denote K_a as the ionization constant of a weak acid, and K_b as the ionization constant for a weak base.

ACIDITY OF SOLUTIONS

At room temperature, the equilibrium constant for water solutions, K_w, is 1×10^{-14}. In pure water the hydrogen-ion and hydroxide-ion concentrations are equal; at 25°C, each has a value of 1.0×10^{-7}, which is a very small value. In dealing with hydrogen-ion concentrations of aqueous solutions, quite often the values are also very small numbers, usually written in exponential form, which may vary through a wide range of concentration. For example, the numbers 5×10^{-2}, 2×10^{-4}, and 7×10^{-5} might represent common hydrogen-ion concentrations of solutions of weak acids. Numbers of this size are impossible to plot in the form of a graph, and other modes of correlation are difficult to indicate. To avoid the awkwardness involved in comparing small numbers varying through an extended range, Sörensen, a Danish biochemist, devised a system by which such numbers are changed to their negative logarithms.

By the Sörensen system, a geometric progression of small numbers, such as 1×10^{-2}, 1×10^{-3}, 1×10^{-4}, and 1×10^{-5}, is converted to an arithmetical progression of small numbers, such as 2, 3, 4, and 5. By means of this mathematical device of reducing widely varying small numbers to slightly varying positive numbers, ordinary concentration of hydrogen ions may be converted to simple finite numbers, usually ranging from 0 to 14.

3·6 pH and pOH Values

By original definition pH is the negative logarithm, to the base 10, of the hydrogen-ion concentration. The formula for this statement is

$$pH = \log \frac{1}{[H^+]} = -\log [H^+]$$

In 1909, at the time Sörensen proposed this method of expressing hydrogen-ion concentrations, the term *hydronium ion*, which is a more accurate expression than hydrogen ion, was unknown. Also, the concepts of *activity* and *activity coefficient*, which relate actual concentration to effective concentration, had not been developed. Although, the foregoing formula is still widely used for pH conversion, the term pH by interpretation should be the negative logarithm of activity of the hydrated proton. In quantitative chemistry it is assumed that for dilute solutions, formality does not differ greatly from activity; therefore, pH is the negative logarithm of the hydronium-ion concentration, or

$$pH = \log \frac{1}{[H_3O^+]} = -\log [H_3O^+]*$$

* It is sufficient to use the symbol H^+ for the hydrated proton.

To illustrate the use of this formula, suppose a solution has a hydrogen-ion concentration of $2 \times 10^{-9} F$, what is the pH of the solution? The easiest way to produce the conversion is to take the exponent as a whole number, change its sign, and subtract from it the logarithm of the coefficient number. Thus, 10^{-9} becomes 9, and the logarithm of the coefficient number, 2, is 0.3. If 0.3 is subtracted from 9 the result is 8.7, which is the pH of the solution.

To calculate the hydronium-ion concentration from the pH of a given solution, the preceding process is reversed. For example, consider a solution with a pH value of 6.8, what is the hydronium-ion concentration? Since

$$pH = 6.8 = -\log [H^+]$$

it follows that

$$[H^+] = 10^{-6.8} = 10^{-0.8} \times 10^{-6}$$

To evaluate the number, it is necessary to convert the negative decimal exponent from a negative logarithm to a positive logarithm. This is accomplished by borrowing 10^1 from the second portion of the expression, and multiplying the decimal exponent by 10^1 to give the following result:

$$[H^+] = 10^{+0.2} \times 10^{-7}$$

or

$$[H^+] = \text{antilog of } 0.2 \times 10^{-7}$$

and

$$[H^+] = 1.6 \times 10^{-7} F$$

Although rarely used, the hydroxide-ion concentration can be expressed as pOH, which is the negative logarithm of the hydroxide-ion concentration, or

$$pOH = \log \frac{1}{[OH^-]} = -\log[OH^-]$$

3·7 pK Values

Occasionally it becomes convenient to express dissociation constants as pK values, which are analogous to pH values. The pK value of an equilibrium constant for a weak acid or base, is defined as the negative logarithm of the equilibrium constant, or

$$pK = \log \frac{1}{K_e} = -\log K_e$$

To illustrate with an example of a specific equilibrium, pK for the ionization constant of acetic acid, where $K_a = 1.75 \times 10^{-5}$, is

$$pK = -\log K_a = -\log(1.75 \times 10^{-5})$$

or

$$pK = 5.00 - \log 1.75 = 5.00 - 0.24 = 4.76$$

K_w for water represents a special equilibrium constant for dilute solutions, and since K_w has a value of 1×10^{-14}, it is often advantageous to use the following equations in problems involving pH or pOH of aqueous solutions:

$$pH + pOH = pK_w$$

$$pH + pOH = 14$$

and in neutral aqueous solutions

$$pH = pOH = 7$$

pH OF SOLUTIONS OF ACIDS AND BASES

As a matter of convenience acids and bases may be divided into three classes, namely: (1) strong acids and bases; (2) weak monoprotic acids and bases; (3) polyprotic acids.

3·8 Solutions of Strong Acids and Bases

The term "strong" implies that acids and bases of this description are highly dissociated in water solutions. It may be safely assumed that in dilute solutions ionization is practically complete for such compounds, and that any reversibility is negligible. Consequently, aqueous solutions of strong acids and bases do not have measurable equilibrium constants.

There is a structural difference between certain acidic and basic substances. Anhydrous HCl, HBr, and HI are gases, and the pure compounds consist of molecules that are largely covalent in character. The reactions between these molecules with water produce hydronium ions and anions, and the reactions go practically to completion. Thus, the reaction of HCl can be indicated as

$$HCl + H_2O \longrightarrow H_3O^+ + Cl^-$$

In terms of the Brønsted acid–base concept, the chloride ion can be considered as a much weaker base than water when it is a base; in fact it is so weak that the water molecules capture protons to produce an irreversible reaction.

In comparison, the solids NaOH and KOH are largely ionic in the crystalline state. In the process by which NaOH enters solution, water molecules, because of their dipole moments, orient their negative centers toward the sodium ions and their positive ends toward the chloride ions at the surface of the crystal. The dipole forces exerted by the oriented water molecules weaken the interionic attraction within the crystal so that some of the ions are pulled into the solution. The removed ions wander away from the crystal, carrying their attached water molecules with them. Such a solution process cannot be called an acid–base reaction but rather a hydration reaction, which may be indicated as

$$Na^+OH^- + x\,H_2O \rightleftharpoons$$
$$Na(H_2O)_y^+ + OH(H_2O)_z^-$$

The hydration reaction may be considered as reversible, but no additional ionization reaction is involved.

Since strong acids and bases exist largely as ions in dilute solutions, the concentration of the hydrogen ion of a strong acid and the concentration of the hydroxide ion of a strong base are considered to be approximately the same as the formality of each of the substances in solution. Thus, a $0.01\,F$ solution of HCl is also $0.01\,F$ in hydrogen ions, and a $0.001\,F$ solution of NaOH would be $10^{-3}\,F$ in hydroxide ions. As an illustration, consider the problem of obtaining the pH of a solution which is $0.0005\,F$ in HCl.

From the explanation just given, the hydrogen ion concentration can also be considered as 0.0005 F, or

$$0.0005\ F\ HCl \longrightarrow 0.0005\ F\ [H^+]$$

$$[H^+] = 5 \times 10^{-4}$$

$$pH = -\log(5 \times 10^{-4})$$

$$pH = -0.7 + 4.0 = 3.3$$

To obtain the pH of a strong base, a solution which is 0.02 F in $Ba(OH)_2$ may be used for an example. One way of approaching this problem is first to obtain the pOH of the solution.

$$Ba^{++}(OH)_2^- \longrightarrow Ba^{++} + 2\ OH^-$$

$$0.02\ F \qquad\qquad\qquad 0.04\ F$$

Therefore,

$$pOH = -\log(4 \times 10^{-2})$$

$$pOH = -0.6 + 2.0$$

$$pOH = 1.4$$

Since

$$pH + pOH = 14$$

Then

$$pH = 14.0 - 1.4 = 12.6$$

3·9 Solutions of Weak Acids and Weak Bases

All weak acids and weak bases are incompletely dissociated in aqueous solutions. The dissociation of weak acid or base in water is usually viewed as an acid–base reaction. For example, acetic acid reacts with water as follows:

$$HC_2H_3O_2 + H_2O \rightleftharpoons H_3O^+ + C_2H_3O_2^-$$

The equilibrium constant for the reaction is

$$\frac{[H_3O^+][C_2H_3O_2^-]}{[HC_2H_3O_3][H_2O]} = K_e$$

which may be converted to

$$\frac{[H_3O^+][C_2H_3O_2^-]}{[HC_2H_3O_2]} = K_a$$

To obtain the hydrogen-ion concentration for a 0.1 F solution $HC_2H_3O_2$, the following relation is valid

$$HC_2H_3O_2 \rightleftharpoons H^+ + C_2H_3O_2^-$$

$$0.1 - x \qquad x \qquad x$$

Since K_a for the acid is 1.75×10^{-5}, then

$$\frac{x^2}{0.1 - x} = 1.75 \times 10^{-5}$$

The foregoing equation is a quadratic and may be solved as such; however, it may be simplified and the quadratic avoided by discarding the value, x, in the denominator. From a mathematical viewpoint, x may be discarded only if its value is much smaller than the value from which it is subtracted. In this particular illustrative problem, x can be ignored because it does not represent a significant number, as does the concentration of acetic acid, given as 0.1 F. The validity of this approximation may be verified by solving the quadratic, and comparing the answer with that obtained after discarding the x. In this particular case the answer will be the same for each computation.

As a general rule the value "x" is insignificant for weak acids and bases, in which the concentrations of the weak electrolytes are not too dilute. When K_1 of the weak acid or base is 10^{-4} or less, the term $C - x$ (C = concentration) does not differ appreciably from C. By inspection, it can be seen that this estimation is true when C is not less than 0.1 F. On the other hand, this approximation is applicable to lower concentrations of those acids and bases whose ionization constants are somewhat less than 10^{-4}.

If the equilibrium equation for 0.1 $F\ HC_2H_3O_2$ solution is solved after it has been assumed that $C - x$ does not differ appreciably from C, and after x has been discarded in the denominator, then

$$\frac{x^2}{0.1} = 1.75 \times 10^{-5} \quad \text{or} \quad x^2 = 1.75 \times 10^{-6}$$

and

$$x = 1.32 \times 10^{-3}\ F\ H^+$$

The pH of the solution is

$$pH = 3.00 - 0.12 = 2.88$$

The pH of a weak base follows the same pattern as that for a weak acid, except that it is usually convenient to obtain the pOH value, and from that value the pH may be obtained.

The most typical weak base in quantitative chemistry is a solution of NH_3 molecules in water. The reaction is

$$NH_3 + H_2O \rightleftharpoons NH_4^+ + OH^-$$

$$C - x \qquad\qquad x \qquad x$$

The equilibrium expression is

$$\frac{[NH_4^+][OH^-]}{[NH_3]} = K_b$$

in which the concentration of $[H_2O]$ is assumed to be a constant. The K_b value for ammonia is 1.8×10^{-5}.

If the pH of a 0.2 F solution of NH_3 is sought, the above equilibrium becomes

$$\frac{x^2}{0.2} = 1.8 \times 10^{-5}$$

(The x value in the denominator is discarded.)

$$x^2 = 3.6 \times 10^{-6} = 3.6 \times 10^{-6}$$

$$x = 1.88 \times 10^{-3} \, F \text{ in } OH^- \text{ ions}$$

$$pOH = -0.28 + 3.0 = 2.72$$

$$pH = 14.0 - 2.72 = 11.28$$

3·10 Solution of Polyprotic Acids

A polyprotic acid contains more than one proton that may react with water to produce hydronium ions. A detailed study of acid strength for polyprotic acids would be too lengthy for this course, but it seems pertinent to examine some generalizations concerning these compounds. A few characteristics have been fairly well established for inorganic polyprotic acids.

The removal of each proton from a polyprotic acid constitutes a separate equilibrium, and hence, there are always as many equilibrium constants from such an acid as there are protons that can ionize. Consider the dissociation of phosphoric acid,

$$H_3PO_4 \rightleftharpoons H_2PO_4^- + H^+ \qquad pK_{a_1} = 2.22$$

$$H_2PO_4^- \rightleftharpoons HPO_4^{--} + H^+ \qquad pK_{a_2} = 7.21$$

$$HPO_4^{--} \rightleftharpoons PO_4^{3-} + H^+ \qquad pK_{a_3} = 12.67$$

The ionization constant for removing the second proton is usually much smaller than that for the first proton. Phosphoric acid has a very simple structure

$$\begin{array}{c} OH \\ | \\ HO-P=O \\ | \\ OH \end{array}$$

in which the OH groups (from which protons may be derived) are attached to the same atom. This accounts for each successive ionization constant being smaller than the preceding one.

In pyrophosphoric acid, which has a structure of

$$\begin{array}{ccc} OH & & OH \\ | & & | \\ O=P-O-P=O \\ | & & | \\ OH & & OH \end{array}$$

the first two protons are removed from different ends of the molecule and have ionization constants that are not widely different. The third and fourth ionization constants are much smaller than the first two, and are closer together than might be expected. the pK_a values for pyrophosphoric acid, $H_4P_2O_7$, (at 18°C) are

$$pK_{a_1} = 0.85, \qquad pK_{a_2} = 1.49,$$

$$pK_{a_3} = 5.77, \qquad pK_{a_4} = 8.22.$$

For hydride acids derived from the elements of any given family in the periodic table, the strengths of the acids increase with increasing molecular weight. The increasing strength may also be attributed to the decreasing electronegativity* of the central atom. In other words, it is easier to remove a proton attached to a large atom than one attached to a small atom.

---

* Electronegativity is a function of the size and nuclear charge of an atom.

The hydride acids of Group VIB of the periodic table are listed with their pK_a values as Table 3·2.

TABLE 3·2 pK_a values of hydrides of Group VIB of periodic table

Acid	pK_{a_1}	pK_a
H_2O	16.00	—
H_2S	7.00	12.92
H_2Se	3.89	11.00
H_2Te	2.74	11.00

It is not practical to make generalizations concerning the ionization of polyprotic organic acids. Usually, but not always, the protons from such acids are derived from carboxyl ($-COOH$) groups. The simplest polyprotic acid of this type is oxalic acid, which has the following structure;

$$
\begin{array}{c}
\text{O} \\
\| \\
\text{C}-\text{OH} \\
| \\
\text{C}-\text{OH} \\
\| \\
\text{O}
\end{array}
$$

The factors that contribute to the strength of organic polyprotic acids are too numerous and diverse to be considered in our present discussion. A few typical acids of this type are given in Table 3·3 along with their pK_a values.

TABLE 3·3 pK_a values for selected polyprotic organic acids

Acid	pK_{a_1}	pK_{a_2}
oxalic	1.23	4.19
malonic	2.83	5.69
tartaric	2.89	4.34
succinic	4.16	5.61
phthalic	2.89	5.51

The dissociation of a polyprotic acid involves as many dissociation constants as there are protons that may be removed. To calculate the pH of a solution of most polyprotic acids, only the value of K_1 is employed, since the $[H^+]$ is essentially that of a monoprotic acid. This is true of phosphoric acid, which behaves as a monoprotic acid, except in very dilute solutions;

therefore is $[H^+] = x$, and $C = $ concentration of the acid, the quadratic (see Section 3·9) for obtaining the hydrogen-ion concentration is

$$\frac{x^2}{C - x} = K_1$$

However, for a polyprotic acid with ionization constants that are not widely different, the calculation of $[H^+]$ becomes more complex. Consider the structure of succinic acid

$$
\text{HO}-\overset{\overset{\textstyle O}{\|}}{\text{C}}-\text{CH}_2-\text{CH}_2-\overset{\overset{\textstyle O}{\|}}{\text{C}}-\text{OH}
$$

in which the two protons are derived from different ends of a fairly long molecule. The ionization constants are 6.7×10^{-5} and 2.5×10^{-6}, which are fairly close together. In calculating hydrogen-ion concentration, it is not permissible to neglect the second ionization constant, and a more exact equation should be applied. Such an equation* is as follows:

$$[H^+]^3 + [H^+]^2 K_1 - [II^+]$$
$$\times (K_{1c} - K_1 K_2) = 2K_1 K_{2c}$$

Polyprotic acids with ionization constants that do not differ greatly are rare. The few that do exist are organic acids that are seldom used in quantitative chemistry. Consequently, the rigorous equation given above is more of theoretical interest than of practical value. Usually a polyprotic acid can be treated as a monoprotic acid without appreciable error. For rough approximations, the K_1/K_2 ratios for diprotic acids are significant to the following extent: for a ratio of 10,000 or more, K_2 can be neglected without appreciable error; for a ratio of 1000, K_2 can be neglected, but an appreciable error is involved; for a ratio of 100, K_2 cannot be neglected.

pH OF SOLUTIONS OF SALTS

Acids and bases react with each other to form water molecules and ionic compounds known

---

*For the derivation of this equation see *Acid Base Indicators*, by I. M. Kolthoff and R. C. Rosenblum, The Macmillan Company, New York, 1937, p. 10.

as salts. The latter are electrically neutral combinations of cations and anions, which exist as ions in the solid state or in water solution. A salt, resulting from the reaction of equivalent amounts of acid and base, does not contain hydrogen and hydroxide ions in the crystal lattice. However, its components (cation and anion), which, dissolved in water, may produce a shift in the water equilibrium,

$$H_2O + H_2O \rightleftharpoons H_3O^+ + OH^-$$

and, thereby, affect the pH of the resulting solution. Therefore, it is convenient to classify salts into various types with respect to the reaction of their ions with water.

3·11 Salts Whose Solutions Are Aprotic

Salts of this type contain cations and anions, which apparently do not gain or lose protons, and, therefore, do not affect the equilibrium concentrations of H_3O^+ and OH^- in the solvent, water. Examples of such salts are NaCl and KNO_3. The entrance of these salts into solution is caused by forces operating between the electrically charged ions in the crystal and the electrostatic dipoles of water molecules. It is quite probable that Na^+ and Cl^- ions in water solution are loosely hydrated, but the extent of hydration is not sufficient to shift the water equilibrium of H_3O^+ and OH^- ions. If the solvent is pure water, the pH of a solution of NaCl if 7.0. When the solution is in equilibrium with air, containing a normal content of CO_2, the pH is approximately 6.0, as a result of the reaction

$$CO_2 + H_2O \rightleftharpoons H_3O^+ + HCO_3^-$$

3·12 Salts Whose Anions Are Proton Acceptors

This discussion limits these salts into two types:

The Salt of a Monoprotic Anion. Such anions, according to Brønsted interpretations, are defined as bases. In consequence of protons being accepted by the anions, there is a decrease in hydronium-ion concentration of the solution to make it basic. Examples of such salts are $NaC_2H_3O_2$ and KCN. Since the sodium and potassium ions are apparently aprotic, it must be concluded that the acetate and cyanide ions are bases. The reaction of CN^-, acting as a base, is represented by the following reaction:

$$CN^- + H_2O \rightleftharpoons OH^- + HCN$$
$$\text{base}_1 \quad \text{acid}_2 \qquad \text{base}_2 \quad \text{acid}_1$$

The extent to which the reaction proceeds is inversely related to the strength of the weak acid, HCN, formed. The equilibrium constant for the conjugate base, CN^-, is formulated in the same manner as that of other equilibrium constants for bases, and may be indicated as

$$\frac{[OH^-][HCN]}{[CN^-]} = K_b$$

But since K_b is inversely related to K_a for the HCN formed in the reaction, and the OH^- concentration is directly related to K_w, the two relationships are shown as follows:

$$\frac{[H_3O^+][OH^-]}{\dfrac{[H_3O^+][CN^-]}{[HCN]}} = \frac{K_w}{K_a} = K_b$$

which may be simplified to

$$\frac{[HCN]\,[OH^-]}{[CN^-]} = \frac{K_w}{K_a}$$

To obtain the pH for a solution of $0.1\ F$ KCN (where K_a for HCN is 5×10^{-10}), values may be substituted into the above equation to produce

$$\frac{x^2}{0.1} = \frac{10^{-14}}{5 \times 10^{-10}}$$

(where $x = [HCN] = [OH^-]$) or

$$x = \sqrt{\frac{10^{-15}}{5 \times 10^{-10}}} = 1.4 \times 10^{-3}$$

Since

$$[OH^-] = 1.4 \times 10^{-3}$$

Then

$$pOH = 3.00 - 0.15 = 2.85$$

and

$$pH = 14.00 - 2.85 = 11.15$$

The Salt of a Diprotic Anion. A number of diprotic-anion salts, such as Na_2CO_3, are important in analytical chemistry. In a solution of sodium carbonate and water, two equilibria are produced:

$$CO_3^{--} + HOH \rightleftharpoons HCO_3^- + OH^-$$

$$HCO_3^- + HOH \rightleftharpoons H_2CO_3 + OH^-$$

The extent to which the two reactions take place depends upon the relative values of the ionization constants of the acids (HCO_3^- and H_2CO_3) produced, *compared with the ion product of water.* The first reaction takes place to a much greater extent than the second, since the ionization constant for HCO_3^- is 5.6×10^{-11} and that for H_2CO_3 is 4.3×10^{-7}. This fact appears more evident when the ionization constants are divided into the value of K_w, or,

$$\frac{10^{-14}}{5.6 \times 10^{-11}} = 2.4 \times 10^{-4}$$

(for first reaction)

$$\frac{10^{-14}}{4.3 \times 10^{-7}} = 2.3 \times 10^{-8}$$

(for second reaction).

Therefore, to calculate the pH of a solution containing Na_2CO_3, only the first reaction need be considered.

As an example of the foregoing discussion, consider the procedure for determining the pH of a $0.1\ F\ Na_2CO_3$. The equilibria for the first reaction is expressed as

$$\frac{[HCO_3^-][OH^-]}{[CO_3^{--}]} = \frac{K_w}{K_a}$$

(where $K_a = 5.6 \times 10^{-11}$).

After substituting known values into the formula, it becomes

$$\frac{[HCO_3^-][OH^-]}{0.1} = \frac{10^{-14}}{5.6 \times 10^{-11}}$$

and

$$[OH^-] = \sqrt{\frac{10^{-15}}{5.6 \times 10^{-11}}} = \sqrt{1.8 \times 10^{-5}}$$

$$= 4.2 \times 10^{-3}$$

$$pOH = 3.00 - 0.63 = 2.37$$

and

$$pH = 14.00 - 2.37 = 11.63$$

3·13 Salts Whose Cations Are Proton Donors

In accordance with Brønsted definitions, such cations may be considered as acids, inasmuch as they increase the hydronium-ion concentration of a water solution. The most typical example is NH_4Cl. Salts of the ammonium chloride type react with water in the following manner:

$$NH_4^+ + HOH \rightleftharpoons H_3O^+ + NH_3$$

The extent of the reaction is inversely proportional to the strength of the base, NH_3, produced. The K_a for the conjugate acid, NH_4^+, is indicated as

$$\frac{[H_3O^+][NH_3]}{[NH_4^+]} = K_a$$

According to the reasoning given in Section 3·12: K_a is inversely related to the K_b of NH_3, and $[H_3O^+]$ is directly related to K_w. The combination can be formulated as

$$\frac{[H_3O^+][OH^-]}{\dfrac{[NH_4^+][OH^-]}{[NH_3]}} = \frac{K_w}{K_b} = K_a$$

which can be reduced to

$$\frac{[H_3O^+][NH_3]}{[NH_4^+]} = \frac{K_w}{K_b}$$

The $[H_3O^+]$ of a solution of $0.1\,F$ NH_4Cl, where $K_b = 1.8 \times 10^{-5}$, and $[H_3O^+] = [NH_3] = x$, can be expressed as

$$\frac{x^2}{0.1} = \frac{10^{-14}}{1.8 \times 10^{-5}}$$

$$x = \sqrt{\frac{1 \times 10^{-15}}{1.8 \times 10^{-5}}} = 7.4 \times 10^{-6}$$

Thus

$$[H_3O^+] = 7.4 \times 10^{-6}$$

and

$$pH = 6.00 - 0.87 = 5.13$$

3·14 Salts Whose Anions Are Proton Acceptors and Whose Cations Are Proton Donors

A water solution containing this type of salt will be either acidic or basic depending upon the relative strength of the cation as an acid and the anion as a base. An example of such a salt is ammonium cyanide. In the aqueous solution of ammonium cyanide the following equilibria exist:

$$CN^- + H_2O \rightleftharpoons HCN + OH^-$$

$$NH_4^+ + H_2O \rightleftharpoons NH_3 + H_3O^+$$

Whether the solution is acidic or basic depends upon the relative extent of the two ionic reactions. The calculation of the pH of such a solution must take into account both equilibria.

It is not feasible to titrate a salt of this type; consequently, the calculation of pH is of theoretical interest only. If the equations for K_a and K_b, as given in Sections 3·12 and 3·13, are combined with the water equilibrium for K_w, the resulting combination produces the following equation:

$$[H_3O^+] = \sqrt{\frac{K_w \times K_a}{K_b}}$$

This theoretical equation is unique in that the factor of concentration does not appear in it.

Experimental evidence supports the fact that the pH of such a salt is independent of concentration within normal limits. However, the equation does not apply for either very high concentrations of the salt or very low concentrations.

The pH of a $0.1\,F$ NH_4CN solution (when $K_a = 5 \times 10^{-10}$, $K_b = 1.8 \times 10^{-5}$, and $K_w = 10^{-14}$) may be obtained by substituting into the above equation. The substitution produces

$$[H_3O^+] = \sqrt{\frac{10^{-14} \times 5 \times 10^{-10}}{1.8 \times 10^{-5}}}$$

$$[H_3O^+] = 5.3 \times 10^{-10}$$

$$pH = 10.00 - 0.72 = 9.28$$

3·15 Acid Salts

When a diprotic acid is incompletely neutralized, the resulting salt contains the excess hydrogen ions in its structure; typical examples are $NaHSO_3$ and $NaHCO_3$. An acid salt such as $NaHCO_3$ may react with water to give two possible equilibria:

$$HCO_3^- + HOH \rightleftharpoons H_2CO_3 + OH^- \qquad (1)$$

$$HCO_3^- + HOH \rightleftharpoons H_3O^+ + CO_3^{--} \qquad (2)$$

The first reaction produces hydroxide ions, whereas the second gives hydronium ions as a product. Whether the solution is basic or acidic is determined by the relative values of the equilibrium constants for the two reactions. As both reactions proceed, the resulting H_3O^+ and OH^- ions form a third equilibrium in the production of water molecules.

$$H_3O^+ + OH^- \rightleftharpoons 2H_2O \qquad (3)$$

The equilibrium for reaction (1) is

$$\frac{[H_2CO_3][OH^-]}{[HCO_3^-]} = \frac{K_w}{K_{a_1}}$$

(where $K_{a_1} = 4.3 \times 10^{-7}$). $\qquad (4)$

The equilibrium for reaction (2) is

$$\frac{[H_3O^+][CO_3^{--}]}{[HCO_3^-][H_2O]} = K_{a_2}$$

(where $K_{a_2} = 5.6 \times 10^{-11}$).　　(5)

The equilibrium for reaction (3) is

$$\frac{[H_2O]^2}{[H_3O^+][OH^-]} = \frac{1}{K_w}$$

(where $K_w = 10^{-14}$).　　(6)

If the three equilibria, (4), (5), and (6) are combined the result is

$$\frac{[H_2CO_3][CO_3^{--}]}{[HCO_3^-]^2} = 1.3 \times 10^{-4}$$

To determine the pH of a 0.1 F NaHCO$_3$, the following simplification may be assumed:

$$x = [H_2CO_3] = [CO_3^{--}]$$

Then

$$\frac{x^2}{(0.1)^2} = 1.3 \times 10^{-4}$$

and

$$x = 1.14 \times 10^{-3} = [H_2CO_3] = [CO_3^{--}]$$

If known values are substituted into equilibrium (5) the result is

$$\frac{[H_3O^+][CO_3^{--}]}{[HCO_3^{--}]} = \frac{[H_3O^+](1.14 \times 10^{-3})}{0.1}$$

$$= K_{a_1} = 5.6 \times 10^{-11}$$

Then

$$[H_3O^+] = 4.9 \times 10^{-9}$$

and

$$pH = 9.00 - 0.69 = 8.31$$

Therefore, a solution of NaHCO$_3$ is basic, and reaction (1) is predominant.

pH OF BUFFER SOLUTIONS

A buffer solution is one that resists change in pH when an appreciable amount of either a strong acid or base is added to the solution. A buffer is composed of a mixture of a weak acid and its salt, or a mixture of a weak base and its corresponding salt. Typical examples of buffer mixtures are acetic acid with sodium acetate, and ammonia water with ammonium chloride. In other words, a buffer solution contains either a weak acid or a weak base with the conjugate acid or base (in the form of a salt), and, by common-ion effect, it maintains a fairly constant hydrogen-ion concentration, even when a strong acid or base is added.

Buffer solutions are important in many branches of chemistry, particularly in analytical and biochemical investigations, in which close control of pH is necessary. For example, the human blood is buffered at a pH of 7.3–7.5 by a mixture of carbonates, phosphates, and proteins. A change of less than one pH unit from this range is sufficient to cause death.

3·16 Acid Buffers

The operation of an acid buffer depends upon the equilibrium between a weak acid (largely molecular in structure) and its ions. An increase in concentration of one of the ions results in a momentary decrease of the other ion; however, the two ions combine to form the undissociated acid, and the original concentrations of the ions are not greatly changed. We can explain this process more clearly by considering a specific example of a buffer mixture. Assume that a liter of an acid buffer is 0.5 F acetic acid and 0.5 F sodium acetate. Acetic acid is a weak acid and only slightly dissociated, whereas sodium acetate may be considered as completely ionized. These conditions are indicated in the following equations:

$$HC_2H_3O_2 + HOH \rightleftharpoons H_3O^+ + C_2H_3O_2^-$$
$$0.5\,F \qquad\qquad\qquad x \qquad\quad x$$

$$Na^+C_2H_3O_2^- \xrightarrow{\;HOH\;} Na^+ + C_2H_3O_2^-$$
$$0.5\,F \qquad\qquad\qquad 0.5\,F \quad 0.5\,F$$

The sodium ion can be considered as a spectator ion, since it serves only to maintain the electrical-charge-balance of the solution; consequently, the buffer-mixture may be indicated as

$$HC_2H_3O_2 + HOH \rightleftharpoons H_2O^+ + C_2H_3O_2^-$$

$$0.5\,F \qquad\qquad\qquad x \qquad\quad 0.5\,F$$

The ionization equilibrium of the acetic acid-acetate buffer is

$$\underset{\text{acid}}{\overset{\text{salt}}{\frac{[H_3O^+][C_2H_3O_2{}^-]}{[HC_2H_3O_2]}}} = K_a$$

(where $K_a = 1.75 \times 10^{-5}$).

The substitution of known concentrations into the equilibrium results in

$$\underset{\text{acid}}{\overset{\text{salt}}{\frac{[H_3O^+](0.5\,F)}{(0.5\,F)}}} = 1.75 \times 10^{-5}$$

or

$$[H_3O^+] = \frac{0.5\,F}{0.5\,F} \times 1.75 \times 10^{-5} = 1.75 \times 10^{-5}$$

and

$$pH = 4.76$$

If 0.1 gfw of HCl is added to the preceding buffer mixture (assume that the hydrochloric acid is completely ionized and that its addition does not change the volume of the solution), the added hydrogen ions will react with excess acetate ions to increase the total concentration of undissociated acetic acid, and the acetate ion concentration will be decreased correspondingly. The resulting equilibrium can be indicated as follows:

$$\underset{\text{acetic acid}}{\overset{\text{acetate ion}}{\frac{[H_3O^+](0.5\text{ gfw} - 0.1\text{ gfw})}{(0.5\text{ gfw} + 0.1\text{ gfw})}}} = 1.75 \times 10^{-5}$$

or

$$[H_3O^+] = \frac{0.6\,F}{0.4\,F} \times 1.75 \times 10^{-5} = 2.63 \times 10^{-5}$$

and

$$pH = 4.58$$

The change of 0.18 pH units (from 4.76 to 4.58) is fairly small considering the fact that the addition of 0.1 gfw HCl to a liter of pure water produces a change of approximately 6 pH units.

To explain how an acid buffer acts upon the addition of a strong base, suppose the preceding problem is changed to the extent that 0.1 gfw of NaOH is added to the buffer instead of hydrochloric acid. It is assumed that the NaOH is completely ionized, and that its addition does not change the volume of the solution. The added hydroxide ions react with molecular acetic acid to increase the concentration of acetate ions, and the concentration of the undissociated acetic acid is decreased correspondingly. Under these conditions, the buffer equilibrium becomes

$$\underset{\text{acetic acid}}{\overset{\text{acetate ion}}{\frac{[H_3O^+](0.5\text{ gfw} + 0.1\text{ gfw})}{(0.5\text{ gfw} - 0.1\text{ gfw})}}} = 1.75 \times 10^{-5}$$

or

$$[H_3O^+] = \frac{0.4\,F}{0.6\,F} \times 1.75 \times 10^{-5} = 1.17 \times 10^{-5}\,F$$

and

$$pH = 4.93$$

The change of 0.17 pH units (from 4.93 to 4.76) is relatively small, inasmuch as the addition of 0.1 gfw of NaOH to a liter of pure water results in a difference of 6 pH units.

The general formula for computing the hydronium-ion concentration of an acid buffer is

$$[H_3O^+] = \frac{C_{\text{acid}}}{C_{\text{salt}}} \times K_a$$

where C represents formula-weight concentration, and K_a is the ionization constant of the acid in the buffer mixture.

3·17 Base Buffers

A buffer solution produced by a mixture of a weak base and its salt operates in the same manner as an acid buffer, except that it is more convenient to make calculations of hydroxide-ion concentrations. A hydroxide-ion concentration may be converted to hydronium-ion concentration, or pH, when necessary. An example of a buffer of this type is a mixture of ammonia water and ammonium chloride. A general formula for computing the hydroxide-ion concentration of a base buffer is

$$[OH^-] = \frac{C_{base}}{C_{salt}} \times K_b$$

where K_b is the ionization constant of the weak base. The addition of a strong base to an ammonia-ammonium-ion buffer produces a decrease in the ammonium-ion concentration and an increase in the ammonia concentration by the following reaction:

$$NH_4^+ + OH^- \rightleftharpoons NH_3 + H_2O$$

The addition of a strong acid to the same buffer decreases the concentration of the ammonia and increases the ammonium-ion concentration by the reaction

$$H_3O^+ + NH_3 \rightleftharpoons NH_4^+ + H_2O$$

To illustrate the action of a base buffer, consider a liter of solution which is $0.5\ F$ in respect to both NH_3 and NH_4^+, and determine the effect upon the hydroxide concentration by the addition of 0.1 gfw of NaOH. Before the addition of NaOH, the hydroxide concentration of the solution is

$$[OH^-] = \frac{0.5\ F\ NH_3}{0.5\ F\ NH_4^+} \times 1.8 \times 10^{-5}$$

$(K_b$ for $NH_3)$

and

$$[OH^-] = 1.8 \times 10^{-5}\ F$$

or

$$pOH = 4.75$$

After 0.1 gfw of NaOH is added, assuming no volume change, the ammonium-ion concentra-

tion is reduced to $0.4\ F$, and the ammonia concentration is increased to $0.6\ F$. The formality of the hydroxide ion becomes

$$[OH^-] = \frac{0.6\ F\ NH_3}{0.4\ F\ NH_4^+} \times 1.8 \times 10^{-5}$$

$$= 2.7 \times 10^{-5}\ F$$

and

$$pOH = 4.57$$

Therefore, the addition of 0.1 gfw of NaOH increases the hydroxide-ion concentration of this ammonia-ammonium-ion buffer only 0.18 pH unit.

The preparation of buffer systems for approximate pH values is described in various chemistry handbooks.

3·18 Buffer Capacity

The *buffer capacity* of a solution is in its resistance to changes in pH upon the addition of a strong acid or a strong base. Usually a high capacity is desirable, and this characteristic is obtained by means of a high concentration of the buffer's components. It is also true that maximum capacity is obtained when the buffer ingredients have a 1:1 ratio in their concentrations. In theory, any weak acid or base and its respective salt will produce a buffer mixture; however, not all such mixtures have the same buffering effect. The average effective range of a buffer varies approximately tenfold; consequently, the useful pH range may be defined mathematically as

$$pH = pK_a \pm 1$$

In quantitative terms, buffer capacity is frequently defined as the number of gfw of strong acid or base required to produce a change of one pH unit in a liter of buffer solution. Generally, the maximum buffer capacity is obtained when the K_a or K_b of the weak acid or base is numerically equal to the desired hydrogen-ion or hydroxide-ion concentration.

TYPES OF EXERCISES

Type 1. pH and pOH Values

Problem: What is the hydrogen-ion concentration that corresponds to a pH of 5.7?

Solution:

$$pH = -\log[H^+]$$

$$5.7 = -\log[H^+]$$

$$[H^+] = 10^{-5.7} = 10^{-5} \times 10^{-0.7} = 10^{-6} \times 10^{+0.3}$$

$$[H^+] = 2.0 \times 10^{-6}$$

Type 2. pH of Strong Acids and Bases

Problem: What is the pH of a 0.01 F KOH solution?

Solution:

$$[OH^-] = 10^{-2}\,F$$

$$pOH = 2.00$$

$$pH = 14.00 - 2.00 = 12.00$$

Type 3. pH of Weak Acids and Bases

Problem: Calculate the pH of a 0.12 F solution of boric acid ($K_a = 5.8 \times 10^{-10}$).

Solution:

$$H_3BO_3 \rightleftharpoons H^+ + H_2BO_3{}^-$$

$$0.12 - x \qquad x \qquad x$$

$$\frac{[H^+][H_2BO_3{}^-]}{[H_3BO_3]} = K_a \quad \text{and} \quad \frac{x^2}{0.12 - x} = 5.8 \times 10^{-10}$$

Since the x in the denominator is not significant, it may be discarded, and

$$x^2 = 0.12(5.8 \times 10^{-10}) = 70 \times 10^{-12}$$

$$x = \sqrt{70 \times 10^{-12}} = 8.4 \times 10^{-5}\,F \text{ in } H^+ \text{ ions}$$

$$pH = 5.00 - 0.92 = 4.08$$

Type 4. pH of a Salt of a Monoprotic Anion

Problem: Compute the pH of a liter of solution containing 0.2 gfw of sodium acetate ($K_a = 1.75 \times 10^{-5}$, and $K_w = 10^{-14}$).

Solution:

$$C_2H_3O_2{}^- + HOH \rightleftharpoons HC_2H_3O_2 + OH^-$$

$$\frac{[HC_2H_3O_2][OH^-]}{[C_2H_3O_2{}^-]} = \frac{K_w}{K_a}$$

$$(\text{Let } x = [HC_2H_3O_2] = [OH^-])$$

Then,

$$\frac{x^2}{0.2\,F} = \frac{10^{-14}}{1.75 \times 10^{-5}}$$

$$x^2 = \frac{2 \times 10^{-15}}{1.75 \times 10^{-5}} = 1.14 \times 10^{-10}$$

$$x = \sqrt{1.14 \times 10^{-10}} = 1.07 \times 10^{-5} = [OH]$$

$$pOH = 5.00 - 0.03 = 4.97$$

$$pH = 14.00 - 4.97 = 9.03$$

Type 5. pH of a Cation Which is a Proton Donor

Problem: Calculate the pH of a $0.4\,F$ solution of ammonium chloride ($K_b = 1.8 \times 10^{-5}$, and $K_w = 10^{-14}$).

Solution:

$$NH_4^{\,\prime} + HOH \rightleftharpoons NH_3 + H_3O^+$$

$$\frac{[NH_3][H_3O^+]}{[NH_4^+]} = \frac{K_w}{K_b}$$

$$(\text{Let } x = [NH_3] = [H_3O^+])$$

Then,

$$\frac{x^2}{0.4\,F} = \frac{10^{-14}}{1.8 \times 10^{-5}}$$

$$x^2 = 2.2 \times 10^{-10}$$

$$x = \sqrt{2.2 \times 10^{-10}} = 1.48 \times 10^{-5} = [H_3O^+]$$

$$pH = 5.00 - 0.17 = 4.83$$

Type 6. pH of a Weak Diprotic Acid

Problem: What is the pH of a $0.1\,F$ solution of H_2S ($K_1 = 5.7 \times 10^{-8}$)?

Solution:

$$H_2S \rightleftharpoons H^+ + HS^-$$

$$\frac{[H^+][HS^-]}{[H_2S]} = 5.7 \times 10^{-8}$$

$$(\text{Let } x = [H^+] = [HS^-])$$

Then,

$$\frac{x^2}{0.1\,F} = 5.7 \times 10^{-8} \quad \text{and} \quad x^2 = 57 \times 10^{-10}$$

$$x = \sqrt{57 \times 10^{-10}} = 7.5 \times 10^{-5}$$

$$pH = 5.00 - 0.87 = 4.13$$

Type 7. pH of an Acid Buffer

Problem: Calculate the pH of a solution which is $0.2\,F\,HNO_2$ and $0.5\,F\,NaNO_2$ ($K_a = 4.6 \times 10^{-4}$).

Solution:

$$[H^+] = \frac{C_{\text{acid}}}{C_{\text{salt}}} \times K_a$$

$$[H^+] = \frac{0.2}{0.5} \times 4.6 \times 10^{-4}$$

$$[H^+] = 1.84 \times 10^{-4}$$

$$pH = 4.00 \times 0.26 = 3.74$$

Type 8. pH of a Base Buffer

Problem: Approximately how many grams of NH_4Cl should be dissolved in a liter of $0.3\,F\,NH_3$ to reduce the hydroxide-ion concentration to one-tenth of its original value ($K_b = 1.8 \times 10^{-5}$)?

Solution:
$$NH_3 + HOH \rightleftharpoons NH_4^+ + OH^-$$

$$0.3\,F \qquad\qquad x \qquad x$$

$$\frac{x^2}{0.3\,F} = 1.8 \times 10^{-5}$$

$$x = \sqrt{5.4 \times 10^{-6}} = 2.33 \times 10^{-3} = [OH^-]\ \text{originally}$$

(one-tenth of $2.33 \times 10^{-3} = 2.33 \times 10^{-4}$)

$$[OH^-] = \frac{C_{base}}{C_{salt}} \times K_b$$

$$2.33 \times 10^{-4} = \frac{0.3}{C_{salt}} \times 1.8 \times 10^{-5}$$

$$C_{salt} = \frac{0.3 \times 1.8 \times 10^{-5}}{2.33 \times 10^{-4}} = 2.3 \times 10^{-2}$$

$$g = 2.3 \times 10^{-2} \times \text{fw of } NH_4Cl$$

$$g = 2.3 \times 10^{-2} \times 53.49 = 1.23\ g$$

EXERCISES

1. Convert the following to pH values:

   (a) $[H^+] = 0.0005$ (g) $[OH^-] = 12$

   (b) $[H^+] = 0.075$ (h) $[OH^-] = 8.74 \times 10^{-6}$

   (c) $[H^+] = 7.42 \times 10^{-5}$ (i) $[OH^-] = 0.000085$

   (d) $[H^+] = 1.50$ (j) $[OH^-] = 10^{-13}$

   (e) $[H^+] = 10$ (k) $[OH^-] = 0.50$

   (f) $[H^+] = 0.20$ (l) $[OH^-] = 7.4$

2. Designate the $[H^+]$ which is equivalent to the following:

   (a) $pH = 0.00$ (e) $pOH = 5.60$

   (b) $pH = 7.45$ (f) $pOH = 0.0004$

   (c) $pH = 13.00$ (g) $pOH = 0.20$

   (d) $pH = -1$ (h) $pOH = 15.00$

3. Calculate the pH of the following solutions:

   (a) $0.002\ F$ HCl   (c) $0.06\ F$ KOH

   (b) $0.4\ F$ $HClO_4$ (d) $0.01\ F$ $Ba(OH)_2$

4. Determine the pH of the following solutions:

   (a) $0.01\ F$ acetic acid   (d) $0.3\ F$ ammonia

   (b) $0.005\ F$ benzoic acid   (e) $0.01\ F$ ethylamine

   (c) $0.04\ F$ formic acid   (f) $0.1\ F$ phosphoric acid (quadratic)

5. Compute the pH of the following solutions:

   (a) $0.01\ F$ $NaNO_2$ (c) $0.7\ F$ NH_4Cl

   (b) $0.01\ F$ KCN   (d) $0.05\ F$ $C_2H_5NH_3Cl$

6. Calculate the pH of the following solutions:

   (a) $0.05\ F$ H_2S (c) $0.01\ F$ K_2CO_3

   (b) $0.1\ F$ H_3AsO_4 (quadratic)   (d) $0.01\ F$ $KHCO_3$

7. Find the pH of the following solutions:

   (a) $1\,g$ $NaC_2H_3O_2$ in 100 ml   (d) $1\,g$ HCl in 100 ml

   (b) $1\,g$ NH_4NO_3 in 100 ml   (e) $1\,g$ ammonium sulfate in 100 ml

   (c) $1\,g$ NaOH in 100 ml

8. Determine the pH of the following solutions:
   (a) $0.11\ F\ NH_3$ and $0.2\ F\ NH_4Cl$
   (b) $0.11\ F\ HC_2H_3O_2$ and $0.2\ F\ NaC_2H_3O_2$
   (c) 0.2 gfw of HCl added to a liter of solution which is $0.3\ F\ NH_3$ and $0.3\ F\ NH_4Cl$
   (d) 0.1 gfw of KOH added to a liter of solution which is $0.4\ F\ HC_2H_3O_2$ and $0.6\ F\ NaC_2H_3O_2$

9. Calculate the pH of the following mixtures:
   (a) 100 ml of $0.02\ F\ Ca(OH)_2$ and 50 ml of $0.02\ F$ HCl
   (b) 100 ml of $0.02\ N\ Ca(OH)_2$ and 100 ml of $0.01\ N$ HCl
   (c) 100 ml of $0.02\ N\ Ca(OH)_2$ and 200 ml of $0.01\ N$ HCl
   (d) 100 ml of $0.1\ N$ propionic acid
   (e) 100 ml of $0.1\ N$ propionic acid and 10 ml of $0.1\ N$ KOH
   (f) 100 ml of $0.1\ N$ propionic acid and 100 ml of $0.1\ N$ KOH
   (g) 100 ml of $0.1\ N$ propionic acid and 200 ml of $0.1\ N$ KOH
   (h) 100 ml of $0.1\ F$ potassium propionate

10. What concentration of nitrous acid has a hydrogen-ion concentration of $0.01\ F$?

11. What weight of lactic acid must be dissolved in a liter of solution to produce a hydrogen-ion concentration of $1 \times 10^{-3}\ F$?

12. How many grams of $NaC_2H_3O_2$ must be added to one liter of $0.1\ F\ HC_2H_3O_2$ to yield a hydrogen-ion concentration of 2×10^{-6} gfw/liter?

13. Approximately how many grams of NH_4Cl should be dissolved in a liter of $0.3\ F\ NH_3$ to reduce the hydroxide-ion concentration to one-hundredth of its original value?

14. If 20 ml of $3\ F\ NH_3$ and 5 g of $(NH_4)_2SO_4$ are added to 200 ml of water, what is the pH of the resulting solution?

15. The pOH of a solution, which is $0.5\ F$ with respect to a weak monoprotic acid, is 9.2. Calculate the ionization constant of the acid.

16. What concentration of nitrous acid is required to produce a solution with a pH of 2.6?

17. How much $0.1\ F$ HCl is required to take 250 ml of solution with a pH of 1.7?

18. How many ml of $0.1\ F\ HC_2H_3O_2$ are required to make 500 ml of solution with a pH of 4.0?

19. How many grams of sodium cyanide must be added to a liter of water to produce a hydroxide-ion concentration equivalent to a $0.05\ F\ NH_3$ solution?

20. Calculate the pH of a $0.01\ F$ solution of ammonium benzoate.

21. A $0.1\ F$ solution of an acid designated as HA has a pH of 3.92. Calculate the K_a of the acid.

22. Formic acid has an ionization constant of 1.77×10^{-4}. What percent of the acid is ionized in a $0.01\ F$ solution?

23. What is the pH of a solution containing 25 mfw of pyridine and 50 mfw of NH_4NO_3?

24. What is the pH of a solution containing 30 mfw of trichloroacetic acid and 20 mfw of KOH per liter?

25. Compute the pH of the following solutions:
   (a) $10\ F$ NaOH
   (b) $0.1\ F\ (NH_4)_2SO_4$
   (c) $2\ F$ HCl
   (d) $0.01\ F$ aniline
   (e) $0.1\ N$ HOCl
   (f) $0.1\ F$ dichloroacetic acid
   (g) 50 ml of $0.1\ F\ Ba(OH)_2$ with 50 ml of $0.6\ N$ HCl
   (h) 50 ml of $0.1\ N$ lactic acid with 50 ml of $0.5\ N$ KOH
   (i) 50 ml of $0.1\ N$ diethylamine with 50 ml of $0.6\ N$ NaOH
   (j) 50 ml of $0.1\ N\ NH_3$ with 50 ml of $0.6\ F\ (NH_4)_2SO_4$
   (k) 50 ml of $0.1\ F$ HCl with 50 ml of $0.6\ F$ NaCl
   (l) 50 ml of $0.1\ F$ NaCl

26. How many meq of NH_3 are necessary to prepare exactly one liter of solution with a pH of 10.0?

CHAPTER 4

Acid–Base Titrations

Neutralization titrimetry involves the quantitative measurement of an acid against a base, either through a volume to volume ratio or by a volume to a weight relationship. It is essential that the titrimetric reaction be rapid and complete; also, the completion of the neutralization should produce a pH change, which can be recognized by means of an indicator.

The actual process of measuring the quantitative reaction between an acid and a base is called a *titration*. Since acid–base equivalents always react in a 1:1 ratio, the titrant is invariably described in terms of normality.

ACID–BASE INDICATORS

4·1 Equivalence Point and End Point

The exact volumetric location at which equivalent quantities of an acid and a base react, is defined as the *equivalence* or the *stoichiometric point*. This is a theoretical position that cannot be determined absolutely, but that may be estimated with some degree of precision. In practice, the estimation is an observation of a physical change occurring during the titration. Actually, the change may be chemical in nature, but the recording of the change is a physical process, involving either an estimation of a color change in a dye, or a series of readings from a scale on a pH meter. The observed change is designated as an end point, and in order to minimize errors in volumetric readings, it is very important that the end point should be as near the equivalence point as possible. Two methods of obtaining end points in acid–base titrations are widely used: namely, a color change in a visual indicator, and the midpoint in a series of pH readings.

4·2 Visual Indicators

A few highly colored organic dyes have the property of changing color when the hydrogen-ion concentration of their solutions is changed through a definite pH range. These particular dyes are known as acid–base indicators. The color change is produced by the gain or the loss of a proton by the indicator; therefore, all such dyes are either weak acids or weak bases.

If an indicator acid is symbolized as HIn, it produces a color change by losing a proton to form the conjugate base In^-, and the equilibrium between the two colored species can be designated as

$$HIn + HOH \rightleftharpoons H_3O^+ + In^-$$
$$\text{(color A)} \qquad \text{(color B)}$$

The expression for the ionization of an acid indicator is the same as that for any weak acid, and may be indicated as

$$\frac{[H_3O^+][In^-]}{[HIn]} = K_a$$

TABLE 4·1   Acid–base indicators

Indicator	Acid color	Base color	pH range
crystal violet	green	blue	0.0–2.0
thymol blue	red	yellow	1.2–2.8
methyl yellow	red	yellow	2.9–4.0
bromphenol blue	yellow	blue	3.0–4.6
methyl orange	red	yellow	3.1–4.4
bromcresol green	yellow	blue	4.0–5.6
methyl red	red	yellow	4.2–6.3
bromcresol purple	yellow	purple	5.2–6.8
bromthymol blue	yellow	blue	6.0–7.6
phenol red	yellow	red	6.8–8.4
thymol blue	yellow	blue	8.0–9.6
phenolphthalein	colorless	red	8.3–10.0
thymolphthalein	colorless	blue	9.4–10.5
alizarin yellow R	yellow	lilac	10.0–12.1
trinitrobenzene	colorless	orange	12.0–14.0

Since the two colors are also in equilibrium, the expression may be expanded to

$$[H_3O^+] = K_a \times \frac{[HIn](\text{color A})}{[In^-](\text{color B})}$$

If the two-color indicator is a weak base, the ionization reaction may be represented as

$$\underset{(\text{color A})}{InOH} \rightleftharpoons \underset{(\text{color B})}{In^+ + OH^-}$$

and the hydroxide-ion concentration as

$$[OH^-] = K_b \times \frac{[InOH](\text{color A})}{[In^-](\text{color B})}$$

Each acid–base indicator has a definite pH range through which a color change takes place. The detection of a color change depends, to some extent, upon the color perception of the observer. The average individual can usually detect 100 parts of one color in the presence of 10 parts of the other color. In terms of pH and pK_i, such a color perception can be designated as

$$pH = pK_i + \log\tfrac{10}{1} = pK_i + 1$$

and

$$pH = pK_i + \log\tfrac{1}{10} = pK_i - 1$$

or

$$pH = pK_i \pm 1$$

The range of color corresponds to a hundredfold change in hydrogen-ion concentration; and, as indicated in Table 4·1, many indicators have a color-transition interval corresponding to approximately 2 pH units (one hundredfold). However, visual perception of color change is better for some indicators than others. If the indicator is phenolphthalein (from colorless to pink), or if it is a mixture of methyl red and bromcresol green (from pink to green), an individual may recognize a color change involving less than one pH unit.

To be of use in an aqueous solution an acid–base indicator should change color within a pH range of 0 to 14. In actual practice, the range may be narrowed to a change of 3 to 10. Table 4·1 lists a set of indicators covering the necessary range for pH determinations.

4·3 The pH Meter as an Acid–Base Indicator

A pH meter is a self-contained, high impedance, potentiometric system, so arranged that the pH of a solution may be measured (by direct reading on a scale) from the difference in potential between two electrodes placed in the solution. One electrode must be pH-sensitive, that is, the potential should be a function of the pH of the solution; the other electrode must have a constant potential, and is designated as a reference electrode.

Because of its greater convenience in handling and its resistance to chemical action, a "glass" electrode is customarily used as the indicator electrode. This pH-sensitive electrode is usually a silver–silver chloride electrode immersed in a dilute solution of hydrochloric acid, and the entire system is encased in a single bulb. The portion of the bulb used for pH measurements in solution is composed of a membrane of special glass. Such a glass electrode is diagrammed in Figure 4·1.

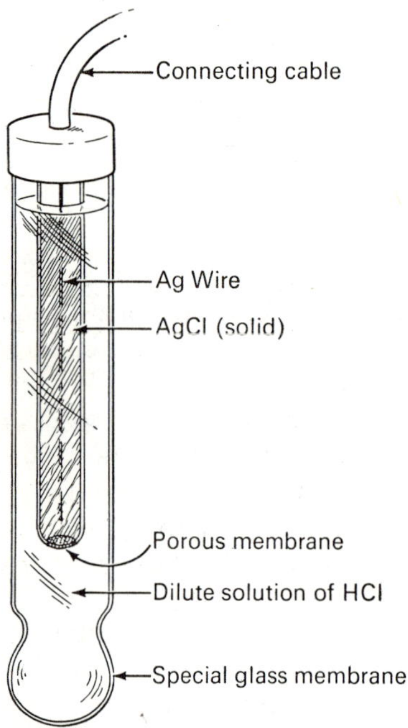

FIGURE 4·1   A glass electrode

A calomel reference electrode is used in conjunction with the glass electrode for pH measurements. Actually, the calomel electrode is connected to a salt bridge, and both components are enclosed in a single unit. A typical calomel electrode is sketched in Figure 4·2.

The emf produced by the combination of the glass electrode and the reference electrode is impressed into the circuit of the pH meter; however, the high resistance of the circuit permits only a small amount of current to be drained from the cell combination. In the circuit, the current is amplified electronically, and then it is passed through an ammeter that is calibrated to

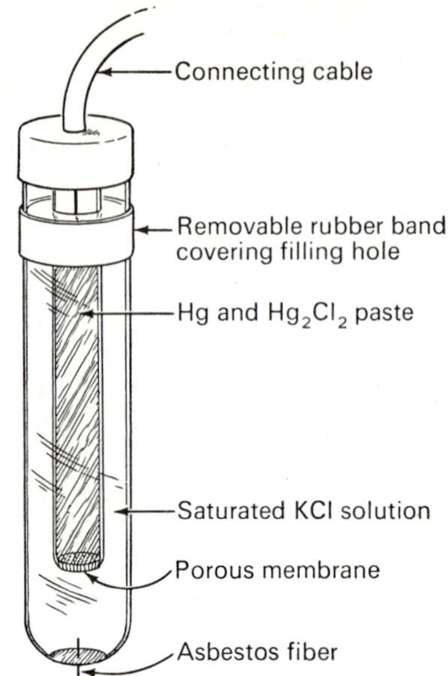

FIGURE 4·2   A calomel reference electrode

register in pH units by means of a movable pointer on a visual scale. (See Section 12·8 for additional explanation concerning use of pH meter.)

The titrating solution is added in small increments, and the pH values are observed and recorded for each addition. When the pH values are plotted against the volume of titrant added, a series of points is obtained, and this can be converted into a solid curve as indicated in Figure 4·3. The shape of the curve is determined by the strengths of the acids and bases involved. Explanations and typical examples of titration curves are given later in this chapter.

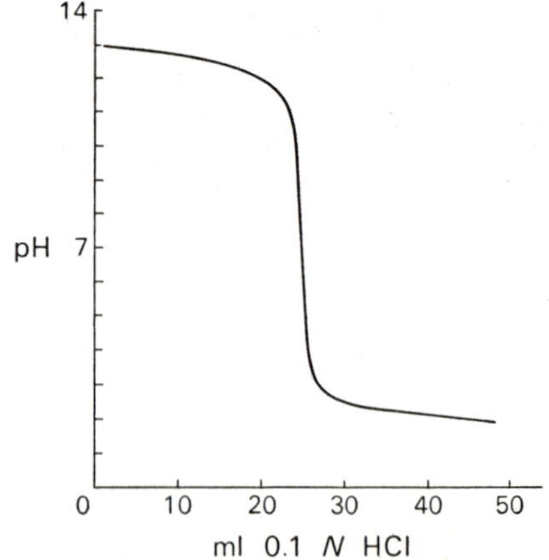

FIGURE 4·3   Potentiometric titration of approximately 25 ml of 0.1 N NaOH with 0.1 N HCL

When properly carried out, a potentiometric titration is as accurate as a titration utilizing a visual indicator. However, the actual titration is more laborious, and subject to a number of possible errors. A description of such errors is not pertinent to this discussion, but may be found in any text concerned with instrumental analysis.

TITRATION CURVES

A graphical representation of the pH change versus the volume of reagent added throughout a titration is termed a *titration curve*. Although many types of titration curves are possible, we shall discuss here only those produced from acid–base reactions. A curve of this type may be obtained experimentally through the use of a pH meter, or it may be calculated by means of a series of pH points, if the initial concentrations of the reaction species are known (as well as their ionization constants). Computed pH values for two titrations are given in Tables 4·2 and 4·3, and the resulting titration curves are plotted in Figures 4·4 and 4·5.

4·4 Titration of a Strong Acid by a Strong Base

Experimental evidence indicates that strong monoprotic acids and bases are almost completely ionized in dilute water solutions. Therefore, the concentrations of the hydrogen ions and the hydroxide ions are approximately the same as the acid and base from which they are derived If no other solute is present, the essential titration reaction for a strong base versus a strong acid is

$$H^+ + OH^- \rightleftharpoons HOH$$

The pH at the equivalence point is 7.0 at room temperature if the water is absolutely pure. The calculations of the pH values for a hypothetical strong-acid titration versus a strong-base titration are relatively simple. If the reacting species are not too dilute, the curve produced from calculated pH values is always $\diagup$ or $\diagdown$ in

shape, with a sharp break at the equivalence point. The salt produced in the reaction is aprotic and, consequently, does not affect the pH of the solution.

A typical curve may be obtained from pH values calculated for the titration of 0.1 N HCl with 0.1 N NaOH. The pH values for 35 ml of acid titrated with selected increments of base are given in Table 4·2.

The calculation of five typical points, in the course of the titration are given as follows:

(1) Before the addition of any NaOH, the solution contains only hydrochloric acid, which is assumed to be completely dissociated into its ions; therefore,

$$[H^+] = 0.1\, F = 1 \times 10^{-1}$$

$$pH = 1.00$$

(2) After the addition of 20 ml of base, the solution contains 1.5 meq of free acid (the equivalent of 15 ml $\times$ 0.1 N of the original acid) plus 2.0 mfw of salt, diluted to a volume of 55 ml. The salt does not appreciably affect the ionization of the excess acid; consequently,

$$[H^+] = 1.5 \text{ meq of HCl}/55 \text{ ml} = 2.73 \times 10^{-2}\, N$$

$$pH = 2.00 - \log 2.73 = 2.00 - 0.43 = 1.57$$

(3) After the addition of 34 ml of base the solution contains only 0.1 meq of free acid (the equivalent of 1 ml $\times$ 0.1 N of the original acid) plus its salt diluted to 69 ml. At this point the slope of the titration curve is much steeper as a consequence of a higher pH value.

$$[H^+] = 0.1 \text{ meq of acid}/69 \text{ ml of solution}$$

$$= 1.45 \times 10^{-3}\, N$$

$$pH = 3.00 - 0.16 = 2.84$$

(4) At the stoichiometric point (after the addition of 35 ml of base) the solution contains only NaCl from the reactants. If the water is pure, the pH is that of the solvent; thus, the pH = 7.00.

(5) When excess base is added (beyond the stoichiometric point), the solution contains NaCl and the excess of hydroxide ions. The salt has no appreciable effect upon the concentration

TABLE 4·2   Titration of 35.00 ml of 0.1000 N HCl with 0.1000 N NaOH

ml of NaOH	Volume of solution	[H$^+$]	pH of solution
0.00	35.00	0.1000	1.00
5.00	40.00	0.0750	1.13
10.00	45.00	0.0556	1.26
15.00	50.00	0.0400	1.40
20.00	55.00	0.0273	1.57
25.00	60.00	0.0167	1.78
30.00	65.00	7.7×10^{-3}	2.11
34.00	69.00	1.45×10^{-3}	2.84
35.00	70.00	1.0×10^{-7}	7.00
36.00	71.00	7.1×10^{-12}	11.15
40.00	75.00	1.5×10^{-12}	11.82
45.00	80.00	8.0×10^{-13}	12.10
50.00	85.00	5.65×10^{-13}	12.25

of hydroxide ions, which depends upon the amount of base added. For example, after 40 ml of NaOH has been added, the resulting solution contains 0.5 meq of free base (the equivalent of 5 ml $\times$ 0.1 N of the original base) plus NaCl, diluted to a volume of 75 ml. The pH is calculated as follows:

$$[OH^-] = 0.5 \text{ meq of base}/75 \text{ ml} = 6.7 \times 10^{-3} \, N$$

$$pOH = 3.00 - 0.82 = 2.18$$

$$pH = 14.00 - 2.18 = 11.82$$

The data for the construction of a hypothetical strong-acid titration versus a strong-base titration are given in Table 4·2, and the pH values obtained are plotted in a solid curve against the volume of titrant as shown in Figure 4·4. Also, shown on this curve are the pH transition ranges of two widely used indicators: namely, phenolphthalein and a mixture of methyl red with bromcresol green. Inasmuch as the steep slope, shown in Figure 4·4, extends across a wide range in pH, it is possible to use either of the two indicators in this particular titration. The difference in using the two indicators (one changing color before the stoichiometric point, and the other after the stoichiometric point has been attained) is approximately 0.1 ml, or about 2 drops of the titrant.

Figure 4·4 also gives a second curve (dotted curve) for the approximate titration of 0.001 N HCl with 0.001 N NaOH. The curve shows the effect of dilution for a strong-acid titration

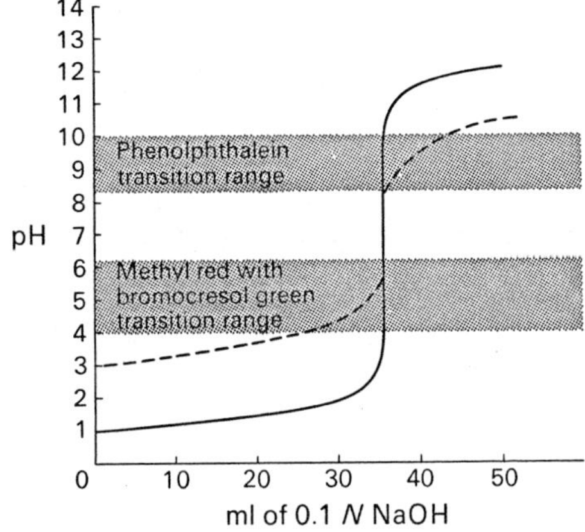

FIGURE 4·4   Titration curve for 35 ml of 0.1 N NaOH. (The dotted curve is for 0.001 N HCl versus 0.001 N NaOH)

versus a strong-base titration. On this particular curve, the pH range at the stoichiometric point is so narrow that neither phenolphthalein nor the mixed indicators are satisfactory for the titration. If we refer to Table 4·1, we see that a suitable indicator (at least theoretically) for a very dilute acid and base titration is bromthymol blue.

4·5 Titration of a Weak Acid by a Strong Base

The construction of a titration curve for a weak acid versus a strong base is more complicated than that (strong base versus a strong acid) given in Section 4·4, because of three factors: the acid is incompletely dissociated and exists as both the molecular and ionic species; during the titration, a salt is produced that buffers the hydrogen-ion concentration of the acid; the equivalence point occurs in a basic solution, and the anion at this point is a proton acceptor.

As explained in Section 3·9, the magnitude of the ionization constant of a weak acid and the dilution of the acid, determine the necessity of the use of quadratics to compute exact pH points. To simplify present calculations, it is assumed that quadratic equations are necessary neither for the approximate data given in Table 4·3 nor for the construction of the titration curve shown in Figure 4·5.

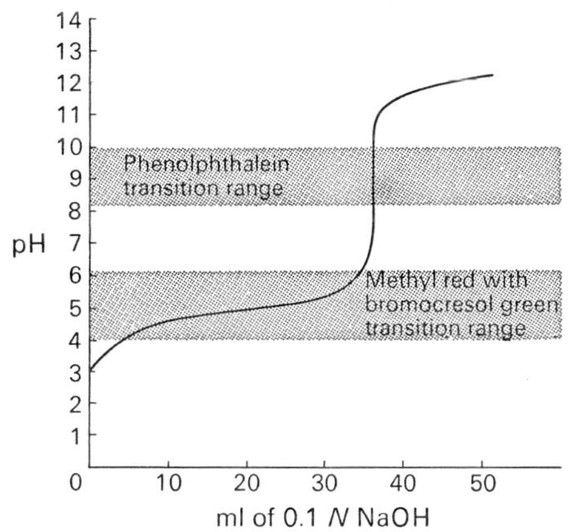

FIGURE 4·5 Titration curve of 35 ml of 0.1 N $HC_2H_3O_2$ titrated with 0.1 N NaOH

Table 4·3 lists selected data obtained in the hypothetical titration of 35 ml of 0.1 N $HC_2H_3O_2$ with 0.1 N NaOH. The calculations for computing five typical pH points are given as follows:

(1) Before the addition of any NaOH, the solution contains only acetic acid and its ions. The equation for obtaining the hydrogen-ion concentration of this solution is

$$[H^+] = \sqrt{\text{conc of acid} \times K_a}$$

$$[H^+] = \sqrt{1.75 \times 10^{-6}}$$

$$[H^+] = 1.32 \times 10^{-3}$$

$$pH = 3.00 - 0.12 = 2.88$$

(2) After the addition of 20 ml of base the solution contains 1.5 meq of free acid (the equivalent of 15 ml × 0.1 N of the original acid) plus 2.0 mfw of salt ($NaC_2H_3O_2$), diluted to a volume of 55 ml. The hydrogen-ion concentration for the buffer mixture is

$$[H^+] = \frac{C_{acid}}{C_{salt}} \times K_a$$

$$[H^+] = \frac{1.5 \text{ meq of acid/55 ml}}{2.0 \text{ mfw of salt/55 ml}} \times 1.75 \times 10^{-5}$$

$$[H^+] = 1.32 \times 10^{-5}$$

$$pH = 5.00 - 0.12 = 4.88$$

(3) The addition of 34 ml of base produces a solution containing 0.1 meq of free acid ($HC_2H_3O_2$), plus 3.4 mfw of salt ($NaC_2H_3O_2$), in a volume of 69 ml. This too is a buffer solution, with a hydrogen-ion concentration of

$$[H^+] = \frac{0.1 \text{ meq of acid/69 ml}}{3.4 \text{ mfw of salt/69 ml}} \times 1.75 \times 10^{-5}$$

$$[H^+] = 5.15 \times 10^{-7}$$

$$pH = 7.00 - 0.71 = 6.29$$

(4) At the stoichiometric point the solution contains 3.5 mfw of salt ($NaC_2H_3O_2$) in 70 ml of solution. Since the anion ($C_2H_3O_2{}^-$) is a Brønsted base, it accepts protons as follows:

$$C_2H_3O_2{}^- + HOH \rightleftharpoons HC_2H_3O_2 + OH^-$$

and the hydroxide-ion concentration may be designated as

$$[OH^-] = \sqrt{C_{salt} \times \frac{K_w}{K_a}}$$

$$[OH^-] = \sqrt{\frac{3.5}{70} \times \frac{10^{-14}}{1.75 \times 10^{-5}}} = 5.3 \times 10^{-6} N$$

$$pOH = 6.00 - 0.72 = 5.28$$

$$pH = 14.00 - 5.28 = 8.72$$

(5) The addition of any strong base beyond the stoichiometric point results in a concentration of excess OH^- ions in the presence of $NaC_2H_3O_2$. However, the presence of the salt

TABLE 4·3   Titration of 35.00 ml of 0.1000 N HC$_2$H$_3$O$_2$ with 0.1000 N NaOH

ml of NaOH	Volume of solution	$[H^+]$	pH of solution
0.00	35.00	1.32×10^{-3}	2.88
5.00	40.00	1.05×10^{-4}	3.98
10.00	45.00	4.77×10^{-5}	4.36
15.00	50.00	2.33×10^{-5}	4.63
20.00	55.00	1.32×10^{-5}	4.88
25.00	60.00	7.0×10^{-6}	5.15
30.00	65.00	2.92×10^{-6}	5.53
34.00	69.00	$5.15 \times 10^{\,7}$	6.29
35.00	70.00	1.88×10^{-9}	8.72
36.00	71.00	$7.1 \times 10^{\,12}$	11.15
40.00	75.00	$1.5 \times 10^{\,12}$	11.82
45.00	80.00	$8.0 \times 10^{\,13}$	12.10
50.00	85.00	$5.65 \times 10^{\,13}$	12.25

has no appreciable effect on the hydroxide-ion concentration. If 10 ml of 0.1 N NaOH are added beyond the stoichiometric point, the total volume is 80 ml, and the hydroxide-ion concentration is

$$[OH^-] = 1 \text{ meq of NaOH}/80 \text{ ml}$$

$$= 1.25 \times 10^{-2}\, N$$

$$pOH = 2.00 - 0.10 = 1.90$$

$$pH = 14.00 - 1.90 = 12.10$$

The titration curve for acetic acid versus NaOH, diagrammed in Figure 4·5, shows that of the two indicators, phenolphthalein and the mixture of methyl red and bromcresol green, only the first is satisfactory for detecting the end point of the titration. In the acid region (below a pH of 7) the slope is gradual and no indicator that changes in a pH range between 3 and 7 can produce a sharp end point. On the other hand, the slope of the curve in the strongly buffered region (between a pH of 7.5 and 9) is actually steeper than that obtained in the titration of a strong acid versus a strong base. Consequently, in the transition range of phenolphthalein, there is a sharp end point –even sharper than that for HCl with NaOH but the range is narrow, and cannot be detected by indicators that change color outside of this pH range.

The slope of a titration curve for a weak acid with a strong base is dependent upon the ionization constant of the acid undergoing titration, and the concentration of the solution. The sharpness of the pH break at the equivalence point decreases as the ionization constant of the acid decreases. Figure 4·6 shows approximate titration curves of weak acids versus 0.1 N NaOH. These curves indicate that acids with a K_a of 10^{-5} can be titrated accurately, but acids weaker than $K_a = 10^{-6}$ do not give sharp end points with visual indicators.

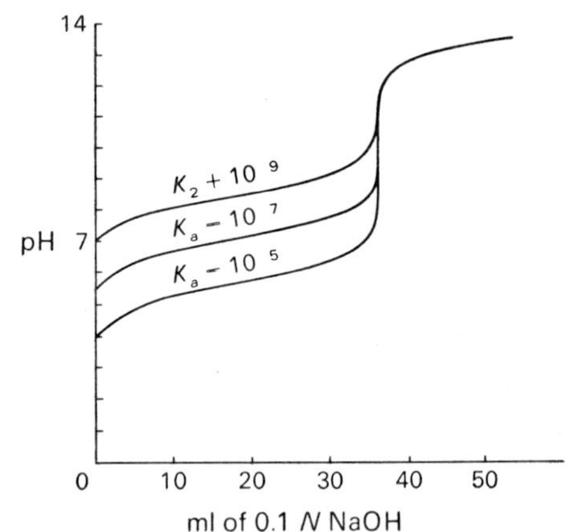

FIGURE 4·6   Titration curves for weak acids with 0.1 N NaOH

4·7 Titration of a Weak Base by a Strong Acid

The titration curve for a weak base versus a strong acid (given approximately in Figure 4·7) is a mirror image of the curve for a weak acid titrated with a strong base (Figure 4·5).

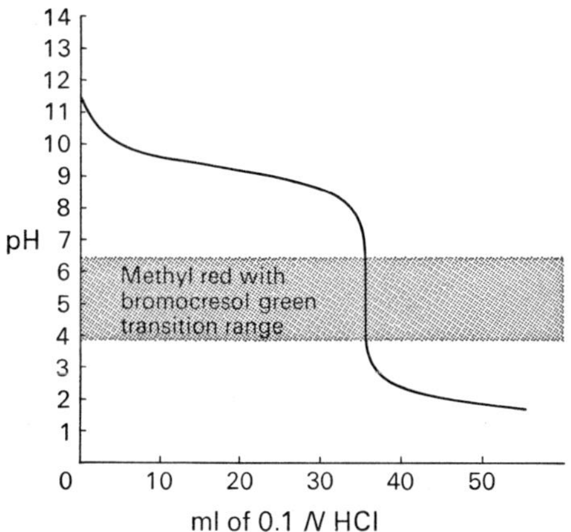

FIGURE 4·7   Titration curve for 0.1 *N* NH₃ with 0.1 *N* HCl

There is likewise an inversion of the equivalence point, which occurs in an acidic pH range. As an example, a 0.1 *N* solution of NH_3 ($K_b = 1.8 \times 10^{-5}$) can be titrated satisfactorily with 0.1 *N* HCl (Figure 4·7) if a mixture of methyl red and bromcresol green is used as the indicator. However, a weak base with an ionization constant smaller than 10^{-6} does not give a sharp end point with a visual indicator if it is titrated with 0.1 *N* HCl. The titration curves for very weak bases (Figure 4·8) versus 0.1 *N* HCl are inverted images of those given for the titration of weak acids in Figure 4·6.

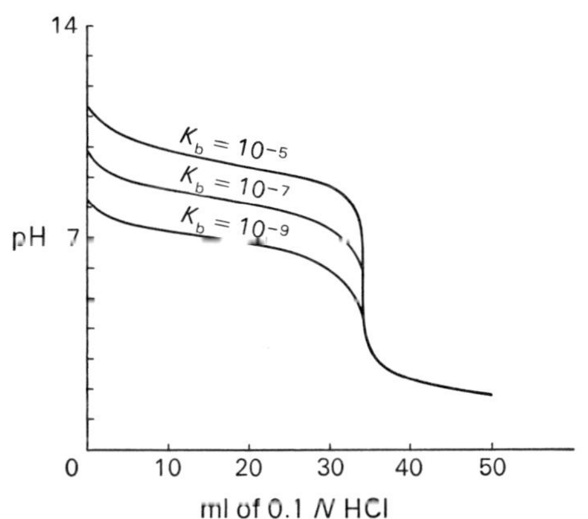

FIGURE 4·8   Titration curves for weak bases with 0.1 *N* HCl

As the K_b of a weak base decreases, the pH break in the titration curve with 0.1 *N* HCl becomes smaller, and for bases with ionization constants smaller than 10^{-10}, the pH break is almost imperceptible. This statement applies also to titration curves for weak acids that have ionization constants less than 10^{-10}, when they are titrated with 0.1 *N* NaOH solution.

4·7 Titration of a Weak Acid by a Weak Base

The titration of a weak acid with a weak base, such as 0.1 *N* HAc versus 0.1 *N* NH₃, produces a series of pH values, which as a solid curve shows no abrupt change at, or near, the stoichiometric point. Consequently, any visual indicator is useless for such a titration. The approximate curve for a titration of this nature is diagrammed in Figure 4·9.

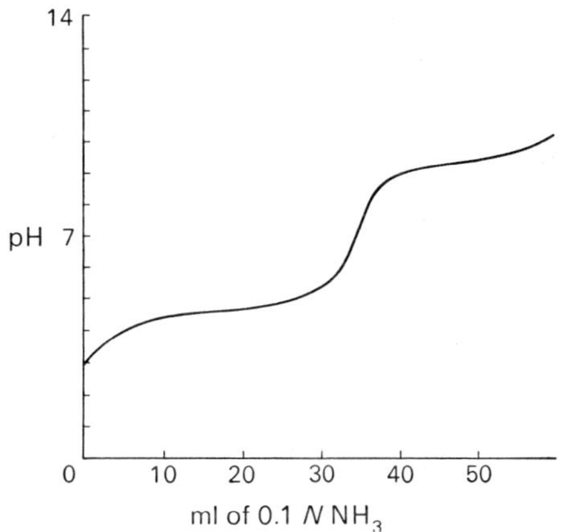

FIGURE 4·9   Titration curve for 0.1 *N* HC₂H₃O₂ with 0.1 *N* NH₃

The first portion of the curve resembles the beginning of the curve for the titration of a weak acid versus a strong base; whereas, the second part is an inverted image of the completion of the titration of a strong acid versus a weak base. Between the two portions there is no pH break sufficient for the use of a visual indicator, see Figure 4·9. At the equivalence point the solution contains a salt (such as $NH_4C_2H_3O_2$) whose anion is a proton-acceptor, and whose cation is a proton-donor. As discussed in Section 3·14, the hydrogen-ion concentration of the solution,

$$[H^+] = \sqrt{\frac{K_w \times K_b}{K_a}}$$

is independent of the concentration of the salt (within normal limits). Although the concentration of the salt and the magnitudes of K_a and K_b do not change the general form of the curve, they do govern the position of the equivalence point in reference to its location on the axes of the plot.

4·8 The Titration of Salts

A complete discussion of all possible types of titrations and titration curves is not possible in an elementary course in quantitative chemistry. On the other hand, the titration of a solution of sodium carbonate with a strong acid is of wide interest and of practical importance.

If a solution of $0.1\ F\ Na_2CO_3$ is titrated with $0.1\ N$ HCl, two successive reactions occur:

$$CO_3^{--} + H^+ \rightleftharpoons HCO_3^-$$

$$HCO_3^- + H^+ \rightleftharpoons H_2CO_3$$

The second reaction does not occur in the presence of CO_3^{--}; consequently, when sufficient acid is added to satisfy the first reaction, a break in the titration curve occurs. Continued addition of acid completes the second reaction, and produces a second end point, as is indicated in Figure 4·10. The pH breaks for the two end points are not easily detected by visual indicators, but they may be calculated, or experimentally determined by means of a potentiometric titration. The first end point occurs in a basic solution, and is the titrimetric midpoint between the beginning of the titration and the second end point. The second end point occurs in an acid solution, and the inflection on the titration curve is somewhat sharper than that caused by the first end point. The two pH inflections constitute experimental evidence that the CO_3^{--} ion is a stronger Brønsted base (see Table 2·1) than is the HCO_3^- ion.

The first end point may be determined approximately with phenolphthalein as an indicator, and the second end point can be observed with a mixture of methyl red and bromcresol green. However, in determining the total basicity of Na_2CO_3, it is more practical to add excess HCl; then, heat the solution to drive off

CO_2, and back titrate with standard $0.1000\ N$ HCl (using the mixed indicator for detecting the end point). The back titration is indicated by a dotted line in Figure 4·10.

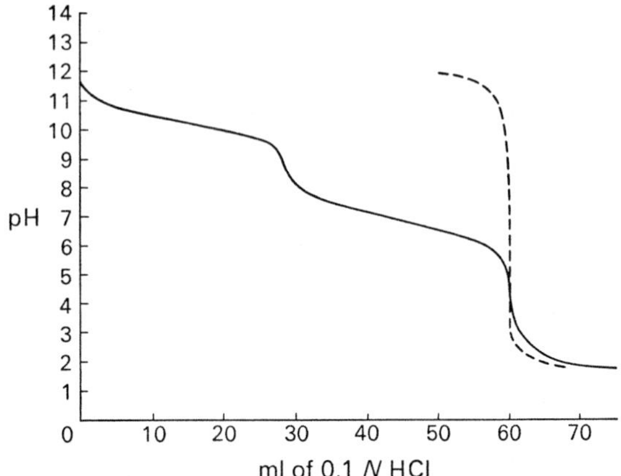

FIGURE 4·10   Titration curve for 30 ml of 0.1 F Na_2CO_3 titrated with 0.1 N HCl. (The dotted curve shows back titration with 0.1 N NaOH after CO_2 has been expelled)

4·9 Feasibility of a Titration

The majority of acid–base indicators change color within a range of about 2 pH units (see Section 4·2). Therefore, a neutralization titration (using a visual indicator) must exhibit an inflection of about 2 pH units near the equivalence point. Also true is the fact that the precision of buret (50 ml) readings is limited to approximately 1–2 parts per thousand. In practical terms, a visual titration must produce a color change within 1–2 drops of the stoichiometric point.

Figure 4·6 shows that the titration of a weak acid with a K_a of 10^{-5} (with a solution of 0.1 N NaOH) is feasible; the same series of curves indicate that a weak acid with a K_a of 10^{-7}, or less, is not feasible.

Figure 4.5 suggests that the titration of any strong acid versus a strong base is feasible, except in very dilute solutions.

In summation, a few generalities concerning the feasibility of titrations are apparent:

(1) When the visual-indicator curve shows an abrupt break of 2 pH units (near the equivalence point) for 1 2 drops of titrant, the titration is

feasible. On the other hand, if the pH inflection is less than 2 pH units, or if more than 2 drops of titrant are required, the feasibility of the titration is doubtful.

(2) A weak acid (0.1 N) with a K_a of less than 10^{-6} cannot be accurately titrated with 0.1 N NaOH. Similarly, a 0.1 N weak base with a K_b of less than 10^{-6} cannot be accurately titrated with a 0.1 N strong acid.

(3) If the dilution of the titrant and solution to be titrated are below 10^{-3} N, the titration is not practical.

TYPES OF EXERCISES

Type 1. Color of a Visual Indicator

Problem: A solution has a hydrogen-ion concentration of 2×10^{-7}. What color is imparted to the solution by 2 drops of methyl red?

Solution:

$$pH = 7.00 - \log \text{ of } 2 = 7.00 - 0.30 = 6.70$$

Methyl red changes from red to yellow in a pH range of 4.2 to 6.3; therefore, the color of the solution is yellow.

Type 2. Suitability of a Visual Indicator

Problem: Is litmus solution a suitable indicator for the titration of 0.1 N $HC_2H_3O_2$ versus 0.1 N NaOH?

Solution: In general, a suitable indicator for a particular titration changes approximately 2 pH units upon the addition of 1 or 2 drops of titrant. The color range of litmus (see any chemical handbook) from red to blue is approximately 4 pH units; consequently, it is not a suitable indicator.

Type 3. Color of an Indicator within a Definite pH Range

Problem: Indicate the color of each of the following indicators (insert numbers in brackets from the colors listed) in a solution with a pH of 3.00.

[ ]	Methyl red	1. Colorless
[ ]	Methyl orange	2. Red
[ ]	Phenolphthalein	3. Blue
[ ]	Phenol red	4. Yellow
[ ]	Bromthymol blue	5. Green
[ ]	Thymol blue	6. Orange

Solution: Correct answers may be obtained from Table 4·1.

Type 4. Selection of a Suitable Indicator for a Definite Titration

Problem: Select the proper indicator for each of the following titrations:
(a) 25 ml of 0.1 N HCl with 0.1 N NaOH
(b) 25 ml of 0.001 N HCl with 0.001 N NaOH
(c) 25 ml of 0.1 N NH_3 with 0.1 N HCl
(d) 25 ml of 0.1 N $HC_2H_3O_2$ with 0.1 N NaOH

Solution:
(a) Any indicator with a pH transition range between a pH of 4 and a pH of 10 (see Figure 4·4)
(b) Bromthymol blue (see dotted curve in Figure 4·4)

(c) Methyl red (see Figure 4·7)
(d) Phenolphthalein (see Figure 4·5)

Type 5.   Calculation of Equivalence Point for the Titration of a Weak Acid by a Strong Base

Problem: What is the pH at the equivalence point if 25 ml of 0.1 N HNO_2 ($K_a = 4.6 \times 10^{-4}$) is titrated with 0.1 N KOH?

Solution: At the equivalence point 50 ml of solution contain 2.5 mfw of KNO_2. Since the NO_2^- ion is a proton acceptor, the following equilibrium exists at the stoichiometric point:

$$NO_2^- + HOH \rightleftharpoons HNO_2 + OH^-$$

Therefore, the hydroxide-ion concentration is

$$[OH^-]^2 = C_{salt} \times \frac{K_w}{K_a}$$

$$[OH^-]^2 = \frac{2.5}{50} \times \frac{10^{-14}}{4.6 \times 10^{-4}}$$

$$[OH^-] = \sqrt{1.09 \times 10^{-12}} = 1.04 \times 10^{-6}$$

$$pOH = 6.00 - 0.02 = 5.98$$

$$pH = 14.00 - 5.98 = 8.02$$

Type 6.   Calculation of the Equivalence Point for the Titration of a Weak Base by a Strong Acid

Problem: What is the pH at the equivalence point when 25 ml of 0.1 N pyridine ($K_b = 1.4 \times 10^{-9}$) is titrated with 0.1 N HCl?

Solution: At the equivalence point, 50 ml of solution contains 2.5 mfw of pyridinium chloride ($C_6H_5NH_3Cl$). The cation of the salt acts as a proton donor, as follows:

$$C_6H_5NH_3Cl + HOH \rightleftharpoons H_3O^+ + C_6H_5NH_2 + Cl^-$$

Therefore, the hydrogen-ion concentration at the equivalence point is

$$[H^+]^2 = C_{salt} \times \frac{K_w}{K_b}$$

$$[H^+]^2 = \frac{2.5}{50} \times \frac{10^{-14}}{1.4 \times 10^{-9}} = 36 \times 10^{-8}$$

$$[H^+] = \sqrt{36 \times 10^{-8}} = 6 \times 10^{-4}$$

$$pH = 4.00 - 0.78 = 3.22$$

Type 7. Affect on pH by Dilution

Problem: One liter of 1.000 N solution of lactic acid ($K_a = 1.38 \times 10^{-4}$) is diluted 1000-fold. What is the change in pH? (Assume no change in activity)

Solution: Before dilution, the hydrogen-ion concentration is

$$[H^+]^2 = C_{acid} \times K_a$$

$$[H^+] = \sqrt{1.38 \times 10^{-4}} = 1.17 \times 10^{-2}$$

$$pH = 2.00 - 0.70 = 1.30$$

After dilution, the hydrogen-ion concentration is

$$[H^+]^2 = 1\,mfw/1000\,ml \times 1.38 \times 10^{-4}$$

$$[H^+] = \sqrt{13.8 \times 10^{-8}} = 3.7 \times 10^{-4}$$

$$pH = 4.00 - 0.57 = 3.43$$

EXERCISES

1. From data in Table 4·1, state what indicator may be used for the following titrations:
   (a) 0.1 N lactic acid with 0.1 N NaOH ($K_a = 1.4 \times 10^{-4}$)
   (b) 0.1 N aniline with 0.1 N HCl ($K_b = 4 \times 10^{-10}$)
   (c) 0.01 N H_2SO_4 with 0.1 N ethylamine ($K_b = 6 \times 10^{-4}$)
   (d) 0.1 N trichloroacetic acid with 0.1 N methyl amine ($K_a = 1.3 \times 10^{-1}$ and $K_b = 4.4 \times 10^{-4}$)

2. An acid–base indicator, InOH, gives a yellow color in a 0.1 N NaOH solution, and a blue color in 0.1 N HCl. If the K_b for the monobasic indicator is 10^{-5}, predict the color of the indicator in solutions having the following pH values: (a) 4.00; (b) 5.00; (c) 6.00; (d) 7.00; (e) 8.00; (f) 9.00; (g) 10.00; (h) 11.00.

3. If 25 ml of 0.500 N $HC_2H_3O_2$ is diluted to 100 ml and titrated with 0.500 N NaOH, calculate the pH of the solution when one-half of the acid has been neutralized.

4. How many gfw/liter of $KC_2H_3O_2$ must be added to a solution to produce a hydroxide concentration of 1×10^{-4}?

5. Calculate the pH of the equivalence for each of the following titrations:
   (a) 35 ml of 0.1000 N trichloroacetic acid ($K_a = 1.3 \times 10^{-1}$) with 0.1000 N KOH solution;
   (b) 25 ml of 0.1000 N pyridine ($K_b = 1.4 \times 10^{-9}$) with 0.1000 N HCl

6. An aqueous solution of 0.500 N NH_3 is partially neutralized with 0.500 N HCl solution. Compute the pH value of the solution when 60 percent of the NH_3 has been neutralized.

7. A 0.500 N $HC_2H_3O_2$ solution is titrated with 0.500 N NH_3 until the titration mixture has a pH of 3.5. What percent of the acid has been neutralized?

8. Calculate the volume, in ml, to which 10 ml of 4.00 F NH_3 must be diluted to produce a solution with a pH of 12.00.

9. A 35 ml sample of 0.2000 N NaOH is titrated with 0.2000 N HCl. Calculate the pH of the solution after the addition of the following amounts of acid: (a) 10.00 ml; (b) 20.00 ml; (c) 30.00 ml; (d) 40.00 ml.

10. Calculate the pH of a solution prepared by adding 100 ml of H_2O to 100 ml of a solution, which is 0.2000 N acetic acid and 0.1000 F sodium acetate.

11. If a 150 ml solution containing NH_3 is titrated with 0.1000 N HCl, and 50 ml of the acid is required to reach the equivalence point, what is the pH of the original NH_3 solution?

12. Given 100 ml of 0.2000 F Na_2CO_3 solution, (a) calculate the pH after 50 ml of 0.1000 N HCl has been added; (b) compute the pH after 90 ml of 0.1000 N HCl has been added.

13. What volume of 2.000 F $Ba(OH)_2$ must be added to a solution containing 12.00 g of ammonium chloride to yield a pH of 10.00?

14. Calculate the hydrogen-ion concentration of a solution consisting of 30.00 ml of 0.1000 N acetic acid, 15.00 ml of 0.2000 F sodium acetate, and sufficient water to give a total volume of 50.00 ml.

15. If 150 ml of 0.2000 N solution of a weak acid has a pH of 3.5, calculate the pH at the equivalence point when the solution is titrated with 0.1000 N NaOH.

16. If 35.00 ml of 0.1000 N NaOH is required to titrate 0.4000 g of a monoprotic organic acid, what is the equivalent weight of the acid?

17. A sodium hydroxide solution is standardized with potassium biiodate. If a 1.095 g sample of the acid salt is equivalent to 26.50 ml of the base, what is the normality of the NaOH solution?

18. Calculate the pH, 0.1 ml before and 0.1 ml after the equivalence point, for solutions involved in the following titrations:
    (a) 30.00 ml of 0.1000 N $HClO_4$ with 0.1000 N NH_3
    (b) 30.00 ml of 0.1000 N $HC_2H_3O_2$ with 0.1000 N NaOH
    (c) 30.00 ml of 0.1000 N $HC_2H_3O_2$ with 0.1000 N HN_3
    (d) 30.00 ml of 0.1000 N NaOH with 0.1000 F boric acid

19. A weak acid has an ionization constant of 1×10^{-6}. What will be the pH at the equivalence point if 25 ml of a 0.1000 N solution of the acid is titrated with 0.1000 N NaOH?

20. Calculate the pH at the equivalence point in titrating 40.00 ml of 0.1000 F benzoic acid with 0.1000 F KOH.

21. If 50.00 ml of a 0.1000 F solution of a weak acid has a pH of 4.00, calculate the pH at the equivalence point when the solution is titrated with 0.1000 F NaOH.

22. A sample of soda ash weighing 0.4000 g was titrated with excess 0.1000 F hydrochloric acid totaling 36.50 ml of acid. A volume of 1.75 ml of NaOH was used in arriving at the mixed indicator end point. The same base required 34.00 ml to titrate a 0.700 g sample of potassium acid phthalate, using a phenolphthalein end point. The ratio of the acid to base at a mixed indicator end point was 1.008. Calculate the percentage of sodium carbonate in the soda ash.

CHAPTER 5

Introduction to Oxidation–Reduction Relations and Calculations

Analytical applications which involve oxidation–reduction titrations are more widely used than all other titrimetric methods combined. Because a complete discussion of oxidation–reduction systems would be so extensive that it alone would fill one whole volume, it is obviously beyond the scope of an elementary course in quantitative chemistry. Consequently, the topic will be confined to a limited survey of redox relationships in selected reactions of inorganic materials in aqueous solutions.

DEFINITIONS

5·1 Oxidation and Reduction

The term *oxidation* originally referred to a process by which substances acquired oxygen in chemical reactions; and *reduction*, to a chemical action in which oxygen was removed from its compounds. In a restricted sense these meanings still hold true, but the two terms now include many reactions in which oxygen plays no part.

A broader concept of oxidation and reduction describes these changes in terms of *electron gain* and *electron loss*. Accordingly, oxidation may be defined as a loss of one or more electrons by an atom, ion, or molecule; and reduction as the reverse process, in which electrons are added to a substance. More broadly, a chemical reaction resulting in a change in the electric charges on the reacting particles may be designated as a *reduction–oxidation reaction*, or *redox* for short. The substance that furnishes the electrons is oxidized, and the material that receives electrons is reduced. Consequently, those particles which donate electrons easily are called *reducing agents*, and those which accept electrons readily are termed *oxidizing agents*. The two processes, oxidation and reduction, cannot take place separately but must occur simultaneously.

5·2 Half-reactions

Reactions that involve the donation or acceptance of electrons are called *electrode reactions* or *electron reactions*. The interactions can take place at electrodes, which supply or remove electrons, or they can result from direct contact of atoms, ions, or molecules, if the contact produces an exchange of electrons.

The majority of oxidation–reduction processes, which occur in aqueous solution, are ionic in nature. The net over-all reaction can frequently be written as the sum of two half-reactions. Consequently, the equation for one half-reaction indicates the release of electrons by a reacting substance, whereas the other half-reaction shows the capture of electrons by the conjugate reactant. For example, the interaction between stannous ions and ferric ions may be indicated (by equations) as the sum of two half-reactions.

$$2Fe^{3+} + 2e \rightleftharpoons 2Fe^{++}$$
$$Sn^{++} - 2e \rightleftharpoons Sn^{4+}$$
$$\overline{2Fe^{3+} + Sn^{++} \rightleftharpoons 2Fe^{++} + Sn^{4+}}$$

The use of half-reactions is convenient for balancing oxidation–reduction equations. However, as shown in Appendix 5, it is conventional to write the equation for a half-cell reaction with the electron, or electrons, on the left-hand side of the equation. The various electrochemical conventions for indicating redox processes will be discussed more completely in Section 6·5.

5·3 Oxidation State or Oxidation Number

In an electrovalent compound, composed of simple ions, the oxidation state of a component is the same as the charge on the particular ion. Thus, when sodium reacts with chlorine to produce sodium chloride, the oxidation state (or oxidation number) of the sodium ion is $+1$, and that for the chloride ion is -1. In some respects the oxidation number is similar to valence, but whereas valence is a specific term defining the combining power of an element, the meaning of oxidation number may be more obscure. For example, the oxidation number of an element in a covalent compound, or in a complex ion, may be defined as the charge that an atom of the element would bear if it were an ion.

Oxidation numbers are usually assigned according to certain conventions. Atoms in an elementary state have an oxidation number of zero. Hydrogen has an oxidation number of $+1$ except in elemental hydrogen and in hydrides. Oxygen is assigned an oxidation number of -2 except in elemental oxygen, peroxides, and the fluorine oxides. The algebraic sum of the oxidation states of all atoms in a compound must equal zero. The oxidation number of an element in a compound of uncertain structure may be calculated by assigning oxidation numbers to the other elements in the compound.

Oxidation numbers may be evaluated by writing balanced hypothetical equations for partial reactions involving the atoms in question. Equations of this type are illustrated as follows:

$$S^0 + 2e \rightleftharpoons S^{--}$$

The oxidation number of the sulfur atom in the sulfide ion is -2, since two electrons are gained in producing the ion from elementary sulfur. On the other hand, if electrons are lost in a partial reaction involving sulfur, such as

$$S^0 + 3H_2O \rightleftharpoons SO_3^{--} + 6H^+ + 4e$$

the oxidation number of the sulfur atom is positive. In the sulfite ion the oxidation number of sulfur is $+4$, since four electrons are released in the formation of the ion. The partial reaction by which the permanganate is hypothetically produced from elementary manganese indicates that seven electrons are lost in the process,

$$Mn^0 + 4H_2O \rightleftharpoons MnO_4^- + 8H^+ + 7e$$

Consequently, the oxidation number of the manganese atom in the permanganate ion is $+7$.

BALANCING REDOX EQUATIONS

5·4 Methods for Balancing Redox Equations for Reactions in Acid Solutions

Simple redox equations, such as those between ions that do not contain oxygen, often may be balanced by inspection, and require no explanation. However, redox reactions involving ions that contain oxygen are more complex, and balancing them requires a series of steps in which an accounting is kept of the electrons lost and electrons gained, as well as of the rearrangements of atoms and ions from the reacting substances that form the reaction products. The two most widely used methods used for balancing ionic redox equations are designated as the *oxidation number-change method*, and the *ion–electron method*. Since these methods are usually covered thoroughly in general chemistry, we shall review them only in an outline form.

But first it should be noted that, in general, oxygen-bearing ions (either as oxidizing or reducing agents) in an acid solution produce either hydrogen ions or water molecules according to the following relationships:

The oxidizing agent loses one or more oxygen particles to form water molecules:

$$O^{--} + 2H^+ \rightleftharpoons H_2O$$

The reducing agent usually gains one or more oxygen particles to form H^+ ions:

$$H_2O \rightleftharpoons O^{--} + 2H^+$$

Oxidation number-change method:

$$SO_3^{--} + MnO_4^- \rightleftharpoons SO_4^{--} + Mn^{++}$$
(skeletal equation)

$$\overset{+7}{} \text{ (Mn gains 5 electrons) } \overset{+2}{}$$

$$SO_3^{--} + MnO_4^- \rightleftharpoons SO_4^{--} + Mn^{++}$$

$$\overset{+4}{} \text{ (S loses 2 electrons) } \overset{+6}{}$$

Thus,

$$5SO_3^{--} + 2MnO_4^- \rightleftharpoons 5SO_4^{--} + 2Mn^{--}$$
(electrically unbalanced)

$$(-10) \quad + (-2) \quad \neq \quad (\ 10) \quad + (+4)$$
$$(-12) \quad \neq \quad (-6)$$

The electrical unbalance is corrected by adding positive charges in the form of H^+ ions on the left, which are converted to water molecules to the left. The balanced equation becomes

$$5SO_3^{--} + 2MnO_4^- + 6H^+ \rightleftharpoons$$
$$5SO_4^{--} + 2Mn^{++} + 3H_2O$$

Ion–electron method:

$$SO_3^{--} + MnO_4^- \rightleftharpoons SO_4^{--} + Mn^{++}$$
(skeletal equation)

Separate the skeletal reaction into two half-reactions: the oxidizing half-reaction, and the reducing half-reaction; then balance total atoms, by adding protons and water molecules, as follows:

oxidizing half-reaction
$$MnO_4^- + 8H^+ \rightleftharpoons Mn^{++} + 4H_2O$$

reducing half-reaction
$$SO_3^{--} + H_2O \rightleftharpoons SO_4^{--} + 2H^+$$

The two half-reactions are unbalanced electrically. Add electrons to produce proper balance.

$$5e + MnO_4^- + 8H^+ \rightleftharpoons Mn^{++} + 4H_2O$$
$$SO_3^{--} + H_2O \rightleftharpoons SO_4^{--} + 2H^+ + 2e$$

Multiply the two half-reactions by such numbers as will make the total electron change the same for both half-reactions, as follows:

oxidizing half-reaction
$$2(5e + MnO_4^- + 8H^+ \rightleftharpoons Mn^{++} + 4H_2O)$$

reducing half-reaction
$$5(SO_3^{--} + H_2O \rightleftharpoons SO_4^{--} + 2H^+ + 2e)$$

Add the two half-reactions to give the following:

$$10e + 2MnO_4^- + 16H^+ +$$
$$5SO_3^{--} + 5H_2O \rightleftharpoons 2Mn^{++} + 8H_2O +$$
$$5SO_4^{--} + 10H^+ + 10e$$

Cancel any common electrons, ions, or molecules that appear on both sides of the equation.

$$2MnO_4^- + 5SO_3^{--} + 6H^+ \rightleftharpoons$$
$$5SO_4^{--} + 2Mn^{++} + 3H_2O$$

5·5 Balancing Redox Equations for Reactions in Basic Solutions

As a general rule, redox reactions taking place in basic solutions are not sufficiently exact for quantitative determinations. The few that are of titrimetric importance are those occurring between low molecular weight, organic molecules, and alkaline oxidants, for example, the reaction between CH_3OH and MnO_4^- in a basic solution to produce CO_2 and MnO_2. The balanced equation for the reaction is

$$2MnO_4^- + CH_3OH \rightleftharpoons$$
$$CO_2 + 2MnO_2 + H_2O + 2OH^-$$

The equation may be balanced by the ion–electron method, but the oxidation number change method is not satisfactory for this purpose.

Redox reactions in basic solutions are frequently made use of in qualitative analysis. In the case of oxygen-bearing ions reacting in alkaline media (either as oxidizing or reducing agents) it is necessary to add either the hydroxide ion or water molecules for balancing purposes. Two general rules usually prevail:

The oxidizing agent loses one or more oxygen particles to form hydroxide ions:

$$O^{--} + H_2O \rightleftharpoons 2OH^-$$

The reducing agent gains one or more oxygen particles to form water molecules:

$$2OH^- \rightleftharpoons H_2O + O^{--}$$

Consider the following skeletal equation:

$$CrO_2^- + ClO^- \rightleftharpoons Cl^- + CrO_4^{--}$$

oxidizing half-reaction
$$3(2e + ClO^- + H_2O \rightleftharpoons Cl^- + 2OH^-)$$

reducing half-reaction
$$2(CrO_2^- + 4OH^- \rightleftharpoons CrO_4^{--} + 2H_2O + 3e)$$

Add the two half-reactions, and cancel any common electrons, ions, or molecules that appear on both sides of the total equation for the reaction. Thus, the balanced equation becomes

$$2CrO_2^- + 3ClO^- + 2OH^- \rightleftharpoons$$
$$2CrO_4^{--} + 3Cl^- + H_2O$$

5·6 Redox Reactions in Neutral Solutions

Certain redox reactions of titrimetric precision are best suited to neutral media. In general, iodimetric processes belong to this category. One particular example, widely used, is the titration of thiosulfate ion with iodine.

$$2S_2O_3^{--} + I_2 \rightleftharpoons S_4O_6^{--} + 2I^-$$

In a neutral solution the reaction is quantitative; on the other hand, the reactants and products are highly sensitive to any pH change in the solution. The foregoing equation represents the key reaction to most iodometric processes (Section 5·11).

STOICHIOMETRY

The calculations of redox processes have two similarities to acid–base reactions:

(1) All redox reactions involve loss and gain of negatively charged particles in the form of electrons. In acid–base reactions, according to the Brønsted concept, there is an interchange of positively charged particles usually designated as protons.

(2) Each of the two types of reactions (acid–base and redox) may theoretically be subdivided into half-reactions, although no single half-reaction may take place without an accompanying conjugate half-reaction.

5·7 Equivalent Weights of Redox Reagents

The units of concentration for redox reagents are the same as those used for acids and bases. As was explained in Section 2·3, such units may be expressed, without ambiguity, in terms of formality or normality. For redox reagents, concentration is most conveniently expressed in terms of normality.

The equivalent weight of an oxidizing or reducing agent is the formula weight of the substance divided by the number of electrons lost or gained by the material in its half-reaction. Similar to the stoichiometry of acidimetry and alkalimetry, the amount of any oxidizing agent which loses one electron is always equivalent to that of the amount of reducing agent which gains one electron. The relationships may be expressed as

$$eq_1 = eq_2$$

or

$$meq_1 = meq_2$$

Consequently,

$$eq \, wt \, (N) = \frac{fw}{electron \, change}$$

As examples, the equivalents for some common oxidizing and reducing agents, as determined from their half-reactions, are indicated as follows:

half-reaction   eq wt of reactant
$$5e + MnO_4^- + 8H^+ \rightleftharpoons$$
$$Mn^{++} + 4H_2O \qquad\qquad fw(salt)/5$$

$$Fe^{++} \rightleftharpoons Fe^{3+} + 1e \qquad \text{fw(salt)/1}$$

$$6e + Cr_2O_7^{--} + 14H^+ \rightleftharpoons$$
$$2Cr^{3+} + 7H_2O \qquad \text{fw(salt)/6}$$

$$2S_2O_3^{--} \rightleftharpoons S_4O_6^{--} + 2e \qquad \text{fw(salt)/1}$$

The equivalent weight of an oxidizing agent or a reducing agent may vary with the conditions under which they react. A specific case is potassium permanganate, which can exhibit three different oxidation changes depending upon the pH of its half-reactions. The conditions for the three different equivalencies are these:

(1) In strongly acid solution, the half-reaction and equivalent weight are

$$5e + MnO_4^- + 8H^+ \rightleftharpoons Mn^{++} + 4H_2O$$
$$eq = fw/5$$

(2) In weakly acid, neutral, or weakly alkaline solution, the half-reaction and equivalent weight are

$$3e + MnO_4^- + 2H_2O \rightleftharpoons MnO_2\downarrow + 4OH^-$$
$$eq = fw/3$$

(3) Alkaline permanganate oxidations are usually carried out in the presence of excess barium ions, which precipitate the reduction product in the form of insoluble barium manganate. The half-reaction and the reduction product are

$$1e + MnO_4^- + Ba^{++} \rightleftharpoons BaMnO_4\downarrow$$
$$eq = fw/1$$

Stoichiometric and equivalent relationships, in redox reactions, can be expressed by any of the following methods:

$$eq = \frac{wt\ in\ g}{eq\ wt} \quad or \quad meq = \frac{g}{meq\ wt}$$

and

$$normality = \frac{eq}{liter} \quad or \quad normality = \frac{meq}{ml}$$

TYPES OF INDICATORS

At least four general methods for detecting end points in redox reactions have been devised. Examples of these procedures are briefly discussed.

5·8 Specific Indicators

(1) A few highly colored oxidizing substances may, under certain conditions, be self-indicating. The best example is that of a potassium permanganate solution (0.1 to 1 N), which is so intensely colored in the MnO_4^- oxidation state, that a slight excess (usually one drop) is a visual indicator of the titrimetric end point.

(2) A dilute dispersion of "soluble" starch forms a deep blue color with the triiodide ion (I_3^-). The extreme sensitivity of the colored complex makes possible a variety of reactions in which iodine is either consumed (iodimetry) or produced (iodometry).

(3) If no internal indicator is available, the use of an external indicator is sometimes possible. The procedure involves the removal of a drop of the titration solution near the end point, and testing the drop with a few drops of the external indicator. In the titration of $Cr_2O_7^{--}$ solution versus Fe^{++} solution, the ferricyanide ion was formerly used as an external indicator to test for a bare excess of ferrous ions. This particular titration is of historical interest only. Since 1924 satisfactory internal redox indicators have been available.

The use of an external indicator is a tedious process, and error results from removing drops of the titration solution.

5·9 Oxidation–Reduction Indicators

In Section 4·2, it was stated that certain highly colored acid–base indicators were weak acids or weak bases, which exhibited color changes through specific pH ranges. In like manner, redox indicators are colored organic substances which change color when oxidized or reduced. Just as certain acid–base indicator changes are subject to certain pH ranges, redox color changes are sensitive to specific redox potentials. A more complete discussion of redox indicators is given in Section 6·12.

TYPICAL REDOX PROCESSES

Selected methods for permanganate and iodine processes are given in Chapter 13; however, the discussions pertain only to the experimental uses of the reagents. Thus it seems pertinent to present in this chapter some of the more common oxidizing agents and their advantages, disadvantages, and general uses.

5·10 Permanganate Oxidations

The behavior of potassium permanganate as an oxidizing agent is unusual in many respects.

Advantages. Potassium permanganate has been used as an oxidizing agent for more than a century, and it is still the most widely used of the various titrimetric oxidizing agents. It is readily available and inexpensive, and it serves as its own indicator in concentrations ordinarily used. It has an oxidation potential of about 1.5 volts in acid solutions, and will quantitatively oxidize most reducing reagents. Inasmuch as the permanganate ion has several oxidation states (see Section 5·7), it has considerable versatility, provided the conditions of the titrimetric reactions are carefully controlled.

Disadvantages. Permanganate solutions are not stable for a long period of time. Special precautionary procedures are used in the preparation of solutions that are reasonably free of reducing agents. Since light seems to facilitate permanganate decomposition, the solutions must be stored in the dark. The tendency of permanganate ions to oxidize chloride ions requires additional precautionary measures when HCl is used as a solvent. The high oxidation potential and reactivity of the permanganate ion may produce a multiplicity of reaction products, which can cause the stoichiometry of a titration to be uncertain.

Uses. The versatility of permanganate titrations is both an advantage and also a disadvantage. A partial list of some of the most widely used permanganate analyses is given in Tables 5·1, 5·2, and 5·3.

5·11 Iodine Titrations

Iodimetric processes are extensively used for a small number of determinations.

Advantages. Iodine, in the form of the tri-iodide ion, is a relatively weak oxidizing agent with an oxidation potential of approximately 0.54 volt. The redox half-reaction is

$$I_3^- + 2e \rightleftarrows 3I^-$$

The low oxidation potential permits the use of iodine as a selective oxidizing agent, since only easily oxidizable substances react with the oxidant. As a consequence, it is sometimes possible to carry out specific redox analyses that would be impossible with stronger oxidizing agents. However, the low oxidation potential limits the number of practical titrations with iodine as the oxidizing agent. The direct use of iodine redox determinations is usually designated as an iodimetric method. On the other hand, iodine may be used indirectly, in a different type of procedure, called an iodometric process, which is extremely accurate and widely used. This method makes use of other oxidizing agents to liberate free iodine, which is back-titrated (in a buffered solution) with standard sodium thiosulfate.

Occasionally the color of iodine can be used as its own indicator in an iodimetric titration; on the other hand, the "soluble starch" indicator (β-amylose) can be used in both iodimetric and iodometric methods. The end point in such titrations is due to the adsorption of the I_3^- ion on the surface of the starch molecule, thereby producing an intense blue color. Under proper conditions, it is considered to be the best visual indicator in titrimetric analysis.

Although iodine is fairly expensive, the reagent-grade purity of the chemical may be used as a primary standard. Standard solutions may be prepared by dissolving a weighed sample in

TABLE 5·1   **Determinations by permanganate titration in acid solution**

$$5e + MnO_4^- + 8H^+ \rightleftarrows Mn^{++} + 4H_2O$$

$$Fe^{++} \rightleftarrows Fe^{3+} + 1e$$

$$Sn^{++} \rightleftarrows Sn^{4+} + 2e$$

$$Ti^{3+} + H_2O \rightleftarrows TiO^{++} + 2H^+ + 1e$$

$$Mo^{3+} + 4H_2O \rightleftarrows MoO_4^{--} + 8H^+ + 3e$$

$$H_2O_2 \rightleftarrows O_2 + 2H^+ + 2e$$

$$H_3SbO_3 + H_2O \rightleftarrows H_3SbO_4 + 2H^+ + 2e$$

$$H_3AsO_3 + H_2O \rightleftarrows H_3AsO_4 + 2H^+ + 2e$$

$$H_2C_2O_4 \rightleftarrows CO_2 + 2H^+ + 2e$$

$$U^{4+} + 2H_2O \rightleftarrows UO_2^{++} + 4H^+ + 2e$$

$$Fe(CN)_6^{4-} \rightleftarrows Fe(CN)_6^{3-} + 1e$$

$$HNO_2 + H_2O \rightleftarrows NO_3^- + 3H^+ + 2e$$

$$H_2S \rightleftarrows S^0 + 2H^+ + 2e$$

TABLE 5·2   **Determination by permanganate titration in neutral solution**

$$3e + MnO_4^- + 4H^+ \rightleftarrows MnO_2 + 2H_2O$$

The half-reaction takes place in a buffered solution.

$$Mn^{++} + 2H_2O \rightleftarrows MnO_2 + 4H^+ + 2e \quad \text{(buffered solution)}$$

TABLE 5·3   **Determinations by permanganate titration in alkaline solutions in the presence of barium ions**

$$MnO_4 \rightleftarrows MnO_4 + 1e$$

and

$$Ba^{++} + MnO_4^- \rightleftarrows BaMnO_4 \downarrow$$

$$I^- + 8OH \rightleftarrows IO_4^- + 4H_2O + 8e$$

$$CN^- + 2OH \rightleftarrows CNO^- + H_2O + 2e$$

$$SO_3^{--} + 2OH \rightleftarrows SO_4^{--} + 5H_2O + 8e$$

water containing an excess of potassium iodide. Even so, the element is somewhat volatile through sublimation either as a solid or in a KI solution. Consequently, it is customary practice to prepare approximate solutions and standardize them against a primary-standard grade of arsenious oxide.

Disadvantages. The most serious possibility or error in iodide chemistry is the oxidation of the iodine ion by air. The equation for the reaction may be indicated as

$$4I^- + 4H^+ + O_2 \rightleftarrows 2I_2 + 2H_2O$$

Although the reaction is normally slow in a neutral solution, the speed of the reaction is greatly increased in an acid solution. Direct light also accelerates the oxidation reaction, and in some instances the presence of impurities may catalyze the decomposition.

TABLE 5·4   Applications of iodimetry

$$H_2SO_3 + I_2 + H_2O \rightleftharpoons SO_4^{--} + 2I^- + 4H^+$$

$$H_2S + I_2 \rightleftharpoons 2H^+ + S^0 + 2I^-$$

$$H_2AsO_3^- + I_2 + H_2O \rightleftharpoons H_2AsO_4^- + 2I^- + 2H^+$$

$$HSbOC_4H_4O_6 + I_2 + H_2O \rightleftharpoons HSbO_2C_4H_4O_6 + 2I^- + 2H^+$$

(The above reactions go to completion only in the presence of excess $NaHCO_3$, which removes hydrogen ions formed in the initial reaction.)

$$Sn^{++} + I_2 \rightleftharpoons Sn^{4+} + 2I^-$$

$$2S_2O_3^{--} + I_2 \rightleftharpoons S_4O_6^{--} + 2I^-$$

$$2Fe(CN)_6^{4-} + I_2 \rightleftharpoons 2Fe(CN)_6^{3-} + 2I^-$$

TABLE 5·5   Applications of iodometry

(Each of the following reaction products are neutralized, or buffered, and then back-titrated as follows:

$$(2S_2O_3^{--} + I_2 \rightleftharpoons S_4O_6^{--} + 2I^-)$$

$$IO_3^- + 5I^- + 6H^+ \rightleftharpoons 3I_2 + 3H_2O$$

$$ClO_3^- + 6I^- + 6H^+ \rightleftharpoons Cl^- + 3I_2 + 3H_2O$$

$$Cr_2O_7^{--} + 6I^- + 14H^+ \rightleftharpoons 2Cr^{3+} + 3I_2 + 7H_2O$$

$$MnO_4^- + 10I^- + 16H^+ \rightleftharpoons 2Mn^{++} + 5I_2 + 8H_2O$$

$$2Cu^{++} + 4I^- \rightleftharpoons 2CuI + I_2$$

$$OCl^- + 2I^- + 2H^+ \rightleftharpoons I_2 + Cl^- + H_2O$$

$$BrO_3^- + 6I^- + 6H^+ \rightleftharpoons Br^- + 3I_2 + 3H_2O$$

$$2HNO_2 + 2I^- + 2H^+ \rightleftharpoons 2NO + I_2 + 2H_2O$$

$$H_2O_2 + 2I^- + 2H^+ \rightleftharpoons I_2 + 2H_2O$$

Not only do acids tend to decompose iodide solutions, as indicated in the foregoing reaction, but also bases will produce a reversal of the reaction, and even change the stoichiometry of an iodine titration. As a result of the sensitivity of iodine solution to such factors as pH, light, and oxygen, as well as its tendency to volatilize from solution, it is necessary to keep such a solution stored in the dark in a glass stoppered bottle. The factors of instability necessitate the restandardization of an iodine solution frequently. As a safe rule, it is unwise to rely upon a standardization for a period of more than two or three weeks.

Uses. The use of iodimetric processes is limited to very few titrimetric determinations. The most common analyses are listed in Table 5·4.

The use of iodometric processes has many applications. They all depend upon the libera-tion of iodine by an active oxidizing agent; the reaction solution is then buffered, and back-titrated with standard thiosulfate solution.

5·12 Other Oxidizing Titrants

Oxidizing reagents, other than potassium permanganate and iodine, of use in elementary quantitative chemistry are usually limited to compounds containing the dichromate and the ceric ions.

Potassium Dichromate. The half-cell reaction of the chromic–dichromate system is

$$Cr_2O_7^{--} + 14H^+ + 6e \rightleftharpoons 2Cr^{3+} + 7H_2O$$

Consequently, the equivalent weight of the di-chromate ion is one-sixth of its formula weight.

The principal quantitative use of the dichro-mate ion is as a titrant of the ferrous ion involving a reaction represented by the following equation:

$$6Fe^{++} + 2Cr_2O_7^{--} + 14H^+ \rightleftharpoons$$
$$6Fe^{3+} + 2Cr^{3+} + 7H_2O$$

The reaction was first exploited, quantitatively, in 1850, and was widely used with an external indicator (Section 5·8) for nearly 75 years. The awkwardness of the use of such an indicator was eliminated in 1924 with the discovery that di-phenylamine, and some of its derivatives, could be used as an internal redox indicator (Section 5·8). Sodium diphenylamine sulfonate is the most widely used in this family of indicators.

Although potassium dichromate is not as strong an oxidizing agent as potassium permanganate, it has a number of advantages over the latter:

(1) It is available in a primary-standard purity, at a reasonable cost, so that standard solutions may be prepared by direct weighing.

(2) Potassium dichromate solutions are stable for an indefinite period, and can be boiled without decomposition.

(3) Oxidation states for chromium between $+6$ and $+3$ do not exist; consequently, side reactions are never present to produce a stoichiometric error.

(4) It is unaffected by the presence of the chloride in dilute solutions.

Many of the substances that can be oxidized with permanganate may also be titrated with the dichromate; yet, frequently the results of the latter are not as good. It might be concluded that in spite of the wide exploitation of $K_2Cr_2O_7$ for a period of many years, the advantages of its purity and stability warrant additional investigation.

Ceric Sulfate. A solution of Ce^{4+} (in sulfuric acid) is nearly as efficient as a permanganate solution for oxidizing purposes. The ceric ion has a number of advantages over the permanganate ion:

(1) Ceric solutions are indefinitely stable, and can be boiled without decomposition.

(2) Because there is only one oxidation change (from $+4$ to $+3$), side reactions in its titrimetry are unknown.

(3) It can be used as an oxidation agent in the presence of dilute solutions of chloride ions without error.

Because of its strong oxidizing properties, cerate ion titrations can usually be substituted for permanganate determinations. However, it is usually necessary to use an internal redox indicator for satisfactory accuracy. The redox indicator of greatest acceptance is prepared by adding orthophenanthroline to a ferrous salt. If this indicator is used, the color change is from intense red to faint blue near the redox equivalence point.

Although ceric sulfate, as an oxidizing titrant, has been hailed as the ideal reagent to replace potassium permanganate for a period of several decades, such a replacement has not taken place. Some of the reasons for the reluctance to substitute cerate titrations for permanganate determinations are these:

(1) It is approximately fifty times as expensive as potassium permanganate.

(2) Some ceric titrations require catalysts, whereas the same reactions proceed smoothly, without catalysts, with permanganate as the oxidizing agent.

(3) The color of the ceric ion rarely serves, accurately, as its own indicator.

TYPES OF EXERCISES

Type 1. Oxidation Numbers

Problem: What is the oxidation number of the central atom in the following ions and compounds: (a) ClO^-; (b) BrO_2^-; (c) IO_4^-; (d) MnO_4^-; (e) $K_2Cr_2O_7$.

Solution: (a) $+1$; (b) $+3$; (c) $+7$; (d) $+7$; (e) $+6$.

Type 2. Balancing Redox Equations for Reactions in Acid Solution

Problem: The following unbalanced skeletal equations represent redox reactions which may take place in acid solutions. Convert them to balanced ionic equations, introducing hydrogen ions and water molecules wherever necessary.

(a) $Fe^{++} + ClO_3^- \rightleftharpoons Fe^{3+} + Cl^-$

(b) $Cr_2O_7^{--} + NO_2^- \rightleftharpoons Cr^{3+} + NO_3^-$

(c) $Cr^{3+} + S_2O_8^{--} \rightleftharpoons Cr_2O_7^{--} + SO_4^{--}$

(d) $MnO_4^- + H_2O_2 \rightleftharpoons Mn^{++} + O_2$

(e) $Zn + NO_3^- \rightleftharpoons Zn^{++} + N_2$

Solution:

(a) $6Fe^{++} + ClO_3^- + 6H^+ \rightleftharpoons 6Fe^{3+} + Cl^- + 3H_2O$

(b) $Cr_2O_7^{--} + 3NO_2^- + 8H^+ \rightleftharpoons 2Cr^{3+} + 3NO_3^- + 4H_2O$

(c) $2Cr^{3+} + 3S_2O_8^{--} + 7H_2O \rightleftharpoons Cr_2O_7^{--} + 6SO_4^{--} + 14H^+$

(d) $2MnO_4^- + 5H_2O_2 + 6H^+ \rightleftharpoons 2Mn^{++} + 5O_2 + 8H_2O$

(e) $5Zn + 2NO_3^- + 12H^+ \rightleftharpoons 5Zn^{++} + N_2 + 6H_2O$

Type 3. Equivalent Weights in Redox Reactions

Problem: Each of the following compounds represents the formula weight of a reactant in a redox process, in which the pertinent product is enclosed in a parenthesis. Select the fraction of the formula weight (numbered on the right), which represents the equivalent weight, and insert the correct answer in the bracket to the left of the reactant.

[4] As_2O_3 (H_3AsO_4)	1. one
[4] KIO_3 (ICl_3^{--})	2. one-half
[2] Na_2SeO_4 (SeO_3^{--})	3. one-third
[2] $U(SO_4)_2$ (UO_2^{++})	4. one-fourth
[1] $VOSO_4$ (VO_3^-)	5. one-fifth
[6] Mo_2O_3 (H_2MoO_4)	6. one-sixth
[4] P_2O_3 (H_3PO_4)	7. one-seventh
[4] $KHC_2O_4 \cdot H_2C_2O_4 \cdot 2H_2O$ (CO_2)	8. one-eighth
[6] $K_2Cr_2O_7$ (Cr^{3+})	9. one-ninth
[1] $Na_2S_2O_3$ $(S_4O_6^{--})$	10. one-tenth

Solution: See numbers in brackets.

Type 4. Stoichiometric Relationships

Problem: How much of the substance listed in column *B* will react with the given amount of pure material indicated in column *A*?

A	*B*
50 ml of 0.1000 *F* $Na_2S_2O_3$	__50__ ml 0.1000 *N* Iodine solution
50 ml of 0.1000 *F* $KMnO_4$	__1.4__ g of Fe
50 ml of 0.1000 *N* I_2	__0.24__ g of As_2O_3
0.2000 g of $H_2C_2O_4 \cdot 2H_2O$	__31.7__ ml 0.1000 *N* $KMnO_4$
0.2000 g of Cu	__31.4__ ml of 0.1000 *N* $Na_2S_2O_3$

Solution: Answers are underlined.

EXERCISES

1. What is the oxidation number of each of the elements (other than H and O) in the following ions, salts, and molecules?

(a) HNO_3	(f) N_2O	(k) $U_2O_7^{--}$
(b) $K_2Cr_2O_7$	(g) $B_4O_7^{--}$	(l) $K_2S_4O_6$
(c) H_3AsO_3	(h) NH_3	(m) $H_4P_4O_7$
(d) $H_2C_2O_4$	(i) As_2S_3	(n) H_3PO_3
(e) N_2O_5	(j) AsS_3^{3-}	(o) $Ce(IO_3)_4$

2. The following unbalanced equations represent redox reactions taking place in acid solutions. Convert them to balanced equations, introducing hydrogen ions and water molecules wherever necessary.

   (a) $PbO_2 + Pb + SO_4^{--} \rightleftharpoons PbSO_4$

   (b) $H_2AsO_4^- + Zn \rightleftharpoons AsH_3 + Zn^{++}$

   (c) $NO_3^- + Zn \rightleftharpoons Zn^{++} + NH_4^+$

   (d) $Fe_3P + NO_3^- \rightleftharpoons Fe^{3+} + H_2PO_4^- + NO$

   (e) $IO_3^- + N_2H_4 \rightleftharpoons N_2 + I^-$

3. The following unbalanced equations represent redox reactions taking place in basic solutions. Convert into balanced ionic equations, introducing hydroxide ions and water molecules wherever necessary.

   (a) $CrO_2^- + Na_2O_2 \rightleftharpoons CrO_4^{--} + Na^+$

   (b) $NO_2^- + Al \rightleftharpoons NH_3 + Al(OH)_4^-$

   (c) $Cu^{++} + CN^- \rightleftharpoons Cu(CN)_3^{--} + CNO^-$

   (d) $Cl_2 \rightleftharpoons ClO_3^- + Cl^-$

   (e) $CrO_4^{--} + HSnO_2^- \rightleftharpoons HSnO_3^- + CrO_2^-$

4. Each of the following compounds represents the formula weight of a reactant in a redox process, in which the product is indicated in parenthesis. Select the fraction of the formula weight (numbered on the right), which represents the equivalent weight, and insert the correct answer in the bracket to the left of the reactant.

[ ] MnO_2 (Mn^{++})	1. one
[ ] $Na_2S_2O_8$ (SO_4^{--})	2. one-half
[ ] BrO_3^- (Br^-)	3. one-third
[ ] $Ca(OCl)_2$ (Cl_2)	4. one-fourth
[ ] $Ca(IO_3)_2$ (I^-)	5. one-fifth
[ ] KIO_4 (I^-)	6. one-sixth
[ ] V_2O_5 (V_2O_4)	7. one-seventh
[ ] K_2MnO_4 (Mn^{++})	8. one-eighth
[ ] PbO_2 $(PbSO_4)$	9. one-ninth
	10. one-tenth
	11. one-eleventh
	12. one-twelfth

5. How much of the substance listed in column B will react with the given amount of material in column A?

   (a) 20 ml of 0.1000 F $KMnO_4$ ______________ ml of 0.1000 F H_2S

   (b) 20 ml of 0.1000 F $KMnO_4$ ______________ g of $Na_2C_2O_4$

   (c) 0.2000 g of Fe_2O_3 ______________ ml of 0.1000 F $K_2Cr_2O_7$

   (d) 40.00 ml of 0.1000 N I_2 ______________ g of As_2O_3

   (e) 0.1000 g CaC_2O_4 ______________ ml of 0.0400 F $KMnO_4$

   (f) 0.3500 g $KBrO_3$ ______________ ml of 0.1000 N $Na_2S_2O_3$

6. Potassium permanganate titrations may give a variety of products with varying pH concentrations. Indicate the number of electrons in a balanced half-reaction, involving the permanganate ion, under the following conditions:

   (a) In a strongly acid solution ______________

   (b) In a neutral solution ______________

   (c) In a weakly acid solution ______________

   (d) In a basic solution in presence of Ba^{++} ions ______________

   (e) In a strongly basic solution ______________

7. What is the normality of the following solutions under the specified conditions?

   (a) 0.0500 F $KMnO_4$ in a weakly acid solution   ________

   (b) 0.0500 F $KMnO_4$ when used as a titrant in the presence of Ba^{++} ions   ________

   (c) 0.1000 F $Na_2S_2O_3$ as a titrant of I_3^- in a neutral solution   ________

   (d) 0.1000 F $K_2Cr_2O_7$ as a titrant of Fe^{++} ions in an acid solution   ________

   (e) 0.0500 F KIO_3 as a source of iodine in a redox reaction, in an acid solution   ________

8. Write balanced equations for permanganate oxidation (in acid solution) of the following substances: (a) Sb_2S_3; (b) Mo^{3+}; (c) VO^{++}; (d) NO_2^-; (e) W^{3+}.

9. Write balanced equations for permanganate oxidation (in basic solution) of the following substances: (a) $HCOO^-$; (b) Sn^{++}; (c) HPO_3^{--}; (d) HS^-; (e) CH_3OH.

10. Write balanced equations for iodine oxidation of the following substances: (a) N_2H_2; (b) CNS^-; (c) HN_3; (d) SO_2; (e) OH^-.

11. Write balanced equations for iodide reduction (in acid solution) of the following substances: (a) MnO_2; (b) Cl_2; (c) ClO_2^-; (d) HIO; (e) SO_4^{--}.

12. How much of the substance in column B will react with the given amount of material in

A	B
0.7500 g of As_2O_3	________ ml of 0.1000 F $Ce(SO_4)_2$
0.5000 g of $SnCl_2$	________ ml of 0.1000 N $K_2Cr_2O_7$
100 ml of 0.1000 F H_2SO_3	________ ml of 0.1000 F $KMnO_4$
0.3500 g of $KHC_2O_4 \cdot H_2C_2O_4$	________ ml of 0.1000 N $KMnO_4$
50 ml of 0.0200 F $K_2Cr_2O_7$	________ g of Fe

CHAPTER 6

Oxidation–Reduction Equilibria

FUNDAMENTAL PRINCIPLES

The driving force of an oxidation–reduction reaction is usually measured in terms of voltage (emf). It is also common practice to write equations for redox reactions so that they proceed from left to right, but this is not a necessary convention. Frequently, redox equations are encountered in which the actual reaction proceeds to the left. The direction of a spontaneous redox reaction is discussed in Section 6·5.

6·1 Galvanic Cells

The tendency of a substance to gain electrons may be designated as its reduction potential. When the potential is determined under specified conventional conditions, the resulting value is termed the *standard electrode potential*. Inasmuch as the ionic half-reaction involved in a standard electrode potential indicates an ion, or atom, gaining one or more electrons, such a potential is frequently called the *standard reduction potential*. In Appendix 5 the standard electrode potentials of a large number of half-reactions are tabulated.

The electromotive force of a redox reaction may be evaluated, if the overall reaction can be set up in the form of a galvanic cell. Such a cell consists of electrodes dipping into separate solutions of electrolytes, in which there is electrical connection between the electrodes, and

electrolytic contact between the solutions. Under ideal conditions the electrodes and solutions are so selected that a spontaneous redox reaction takes place with an exchange of electrically charged particles.

A galvanic cell may be arranged by placing a solution of an oxidizing agent in one vessel, and a reducing solution in another vessel. A conducting electrode is immersed in each solution, and electrolytic contact between the two solutions is established by means of a salt bridge. The salt bridge is an inverted U tube containing a solution of an electrolyte, fitted with porous plugs at each end of the tube. A flow of electrons is produced when external contact between the two electrodes is completed by means of a conducting wire. A simple galvanic cell, as diagrammed in Figure 6·1, may be constructed by using a zinc electrode bathed by zinc ions in one vessel, and a copper electrode immersed in a solution of cupric ions in another container. In this particular cell, electrons move along the wire from the zinc electrode to the copper electrode, since zinc loses electrons more easily than does copper. The two half-reactions are

$$Zn \rightleftharpoons Zn^{++} + 2e \quad \text{and} \quad Cu^{++} + 2e \rightleftharpoons Cu$$

and the overall reaction for the cell is expressed in the equation,

$$Zn + Cu^{++} \rightleftharpoons Zn^{++} + Cu$$

If a sensitive voltmeter is inserted in the external circuit, the voltage of the cell, when the reactants

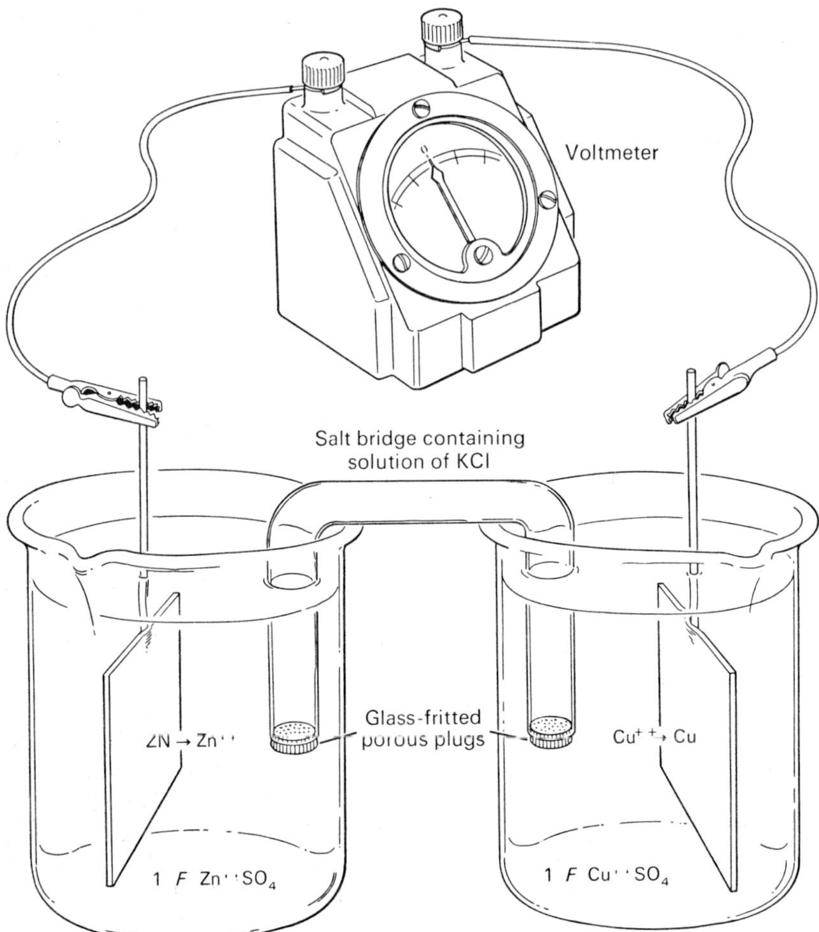

FIGURE 6·1   Diagrammatic representation of a galvanic cell

are at standard state (see Section 6·4 for an explanation of standard state), is found to be 1.10 volts.

The cell cannot operate without a transfer of ions. As the zinc electrode goes into solution, the concentration of Zn^{++} becomes greater, whereas the deposition of copper at the other electrode decreases the Cu^{++} in the second solution. Consequently, an excess of sulfate ions accumulates around the copper electrode and an excess of zinc ions around the zinc electrode. In any ionic solution the sum of the negative charges must equal the total positive charges. Hence, if no provision is made for the migration of ions from one vessel to the other, the current will cease to flow. The salt bridge provides a pathway for the transfer of ions between the two solutions so that the ionic charges within the solutions remain balanced.

It was stated that the potential difference between the zinc and copper electrodes is measured as 1.10 volts, but this potential is obtained only under certain specified conditions. Actually, the magnitude of the potential difference between the electrodes in an electrochemical cell depends upon many factors. Most important of these factors is the nature of the substances which makes up the two half-cells. In other words, the driving force of the overall reaction will depend largely upon the electrode potentials of the two half-reactions. The simplest type of galvanic cell is one in which the electrodes are metals which enter into the chemical reaction. The tendency of a metal to lose electrons and go into solution as ions is called its *electrolytic solution pressure*. This tendency varies from metal to metal. If nickel were used in the reducing half-cell, replacing zinc, the potential of the cell would be lowered because nickel has a lower electrolytic solution pressure than zinc. On the other hand, if the copper–cupric ion half-cell were replaced by the silver–silver ion half-cell, the potential of the cell would be increased.

The concentration, or activity, of the ions of the metal in solution affects the potential of the electrode. In the case of the zinc–zinc ion half-cell, the concentration of the zinc ion is usually arbitrarily fixed at a given value to obtain the standard electrode potential. The half-reaction may be indicated by the equation

$$Zn^{++} + 2e \rightleftharpoons Zn$$

If the concentration of zinc ions is increased, the equilibrium is shifted to the right, and from the principle of Le Châtelier it can be predicted that the charge on the electrode will be less negative. Therefore, when the half-cell is in operation with another half-cell, such as the cupric ion in contact with copper, the increase in zinc ions and decrease in cupric ions will cause a decrease in the potential produced. As the cell reaction proceeds, the cell "runs down" until, at equilibrium, the potential is zero.

Other factors of lesser importance that affect the potential difference between two half-cells are temperature and the emf arising at the liquid–liquid junctions of the two half-cells (produced at the terminals of the salt bridge), the latter being known as the liquid junction potentials. It is customary to describe electrode reactions as taking place at a constant temperature, arbitrarily selected as 25°C. Also, the electrolyte in the salt bridge is usually chosen so that the liquid-junction potentials are reduced to negligible quantities.

6·2 Single Electrode Potentials

When a strip of metal, such as zinc, is placed in a 1 F solution of its ions, an equilibrium reaction develops between the metal and its ion,

as indicated by the equation

$$Zn \rightleftharpoons Zn^{++} + 2e$$

The equilibrium expression may be written as

$$K_e = \frac{[Zn^{++}][e]^2}{[Zn]}$$

The term [Zn] may be omitted since the concentration, or activity, of a solid is constant for a given temperature; therefore,

$$K_e = [Zn^{++}][e]^2$$

No method is known for determining the absolute value of the equilibrium constant of a half-cell reaction. Furthermore, it is impossible to measure directly the electrode potential between a metal and a solution of its ions. Only the potential of a cell, resulting from a combination of two electrodes, may be determined experimentally.

To measure and evaluate potential differences between various half-cells, or potentials developed by various redox half-reactions, it is necessary to designate arbitrarily a particular electrode, and its half-reaction, as a standard. The electrode that has been chosen for this purpose is the standard hydrogen electrode. The electrode is so constructed (see Figure 6·2) that hydrogen molecules are in contact with a 1 F (more correctly, activity = 1) solution of hydrogen ions. Hydrogen gas, at 1 atmosphere pressure and at a temperature of 25°C, is bubbled over a platinum electrode (coated with platinum black), which is immersed in the solution of 1 F hydrogen ions. The finely divided platinum adsorbs hydrogen gas, which exists in contact with the hydrogen ions, resulting in an electrode potential. The equation of the half-cell reaction is

$$2H^+ + 2e \rightleftharpoons H_2$$

The potential of the standard electrode is arbitrarily assigned a value of zero.

Figure 6·2 shows how the potential of the zinc electrode may be measured against that of the hydrogen electrode.

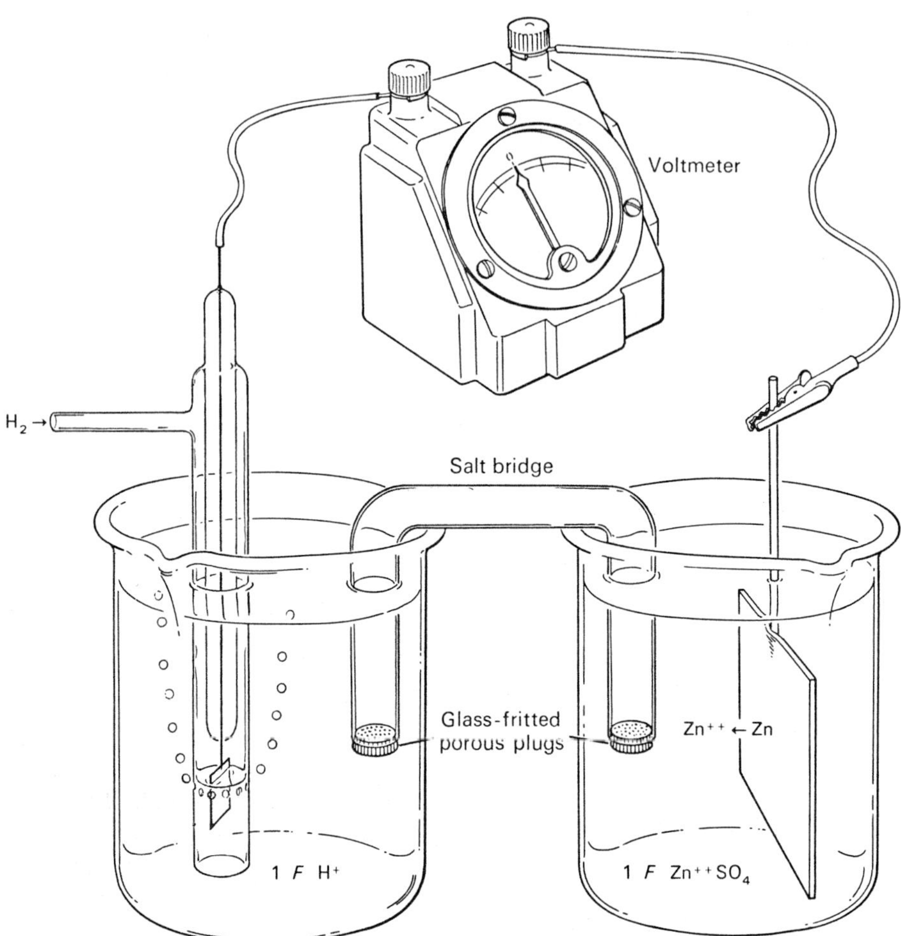

FIGURE 6·2   Diagram of a cell consisting of a hydrogen electrode and a zinc electrode

6·3 Types of Electrodes

Many different types of electrodes, or half-cells, have been devised. If the operating temperature is assumed to be 25°C, and if the electrode's potential is produced with its reactants at known activities, the majority of electrodes can be classified under these categories:

Standard hydrogen electrode. This is the hydrogen electrode diagrammed in Figure 6·2 (with the hydrogen gas at one atmosphere), which operates at a potential defined as zero. With activities at unity, it is the standard against which all electrodes are directly, or indirectly, measured.

Reference electrodes. The standard hydrogen electrode is not easily assembled, owing to fluctuations in the hydrogen pressure, and variations in the consistency of the finely divided platinum on the surface of the electrode. Consequently, several electrodes (usually consisting of a metal in contact with its fairly insoluble salt), which are capable of producing consistent and highly reproducible emf values, have been measured against the standard hydrogen electrode. These electrodes are called reference electrodes, because they produce definite voltages when measured against the hydrogen electrode. The potential of a reference electrode depends upon the composition and structure of the assembly; a typical example is given in

TABLE 6·1　Types of electrodes

Structure	Symbol
metal, metal ion	$Zn\|Zn^{++}$
inert electrode, nonmetal, nonmetal ion	$Pt\|Br_2(1)\|Br^-$
inert electrode, metal ions in different valences	$Pt\|Fe^{++}, Fe^{3+}$
inert electrode, gas, ion	$Pt\|H_2(g)\|H^+$
metal, insoluble salt, ion	$Ag\|AgCl\|Ag^+$
typical calomel electrode	$Hg(1)\|Hg_2Cl_2\|KCl(xF)$

Figure 4·2. The most frequently used reference electrodes are listed as follows:

Calomel electrodes	Potential
$Hg\|Hg_2Cl_2$ (saturated), KCl (saturated)	$+0.242$
$Hg\|Hg_2Cl_2$ (saturated), KCl (1.0 F)	$+0.280$
$Hg\|Hg_2Cl_2$ (saturated), KCl (0.1 F)	$+0.334$
Silver electrode	
$Ag\|AgCl$ (saturated), KCl (saturated)	$+0.197$

Glass electrode. This electrode contains a special glass membrane, which is immersed in the reacting solution when in operation. The electrode is pH sensitive, and is particularly important as a component of a pH meter. A diagram of such an electrode is given in Figure 4·1.

Conducting electrode. An electrode of this type is usually made of platinum. It is inert in itself, and simply serves as a conductor of electrons to the actual half-reaction (see Table 6·1).

Reacting electrode. Any electrode which is part of the half-reaction, and which acts as a conducting medium for the flow of electrons, is a reacting electrode. A typical example is the zinc–zinc ion electrode, listed in Table 6·1 and diagrammed in Figure 6·1.

A partial summary of the various types of electrodes is listed in Table 6·1.

6·4　Standard States

Certain conventions and notations have become fairly well accepted in connection with electrochemical cells. A half-cell composed of a strip of zinc immersed in a solution of 1 F Zn^{++} ions, is indicated as $Zn\|Zn^{++}$ (1 F). The vertical bar denotes a solid phase in contact with a liquid phase. The galvanic cell described in Section 6·1 may be designated as

$$Zn\|Zn^{++} (1\ F)\|Cu^{++} (1\ F)\|Cu$$

in which the liquid junction (salt bridge) is represented by two vertical lines.

For purposes of comparison and reference, electrode potentials must be evaluated at a fixed temperature and under standardized conditions of concentration. Thus, Appendix 5 is a list of standard electrode potentials at 25°C in which the reactants are at unit activity. The term "unit activity" describes the standard state of the substances entering into the electrode reaction. The standard state is arbitrarily chosen as a matter of convenience, to avoid the possibility of ambiguous interpretations.

The following standard states (conditions) are generally accepted for 25°C:

(1) A gas is assumed to be at standard state and unit activity at a pressure of one atmosphere. When a solution is saturated with a gas and the partial pressure of the gas above the solution is one atmosphere, the gas is said to be in its standard state.

(2) The standard state (unit activity) of a liquid or solid is taken as the pure substance under atmospheric pressure.

(3) The standard state (unit activity) of a soluble solute is 1 F, or more strictly, when $a = 1$.

(4) The standard state (unit activity) of a difficulty soluble salt is that of a saturated solution.

Potentials of half-cells, in which the reactants are at standard state are represented as E^0. A cell potential developed by combining two half-cells is indicated as $E^0_{1-2} = E^0_1 - E^0_2$.

6·5 Conventions for Combining Half-Cell Reactions and Standard Electrodes

Until 1953, electrode potentials were written either as oxidation potentials, such as,

$$Zn \rightleftharpoons Zn^{++} + 2e \qquad E^0 = +0.76$$

or as reduction potentials, such as

$$Zn^{++} + 2e \rightleftharpoons Zn \qquad E^0 = -0.76$$

At the XVII Conference of International Union of Pure and Applied Chemistry, held in Stockholm, the Commission on Physico-Chemical Symbols reached an agreement concerning sign conventions of electrode potentials in view of removing the serious confusion existing in regard to the specifications of electrode potentials.*

Certain items of agreement from this conference appear in the following procedures for combining half-cell reactions and single electrode potentials:

Each half-cell reaction is written as a reduction potential. The electrons will appear on the left-hand side of the arrows, for example:

$$Fe^{3+} + 1e \rightleftharpoons Fe^{++}$$

$$I_2 + 2e \rightleftharpoons 2I^-$$

The standard electrode potential, E^0, is evaluated as a reduction potential. The SEP value is placed to the right of the half-reaction.

$$Ag^+ + 1e \rightleftharpoons Ag \qquad E^0 = 0.80\ volt$$

$$Zn^{++} + 2e \rightleftharpoons Zn \qquad E^0 = -0.76\ volt$$

Each half-reaction is multiplied by the correct coefficient so that the number of electrons involved in the half-reactions, to be combined, will be equal. However, the value of E^0 is independent of the coefficient and the number of electrons.

$$2Ag^+ + 2e \rightleftharpoons 2Ag \qquad E^0 = 0.80\ volt$$

$$Pb^{++} + 2e \rightleftharpoons Pb \qquad E^0 = -0.13\ volt$$

* Christiansen, J. A., *J. Am. Chem. Soc.*, **82**, 5517 (1960).

To obtain the potential of the cell, E^0_{1-2}, produced by a combination of the above half-reactions, subtract the second half-reaction from the first half-reaction. It is easier to change all signs in the second half-reaction, and then add it to the first half-reaction. The reactants and products of the second half-reaction are shifted in such a manner to produce a balanced redox reaction.

$$\begin{aligned} 2Ag^+ + 2e &\rightleftharpoons 2Ag \\ -(Pb^{++} + 2e &\rightleftharpoons Pb) \\ \hline 2Ag^+ + Pb &\rightleftharpoons Pb^{++} + 2Ag \end{aligned}$$

$$\begin{aligned} E^0_1 &= 0.80\ volt \\ -(E^0_2 &= -0.13\ volt) \\ \hline E^0_{1-2} &= 0.93\ volt \end{aligned}$$

The sign of the overall standard potential of the cell determines the direction of the spontaneous reaction. If the sign is positive, the reaction proceeds from left to right as written. On the other hand, if the sign is negative, the reaction is not spontaneous as written, but will proceed from right to left. Since the cell potential given in the above example is positive, the spontaneous reaction proceeds from left to right. (Throughout, the ion activities are all considered to be unity.)

If a cell is to be constructed from a redox reaction, which is spontaneous from left to right, and with reactants and products at standard state, then, the half-reactions and SEP values are customarily listed as follows:

(1) The half-reaction with the most positive SEP is indicated as the right-hand electrode.

(2) The half-reaction with the most positive SEP is listed first, and from it is subtracted the half-reaction with the less positive SEP.

Consider the reaction,

$$Sn^{++} + 2Ce^{4+} \rightleftharpoons Sn^{4+} + 2Ce^{3+}$$

The half-reactions and SEP values placed in the following order:

$$\begin{aligned} 2Ce^{4+} + 2e &\rightleftharpoons 2Ce^{3+} \\ -(Sn^{4+} + 2e &\rightleftharpoons Sn^{++}) \end{aligned}$$

$$\begin{aligned} E^0_1 &= 1.23\ (in\ 1\ F\ HCl) \\ -(E^0_2 &= 0.14) \\ \hline E^0_{1-2} &= +1.08\ volt \end{aligned}$$

The electrochemical cell may be depicted as,

$$Pt|Sn^{++}(a = 1), Sn^{4+}(a = 1)\|Ce^{3+}(a - 1),$$

$$Ce^{4+}(a = 1)|Pt$$

Standard electrode potentials may be used to calculate the emf of any cell whose solutes are at unit activity.

Given the cell formulation:

$$Zn|Zn^{++}(a = 1)\|Cu^{++}(a = 1)|Cu$$

The emf is calculated as follows:

$$\begin{aligned} Cu^{++} + 2e &\rightleftarrows Cu \\ -(Zn^{++} + 2e &\rightleftarrows Zn) \\ \hline Cu^{++} + Zn &\rightleftarrows Zn^{++} + Cu \end{aligned}$$

$$\begin{aligned} E_1^0 &= 0.34 \\ -(E_2^0 &= -0.76) \\ \hline E_{1-2}^0 &= 1.10 \end{aligned}$$

The sign $(+ \text{ or } -)$ of E_{1-2}^0 is the polarity of the right-hand electrode as given by the cell formation. The copper electrode is the positive terminal of the cell given above. On the other hand, for a cell written in the reverse direction, as

$$Cu|Cu^{++}(a = 1)\|Zn^{++}(a = 1)|Zn$$

with a cell reaction

$$Cu + Zn^{++} \rightleftarrows Cu^{++} + Zn$$

the emf is given by IUPAC conventions

$$(E_{cell}^0 = E_{right}^0 - E_{left}^0) \text{ as}$$

$$\begin{aligned} Zn^{++} + 2e \rightleftarrows Zn \qquad & E_1^0 = -0.76 \\ -(Cu^{++} + 2e \rightleftarrows Cu) \qquad & -(E_2^0 = 0.34) \\ \hline & E_{1-2}^0 = -1.10 \end{aligned}$$

Although the sign of the cell's emf is changed by a reversal of the cell diagram, the polarities of the electrodes remain unchanged. Thus, Cu is positive and Zn is negative; consequently, the cell reaction is spontaneous from right to left.

To be consistent a cell formulation (diagram) is accepted as given, and the direction of the spontaneous reaction depends upon the sign $(+ \text{ or } -)$ of the emf obtained.

RELATION OF ELECTRODE POTENTIALS TO CONCENTRATION

The symbol E^0 denotes the potential of a half-cell at standard state, and E_{1-2}^0 designates a cell potential (combination of two half-cells) when all the components are in the concentrations that have been selected as standard states. The term, E^0, is a fixed value for a particular ionic half-reaction, which has been determined against the standard hydrogen electrode. Selected values for standard electrode potentials are listed in Appendix 5.

6·6 The Nernst Equation

Frequently, it is impractical to construct half-cells with concentration values that conform with the values for E_{1-2}^0. Consequently, a half cell electrode not at standard state is given the symbol, E, and a combination of two such electrodes is designated as E_{1-2}. One of the notable achievements in electrochemistry was the establishment of the relationship between E_{1-2} and E_{1-2}^0. The Nernst equation, developed in 1889 and named for the German chemist Walther Nernst, contains a quantitative factor that relates the foregoing terms. Thus, Nernst developed a conversion factor for cell potentials, which may be stated as

$$E_{1-2} = E_{1-2}^0 - \text{Nernst factor.}$$

The actual Nernst equation, in toto, is as follows:

$$E_{1-2} = E_{1-2}^0 - \frac{RT}{n\mathscr{F}} \ln Q$$

where

$R = 8.3144$ joules/°K
$T =$ absolute temperature, 298° K, or 25° C
$n =$ the number of electrons transferred in the cell reaction
$\mathscr{F} =$ one faraday = 96,493 coulombs
$\ln =$ the natural logarithm = 2.303 $\log_{10}$
$Q =$ product of the activities of the reaction products, divided by the product of the activities of the reactants, each activity

raised to the numerical power equal to the coefficient of the fw involved in the redox reaction

If the numerical factors, listed above, are substituted into the Nernst equation, it becomes

$$E_{1-2} = E_{1-2}^0 - \frac{8.3144 \times 298}{n \times 96{,}493} \times 2.303 \times \log Q$$

which may be simplified to

$$E_{1-2} = E_{1-2}^0 - \frac{0.059}{n} \log Q$$

6·7 Applications of the Nernst Equation

By means of the Nernst equation the electromotive force of any galvanic cell can be calculated, provided the concentrations of the components in the cell are given. As an example, write the redox reaction involved, and determine the potential which can be obtained from the following cell:

$$Zn|Zn^{++}(0.001)\,F)\|Pb^{++}(0.1\,F)|Pb$$

The half-cell reactions and standard electrode potentials are

$$Pb^{++} + 2e \rightleftarrows Pb$$
$$Zn^{++} + 2e \rightleftarrows Zn$$
$$\overline{Pb^{++} + Zn \rightleftarrows Pb + Zn^{++}}$$

$$E_1^0 = -0.13$$
$$E_2^0 = -0.76$$
$$E_{1-2}^0 = +0.63 \text{ volt}$$

and

$$Q = \frac{[Pb][Zn^{++}]}{[Pb^{++}][Zn]} = \frac{10^{-3}}{10^{-1}} = 10^{-2}$$

therefore,

$$E_{1-2} = E_{1-2}^0 = \frac{0.059}{n} \log Q$$

or

$$E_{1-2} = 0.63 - \frac{0.059}{2} \log 10^{-2}$$

$$E_{1-2} = 0.63 + 0.06 = 0.69 \text{ volt}$$

As another example, write the redox reaction involved, and calculate the potential of the following cell:

$$Pt|Fe^{3+}(1.0\,F),\ Fe^{++}(0.01\,F)\|Ce^{4+}(0.01\,F),$$
$$Ce^{3+}(0.1\,F)|Pt$$

Calculations are given as follows:

$$Ce^{4+} + e \rightleftarrows Ce^{3+}$$
$$Fe^{3+} + e \rightleftarrows Fe^{++}$$
$$\overline{Ce^{4+} + Fe^{++} \rightleftarrows Ce^{3+} + Fe^{3+}}$$

$$E_1^0 = 1.23 \text{ (in 1 } F \text{ HCl)}$$
$$E_2^0 = 0.77$$
$$\overline{E_{1-2}^0 = 0.46 \text{ volt}}$$

$$Q = \frac{[Ce^{3+}][Fe^{3+}]}{[Ce^{4+}][Fe^{++}]} = \frac{(0.1)(1)}{(0.01)(0.01)} = 1000$$

$$E_{1-2} = 0.46 - \frac{0.059}{1} \log 1000$$

$$= 0.46 - 0.17 = 0.29$$

EQUILIBRIUM CONSTANTS FROM HALF-CELL POTENTIALS

The symbol, Q, in the Nernst equation denotes a ratio (relating to activities) of the reactants to the products in a redox reaction. When a cell "runs down" ($E_{1-2} = 0$), the reactants and products are at equilibrium concentrations. Consequently, the value, Q, becomes an equilibrium constant, usually designated as K_e. Various types of equilibrium constants may be calculated, or estimated, from the equilibrium concentrations of components in redox reactions.

6·8 Equilibrium Constant for a Redox Reaction

When the components of a redox reaction are in equilibrium, a combination of the two half-cells, which produces the reaction results in a zero potential. Therefore, the Nernst equation is modified to

$$0 = E^0_{1-2} - \frac{0.059}{n} \log K_e$$

where K_e is the equilibrium constant of the cell reaction. The equation may be simplified to

$$\log K_e = \frac{nE^0_{1-2}}{0.059}$$

The expression permits the calculation of the equilibrium constant for a redox reaction from the electrode potentials of the half-cells, which are components of the overall cell. To illustrate, consider the reaction

$$Cu^{++} + 2Ag \rightleftharpoons Cu + 2Ag^+$$

from which the following cell may be assembled:

$$Ag|Ag^+(1\ F)\|Cu^{++}(1\ F)|Cu$$

The half-cell reactions and electrode potentials are

$$
\begin{array}{ll}
Cu^{++} + 2e \rightleftharpoons Cu & E^0_1 = 0.34 \\
2Ag^+ + 2e \rightleftharpoons 2Ag & E^0_2 = 0.80 \\
\hline
& E^0_{1-2} = -0.46\ \text{volt}
\end{array}
$$

The negative sign of the potential indicates that the reaction is not spontaneous from right to left, as written above, but does proceed from right to left. The magnitude and sign of the cell potential predict that the equilibrium constant for the equation, as written, will be a very small value. The prediction is verified in the following calculations:

$$\log K_e = \frac{nE^0_{1-2}}{0.059} = \frac{2 \times (-0.46)}{0.059} = -15.6$$

$$K_e = 2.5 \times 10^{-16}$$

6·9 Solubility Product Constant from Half-Cell Potentials

The solubility product constant is a special form of an equilibrium constant, and applies to equilibrium concentrations of a fairly insoluble salt with its ions. As an example, consider a saturated solution of silver chloride in equilibrium with its ions, or

$$AgCl_{(solid)} \rightleftharpoons Ag^+ + Cl^-$$

Therefore,

$$K_{sp} = [Ag^+][Cl^-]$$

From Appendix 5, the following potentials are available and may be used for computing the K_{sp} of AgCl:

$$
\begin{array}{l}
AgCl + e \rightleftharpoons Ag + Cl^- \\
Ag^+ + e \rightleftharpoons Ag \\
\hline
AgCl \rightleftharpoons Ag^+ + Cl^-
\end{array}
$$

$$
\begin{array}{l}
E^0_1 = 0.22 \\
E^0_2 = 0.80 \\
\hline
E^0_{1-2} = -0.58\ \text{volt}
\end{array}
$$

Therefore,

$$\log K_{sp} = \frac{nE^0_{1-2}}{0.059} = -\frac{0.58}{0.059} = -9.85$$

$$K_{sp} = 1.4 \times 10^{-10}$$

6·10 Dissociation Constant for a Complex Ion from Half-Cell Potentials

Another form of an equilibrium constant is the dissociation constant (K_d) of a complex ion. For example, the complex ion, $Ag(NH_3)_2^+$, dissociates as follows:

$$Ag(NH_3)_2^+ \rightleftharpoons Ag^+ + 2NH_3$$

and,

$$K_d = \frac{[Ag^+][NH_3]^2}{[Ag(NH_3)_2^+]}$$

If the half-cell reaction of a single electrode contains a complex ion as a component, it is sometimes possible to combine the electrode with a related electrode and determine K_d for the complex ion. For example, calculate the dissociation constant of $Ag(NH_3)_2^+$ from the following data:

$$Ag(NH_3)_2{}^+ + e \rightleftharpoons Ag + 2NH_3$$
$$Ag^+ + e \rightleftharpoons Ag$$

$$\overline{Ag(NH_3)_2{}^+ \rightleftharpoons Ag^+ + 2NH_3}$$

$$E_1^0 = 0.37$$
$$E_2^0 = 0.80$$
$$\overline{E_{1-2}^0 = -0.43}$$

From the Nernst equation, all cell components are at standard state, and

$$0 = -0.43 - \frac{0.059}{1} \log K_d$$

thus,

$$\log K_d = -\frac{0.43}{0.059} = -7.3$$

$$K_d = 10^{-7.3}$$

$$K_d = 5 \times 10^{-8}$$

6·11 Ionization Constant from Half-Cell Reactions and Potentials

If the emf of a cell assembly depends on the hydrogen-ion concentration of a weak acid, it is possible to estimate the K_a of the acid provided the component-concentrations are available. For example, the emf of the following cell is -0.50 volt:

$$Pt|H_2(1\ atm)|HA(0.1\ F)\|KCl(1\ F),$$

$$Hg_2Cl_2\ (saturated)|Hg$$

If HA is a weak monoprotic acid, determine the approximate K_a for the acid. The electrode reactions and potentials are as follows:

$$2H^+ + 2e \rightleftharpoons H_2$$
$$Hg_2Cl_2 + 2e \rightleftharpoons 2Hg + 2Cl^-$$

$$\overline{2H^+ + 2Hg + 2Cl^- \rightleftharpoons H_2 + Hg_2Cl_2}$$

$$E_1^0 = 0.00$$
$$E_2^0 = 0.28$$
$$\overline{E_{1-2}^0 = -0.28}$$

From the Nernst equation,

$$-0.50\ volt = -0.28 - \frac{0.059}{2} \log Q$$

where

$$Q = \frac{[H_2][Hg_2Cl_2]}{[H^+]^2[Hg]^2[Cl^-]^2} = \frac{1}{[H^+]^2}$$

then

$$-0.50 = -0.28 - \frac{0.059}{2} \log \frac{1}{[H^+]^2}$$

$$-0.22 = +\frac{0.059}{2} \log[H^+]^2$$

and

$$\log[H^+] = -\frac{0.22}{0.059} = -3.73$$

or

$$[H^+] = 10^{-3.73} = 1.86 \times 10^{-4}$$

Therefore,

$$K_a = \frac{(1.86 \times 10^{-4})^2}{0.1} = 3.5 \times 10^{-7}$$

REDOX TITRATIONS

The discussion of oxidizing agents in Chapter 5 revealed that there are only four relatively, reliable quantitative-oxidizing ions, namely permanganate, ceric, triiodide, and dichromate (chromate) ions. The advantages and disadvantages characterizing these ions have been discussed.

Many reducing agents are available for redox titrations, and unless they are too unstable or too susceptible to air oxidation, it is possible to select many reducing reagents from Appendix 5 that will react quantitatively with the four oxidizing titrants listed above.

The end point of a redox reaction is usually detected by one of the following three methods: (1) the use of specific indicators (see Section 5·8); (2) the use of redox indicators (see Sections 5·9 and 6·12); (3) an electric method, termed a

TABLE 6·2 A selected list of redox indicators

Indicator	Oxidized form	Reduced form	E^0, volts	Conditions
nitro ferroin	pale blue	red	$+1.25$	$1\,F\,H_2SO_4$
Ferroin	pale blue	red	$+1.06$	$1\,F\,H_2SO_4$
diphenylaminesulfonic acid	purple	colorless	$+0.84$	dil acid
diphenylamine	violet	colorless	$+0.76$	dil acid
methylene blue	blue	colorless	$+0.53$	dil acid

potentiometric titration, which will be discussed in Section 6·13.

6·12 Selection of a Redox Indicator

Redox indicators are highly colored organic substances, which change color when oxidized or reduced. The general half-reaction of a redox indicator may be written as

$$\text{ind}_{\text{oxid}} + ne \rightleftarrows \text{ind}_{\text{red}}$$
$$\text{(color A)} \qquad \text{(color B)}$$

Many redox indicators (in acid solutions) produce half-reactions involving H^+ ions; however, it is usually assumed that the indicator reaction does not change the pH of the solution.

Each redox indicator changes color over a particular potential (emf) range, analogous to the color changes in an acid–base indicator, within a given pH range. Consequently, the selection of a redox indicator depends upon the potential at the equivalence point of the redox titration.

The Nernst equation for an indicator half-reaction may be generalized as

$$E = E_{\text{ind}} - \frac{0.059}{n} \log \frac{[\text{ind}_{\text{red}}]\,(\text{color B})}{[\text{ind}_{\text{oxd}}]\,(\text{color A})}$$

As was stated in Section 4·2, the average individual can detect approximately 100 parts of one color in 10 parts of the other color; consequently,

$$\text{(color B)}\; E = E^0_{\text{ind}} - 0.059 \log 10/1$$
$$= E^0_{\text{ind}} = -0.059$$
$$\text{(color A)}\; E = E^0_{\text{ind}} + 0.059 \log 1/10$$
$$= +0.059$$

(for a detectable change in color) ·
$$\Delta E = \pm 0.12 \text{ volt}$$

Therefore, when $n = 1$, a change in potential of about 0.12 volt is sufficient for detecting the end point of a redox reaction with an indicator, which changes within the desired range. For many indicators, $n = 2$, and a change of approximately 0.06 volt is theoretically sufficient. Table 6·2 lists a few important redox indicators with their characteristics under certain conditions.

In redox titrations (as in acid–base titrations), the visual end point and the equivalence point should be approximately the same. As an illustration of a redox titration, assume that an unknown ferrous iron solution is titrated with a ceric solution of known normality; the equation for the reaction is

$$Ce^{4+} + Fe^{++} \rightleftarrows Ce^{3+} + Fe^{3+}$$

The two half-cells are

$$E_{\text{red}} = 0.77 - \frac{0.059}{1} \log \frac{[Fe^{++}]}{[Fe^{3+}]}$$

$$E_{\text{oxd}} = 1.44 - \frac{0.059}{1} \log \frac{[Ce^{3+}]}{[Ce^{4+}]}$$

and
$$2E = 0.77 + 1.44$$
$$- \frac{0.059}{1} \log \frac{[Fe^{++}][Ce^{3+}]}{[Fe^{3+}][Ce^{4+}]}$$

At the equivalence point,

$$[Fe^{++}] = [Ce^{4+}] \quad \text{and} \quad [Fe^{3+}] = [Ce^{3+}]$$

therefore, the last term becomes equal to log $1 = 0$, and the equation at the equivalence point is modified to

$$E_{\text{eq pt}} = \frac{0.77 + 1.44}{2} = 1.10 \text{ volts}$$

Since only one electron is exchanged ($n = 1$), the end point should be within 0.12 volt of the equivalence point. The most suitable indicator given in Table 6·2 for this particular titration is ferroin, which has a transition voltage of $+1.06$ volts.

6·13 A Redox Potentiometric Titration Curve

The titration curve of a redox reaction can be plotted graphically from measurements, in volts, made with a potentiometer* against volume (in ml) of titrant. Figure 6·3 indicates the essential apparatus for a potentiometric titration. In such a titration, the difference in potential between an indicator electrode and a reference electrode is plotted against the volume of titrant added.

In practice, an electrode of bright platinum metal usually serves as the indicator electrode, inserted into a beaker containing the analate (Fe^{++} ions in solution in this particular setup), and a reference electrode. The potential of the platinum electrode depends on the ratio of oxidizing and reducing agents in solution; consequently, the potential changes during the course of the titration. The calomel electrode is customarily chosen as the reference electrode, inasmuch as its potential remains constant throughout the titration. Since the calomel half-cell maintains a constant potential, the change in potential within the indicator electrode system appears as the value of E measured by the potentiometer. The potential of the indicator electrode, E_{ind}, is dependent upon the ratio of the concentrations of oxidizable and reducible ions in the solution; consequently the following relation exists:

$$E_{ind} = E^0_{ind} - \frac{0.059}{n} \log \frac{[red]}{[oxid]}$$

Figures 6·3 and 6·4 represent the cell setup and the titration curve for a redox reaction. This

*A potentiometer is designed to measure the difference in potential of two electrodes without withdrawing any appreciable electric current from the chemical cell.

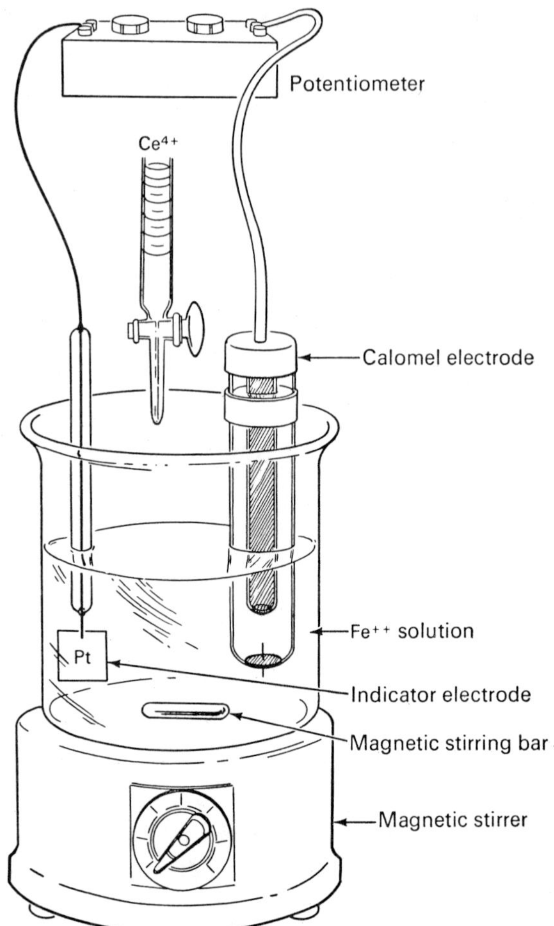

FIGURE 6·3   Potentiometric titration apparatus for titration of Fe^{++} with Ce^{4+} ion

particular curve is obtained by titrating 50.00 ml of $0.1000\,F$ (5.00 mfw) $FeSO_4$ with $0.1000\,F$ $Ce(SO_4)_2$; both solutions are $1\,F$ in H_2SO_4.

The half-cell reactions and standard potentials are as follows:

$$\begin{array}{ll} Ce^{4+} + e \rightleftharpoons Ce^{3+} & E^0_1 = 1.44 \\ Fe^{3+} + e \rightleftharpoons Fe^{++} & E^0_2 = 0.77 \\ \hline Ce^{4+} + Fe^{++} \rightleftharpoons Ce^{3+} + Fe^{3+} & E^0_{1-2} = 0.67 \end{array}$$

The course of the titration is given in Table 6·3.

Initially the 50 ml of solution to be titrated contains only $0.1000\,F$ $FeSO_4$ (in H_2SO_4 solution), although air oxidation produces some ferric ions; therefore, the reducing electrode is arbitrarily assigned a potential of zero. The

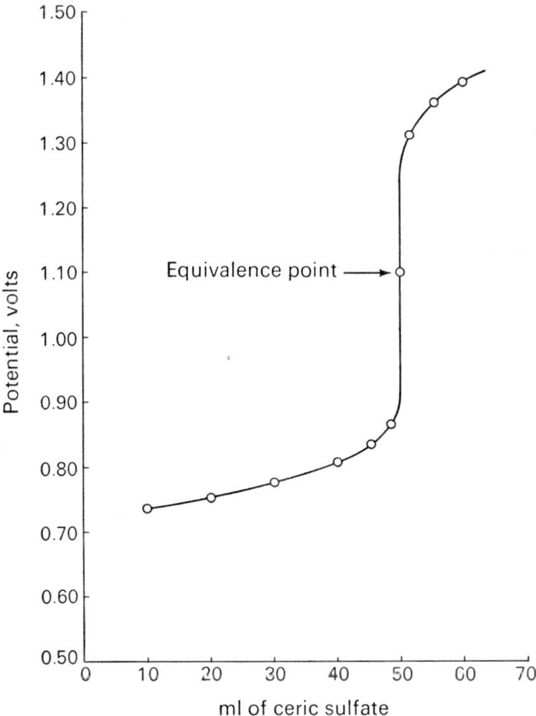

FIGURE 6·4   Titration of ferrous ion with ceric ion

reducing electrode is determined by the ferrous–ferric ion ratio, which is

$$E = 0.77 - \frac{0.059}{1} \log \frac{[\mathrm{Fe}^{++}]}{[\mathrm{Fe}^{3+}]}$$

or

$$0 = 0.77 - \frac{0.059}{1} \log \frac{[0.1000\ \mathrm{Fe}^{++}]}{[0.0000\ \mathrm{Fe}^{3+}]}$$

After 10 ml of 0.1000 F ceric sulfate has been added to the solution (1 mfw of Ce^{4+}), the titrant reacts with 1 mfw of Fe^{++} ions, reducing the Fe^{++} to 4 mfw (from the original total of 5 mfw). At the same time the volume of the solution has been increased to 60 ml. The potential generated by the half-cell may be indicated as

$$E = 0.77 - 0.059 \log \frac{4/60}{1/60}$$

$$E = 0.73$$

When 20 ml of 0.1000 F ceric sulfate has been added (2 mfw of Ce^{4+}) the total amount of Fe^{++} remaining in the solution is reduced to 3 mfw,

and the reducing potential becomes

$$E = 0.77 - 0.059 \log \frac{3/70}{2/70}$$

(70 ml is volume of solution)

$$E = 0.76$$

Upon the addition of 30 ml of ceric sulfate (3 mfw), the potential is

$$E = 0.77 - 0.059 \log \frac{2/80}{3/80}$$

$$E = 0.78$$

The addition of 40 ml of 0.1000 F ceric ions (4 mfw) produces a volume of 90 ml, and the potential of the cell becomes

$$E = 0.77 - 0.059 \log \frac{1/90}{4/90}$$

$$E = 0.81$$

The equivalence point is reached upon the addition of 50 ml of ceric sulfate (see Section 6·12), and the potential is

$$E_{\mathrm{eq\ pt}} = \frac{0.77 + 1.44}{2}$$

$$= 1.10 \text{ volts (in } 1\,F\ \mathrm{H_2SO_4})$$

Beyond the stoichiometric point, it is more convenient to use the oxidizing electrode. Consequently, if 60 ml of 0.1000 F ceric sulfate is added into the titration container, the potential is

$$E = 1.44 - 0.059 \log \frac{[\mathrm{Ce}^{3+}]}{[\mathrm{Ce}^{4+}]}$$

and

$$E = 1.44 - 0.059 \log \frac{5/110}{1/110}$$

$$E = 1.40$$

More complete titration values are given in Table 6·3.

TABLE 6·3   Redox potentials during titration
of 50 ml of 0.1000 F H_2SO_4
with 0.1000 F $Ce(SO_4)_2$

Volume of 0.1000 F ceric sulfate in ml	Total volume of solution	E, volts
0.0	—	—
10.0	60	0.73
20.0	70	0.76
30.0	80	0.78
40.0	90	0.81
45.0	95	0.83
49.0	99	0.87
50.0	100	1.10
51.0	101	1.31
55.0	105	1.37
60.0	110	1.40

The graphic form of the above titration curve is plotted in Figure 6·4. It will be observed that a sharp break in potential occurs in the vicinity of the equivalence point of the titration. In this respect, the curve is analogous to the curves (see Chapter 4) encountered in acid–base titrations.

TYPES OF EXERCISES

Type 1. Cell Assembly and EMF at Standard State

Problem: With all substances at standard state, construct the cell and calculate the voltage for the reaction indicated in the following equation:

$$Fe + 2Ag^+ \rightleftharpoons Fe^{++} + 2Ag$$

Solution: The cell, which corresponds to the above equation, is

$$Fe|Fe^{++}(1\ F)|Ag^+(1\ F)|Ag$$

The half-cell reactions and single electrode potentials are:

$$2Ag^+ + 2e \rightleftharpoons 2Ag \qquad E_1^0 = 0.80$$
$$Fe^{++} + 2e \rightleftharpoons Fe \qquad E_2^0 = -0.44$$
$$\overline{\qquad E_{1-2}^0 = +1.24\ \text{volts}}$$

Type 2. Use of the Nernst Equation

Problem: Write the cell reaction and calculate the voltage that can be obtained from the following cell:

$$Pt|Br_2(aq)|Br^-(10^{-4}\ F)\|Cl^-(0.1\ F)|Cl_2\ (1\ atm)|Pt$$

Solution: The half-cell reactions and electrode potentials are as follows:

$$Cl_2 + 2e \rightleftharpoons 2Cl^- \qquad E_1^0 = 1.36$$
$$Br_2 + 2e \rightleftharpoons 2Br^- \qquad E_1^0 = 1.09$$
$$\overline{Cl_2 + 2Br^- \rightleftharpoons Br_2 + 2Cl^- \qquad E_{1-2}^0 = 0.27\ \text{volt}}$$

$$E_{1-2} = 0.27 - \frac{0.059}{2} \log \frac{[Cl^-]^2}{[Br^-]^2}$$

$$E_{1-2} = 0.27 - 0.0295 \log \frac{(10^{-1})^2}{(10^{-4})^2}$$

$$E_{1-2} = 0.27 - 0.0295 \log 10^6$$

$$E_{1-2} = 0.27 - 0.18 = 0.09 \text{ volt}$$

Type 3. Equilibrium Constant for a Redox Reaction

Problem: Compute the equilibrium constant for the reaction indicated by the following equation (assume that all substances are at standard state).

$$Ce^{4+} + Fe^{++} \rightleftarrows Ce^{3+} + Fe^{3+}$$

Solution:

$$Ce^{4+} + e \rightleftarrows Ce^{3+} \qquad E_1^0 = 1.44 \ (\text{in } 1 \ F \ H_2SO_4)$$
$$Fe^{3+} + e \rightleftarrows Fe^{++} \qquad \underline{E_2^0 = 0.77}$$
$$E_{1-2}^0 = 0.67 \text{ volt}$$

$$\log K_e = \frac{nE_{1-2}^0}{0.059} = \frac{0.67}{0.059} = 11.4$$

$$K_e = 10^{11.4} = 2.5 \times 10^{11}$$

Type 4. Calculation of K_{sp} from Half-Cell Reactions

Problem: Compute the solubility product constant for cuprous iodide, CuI, from the following data:

$$Cu^{++} + e \rightleftarrows Cu^+ \qquad E_1^0 = 0.16$$
$$\underline{Cu^{++} + I^- + e \rightleftarrows CuI} \qquad \underline{E_2^0 = 0.85}$$
$$CuI \rightleftarrows Cu^+ + I^- \qquad E_{1-2}^0 = -0.69$$

Solution:

$$\log K_{sp} = \frac{nE_{1-2}^0}{0.059} = \frac{-0.69}{0.059} = -11.7$$

$$K_{sp} = 2 \times 10^{-12}$$

Type 5. Calculations of K_d for Complex Ion Half-Cell Reactions

Problem: Compute the K_d for the complex ion, $Ag(CN)_2^-$, from the following half-cell reactions and potentials:

$$Ag(CN)_2^- + e \rightleftarrows Ag + 2CN^- \qquad E_1^0 = -0.31$$
$$\underline{Ag^+ + e \rightleftarrows Ag} \qquad \underline{E_2^0 = 0.80}$$
$$Ag(CN)_2^- \rightleftarrows Ag^+ + 2CN^- \qquad E_{1-2}^0 = -1.11$$

Solution:

$$\log K_d = \frac{nE_{1-2}^0}{0.059} = \frac{-1.11}{0.059} = -18.8$$

$$K_d = 1.6 \times 10^{-19}$$

EXERCISES

1. The following equations represent cell reactions in which all reactants and products are in their standard states. Construct the cell and determine the voltage from each reaction.

   (a) $Mg + 2Ag^+ \rightleftharpoons Mg^{++} + 2Ag$

   (b) $2I^- + Cl_2 \rightleftharpoons I_2 + 2Cl^-$

   (c) $2Fe^{3+} + 2Br \rightleftharpoons 2Fe^{++} + Br_2$

   (d) $5Fe^{++} + MnO_4^- + 8H^+ \rightleftharpoons 5Fe^{3+} + Mn^{++} + 4H_2O$

   (e) $5Cl_2 + I_2 + 6H_2O \rightleftharpoons 2IO_3^- + 10Cl^- + 12H^+$

2. Calculate the emf of each of the following cells at 25°C, and write the chemical reaction for each cell.

   (a) $Zn|Zn^{++}(10^{-4}\ F)\|Cd^{++}(0.2\ F)|Cd$

   (b) $Pt|Fe^{++}(0.1\ F),\ Fe^{3+}(0.01\ F)\|Sn^{4+}(0.01\ F),\ Sn^{++}(0.1\ F)|Pt$

   (c) $Cd|Cd^{++}(0.01\ F)\|Cl^-(0.1\ F)|Cl_2(1\ atm)|Pt$

   (d) $Pt|Mn^{++}(0.1\ F),\ MnO_4^-(0.05\ F),\ H^+(0.2\ F)\|Sn^{4+}(0.01\ F),\ Sn^{++}(0.05\ F)|Pt$

   (e) $Pt|Sn^{++}(10^{-2}\ F),\ Sn^{4+}(10^{-6}\ F)\|Cl^-(10^{-2}\ F)|Cl_2(10^{-2}\ atm)|Pt$

3. Calculate the equilibrium constants for each reaction expressed by the following equations (assume that all materials are in their standard states):

   (a) $Sn^{4+} + Pb \rightleftharpoons Pb^{++} + Sn^{++}$

   (b) $2Fe^{3+} + 2I^- \rightleftharpoons 2Fe^{++} + I_2$

   (c) $2I^- + 2HNO_2 + 2H^+ \rightleftharpoons I_2 + 2NO + 2H_2O$

   (d) $2MnO_4^- + 16H^+ + 5Sn^{++} \rightleftharpoons 2Mn^{++} + 5Sn^{4+} + 8H_2O$

   (e) $6Br^- + Cr_2O_7^{--} + 14H^+ \rightleftharpoons 3Br_2 + 2Cr^{3+} + 7H_2O$

4. Calculate the solubility product constant of Ag_2S from the following half-reactions and half-cell potentials:

$$2Ag^+ + 2e \rightleftharpoons 2Ag \qquad E^0 = 0.80$$

$$Ag_2S + 2e \rightleftharpoons S^{--} + 2Ag \qquad E^0 = -0.69$$

5. Calculate the solubility product constant of HgS from the following data:

$$HgS + 2e \rightleftharpoons Hg + S^{--} \qquad E^0 = -0.70$$

$$Hg^{++} + 2e \rightleftharpoons \qquad E^0 = 0.85$$

6. Calculate the solubility product constant of Tl_2S from the following data:

$$Tl_2S + 2e \rightleftharpoons 2Tl + S^{--} \qquad E^0 = -0.93$$

$$Tl^+ + e \rightleftharpoons Tl \qquad E^0 = 0.33$$

7. Calculate the solubility product constant of $Mn(OH)_2$ from the following data:

$$Mn(OH)_2 + 2e \rightleftharpoons Mn + 2OH^- \qquad E^0 = -1.55$$

$$M^{++} + 2e \rightleftharpoons Mn \qquad E^0 = -1.19$$

8. Calculate the solubility product constant of $PbBr_2$ from the following data:

$$Pb^{++} + 2e \rightleftharpoons Pb \qquad E^0 = -0.13$$

$$PbBr_2 + 2e \rightleftharpoons Pb + 2Br^- \qquad E^0 = +0.28$$

9. Compute the dissociation constant of the $AuBr_4^-$ complex ion from the following:

$$Au^+ + e \rightleftharpoons Au \qquad E^0 = 1.68$$

$$AuBr_4^- + 3e \rightleftharpoons Au + 4Br^- \qquad E^0 = 0.87$$

10. Compute the dissociation constant for the $Ni(NH_3)_6{}^{++}$ complex ion from the following data:

$$Ni(NH_3)_6{}^{++} + 2e \rightleftarrows Ni + 6NH_3 \qquad E^0 = -0.49$$

$$Ni^{++} + 2e \rightleftarrows Ni \qquad E^0 = +0.25$$

11. Compute the dissociation constant for the $HgI_4{}^{--}$ complex ion from the following data:

$$HgI_4{}^{--} + 2e \rightleftarrows Hg + 4I^- \qquad E^0 = -0.04$$

$$Hg^{++} + 2e \rightleftarrows Hg \qquad E^0 = +0.79$$

12. If the emf of a cell assembly depends on the hydrogen-ion concentration of a weak monoprotic acid, and the emf of the cell assembly is -0.55 volt, calculate the K_a for the weak acid from the following cell:

$$Pt|H_2(0.5 \text{ atm})|HA(0.5 \; F)\|KCl(1 \; F)|Hg_2Cl_2|Hg$$

13. Calculate the potential of the following cell:

$$Pt|H_2(1 \text{ atm})|H^+(0.2 \; F)\|KCl(1 \; F)|Hg_2Cl_2|Hg$$

14. A certain cell assembly produces a voltage of -0.60 volt. If one of the components of the cell is a weak acid, compute the K_a of the acid from the following set-up:

$$Pt|H_2(0.4 \text{ atm})|HA(0.4 \; F)\{KCl(1 \; F)|Hg_2Cl_2{}'Hg$$

15. Calculate the potential of the following cell:

$$Pt|H_2(1 \text{ atm})|NH_4Cl(1 \; F)\|KCl(1 \; F)|Hg_2Cl_2|Hg$$

16. Calculate the emf of the following cell:

$$Pt|H_2(0.5 \text{ atm})|H^+(1 \times 10^{-5} \; F)\|KCl(1 \; F)|Hg_2Cl_2|Hg$$

17. Calculate the equivalence-point reduction-potential in the titration of $0.1 \; N \; Cr^{++}$ with $0.1 \; N \; I_3{}^-$.
18. Select an appropriate indicator for the titration of 100 ml of $0.1000 \; F \; Sn^{++}$ with $0.1000 \; F \; Fe^{3+}$.

CHAPTER 7

Colorimetry and Spectrophotometry

When a beam of light passes through a medium (usually a solution), some of the energy associated with the light is absorbed. A knowledge of the absorption of energy in this manner has been recognized for more than two centuries, and may account for the fact that colorimetric and spectrophotometric methods are among the most important in analytical chemistry. During this lengthy period, absorptiometric methods have developed from the first colorimetric determinations that were made with the use of simple inexpensive Nessler tubes into spectrophotometric methods of analysis, which utilize elaborate, expensive spectrophotometers of various types. Because it is not practical to trace the development of methods and instruments used for photometric determinations in an elementary course, this discussion will be confined to the fundamental principles of colorimetry and spectrophotometry, with elementary descriptions of certain types of spectrophotometers.

The term *colorimetry* is sometimes very loosely used in reference to absorptimetric methods. Strictly, *colorimetry* should only be applied to the determination of absorptive capacity in the visible region of the spectrum. However, the term may be used to include spectrophotometric procedures that are carried out with visible light. On the other hand, *photometry* carries the connotation of measurements by means of photoelectric instruments. *Spectrophotometric* methods use visible light, and, more important, they are employed in studying the absorption of narrow bands of radiation within the ultraviolet and infrared regions of the electromagnetic spectrum.

THE ELECTROMAGNETIC SPECTRUM

7·1 Radiant Energy, Light, and Color

Radiant energy covers several conveniently defined regions of the electromagnetic spectrum, in which visible light is a very narrow region. A rough division of selected electromagnetic regions is given in Table 7·1; however, it should be emphasized that the figures in the table are arbitrarily chosen, and should be accepted only as rough boundaries.

The electromagnetic spectrum contains all types of radiation and is continuous. But visible light consists only of a narrow region of radiation to which the human eye is sensitive. The visible spectrum occurs in the range between 400 and 800 mμ (millimicrons) wavelengths. When radiation containing all wavelengths, within the visible spectrum, strikes the eye, the sensation recorded is that of white light. However, if white light is divided into separate wavelengths by means of a prism or diffraction grating, the eye can differentiate the visible spectrum into individual colors. The various colors associated with specific regions in the

TABLE 7·1   Arbitrary regions of the electromagnetic spectrum

Types of radiation	Approximate wavelengths	
	Cm	Usual units
X-rays	10^{-7} to 10^{-10}	10^{-2} to 10 Å
ultraviolet	4×10^{-5} to 10^{-7}	1 to 400 mμ
visible light	4×10^{-5} to 8×10^{-5}	400 to 800 mμ
infrared	8×10^{-5} to 3×10^{-2}	0.8 to 300 μ
microwaves	3×10^{-2} to 10^{-2}	0.03 to 100 cm
radio waves	10^2 to 10^5	1 to 1000 m

TABLE 7·2   The visible spectrum

Color	Wavelength, mμ
red	610–750
orange	585–610
yellow	565–585
green	490–565
blue	435–490
violet	400–435

visible spectrum are classified approximately in Table 7·2.

7·2 The Nature of Radiant Energy

Electromagnetic radiation, as indicated in Table 7·1, covers a broad spectrum of energy manifestations. All electromagnetic energy appears to be composed of discrete packets of energy, which are called *photons* or *quanta*. The photon is the unit of electromagnetic energy, and the magnitude of the unit, or quanta, is *hv* where *v* is the frequency of the radiation and *h* is a constant, which is known as *Planck's constant* with a value of 6.62×10^{-27} erg-seconds.

Radiant energy has the dual property of possessing wave motion, and also acting as if it were composed of energy particles. Consequently, electromagnetic radiation may be described in terms of wave motion as well as in terms of energy packets, as follows:

Wave motion. Electromagnetic radiation is presumed to travel at extreme velocities in transverse waves, thereby producing a wave motion. A wavelength, denoted as λ, refers to

the distance between two adjacent crests, as shown in Figure 7·1. The number of complete cycles, or waves, passing a fixed point in unit

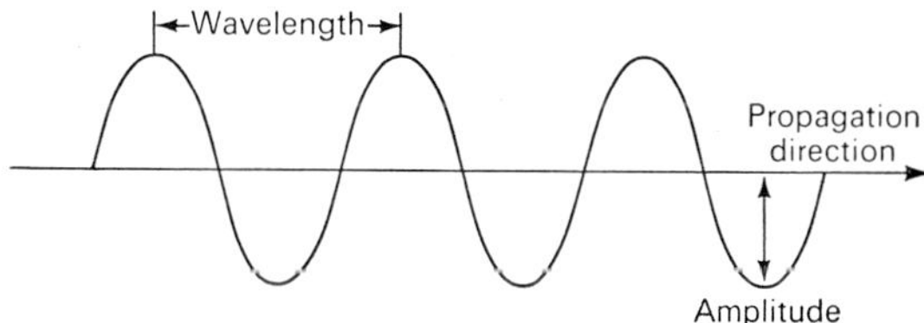

FIGURE 7·1   Propagation of electromagnetic radiation in the form of a wave

time is denoted as frequency (*v*) of the wave motion. The relationship among various units may be expressed as

$$\frac{1}{\lambda} = \frac{v}{c} = \bar{v}$$

or

$$v = \frac{c}{\lambda} = c\bar{v}$$

where

λ = wavelength
v = frequency
$\bar{v}$ = wave number
c = velocity of light (3×10^{10} cm/sec)

Energy packets. The energy content of each photon in a beam of radiation can be expressed as

$$E = hv$$

where *E* represents the energy of a photon in ergs, *h* is Planck's constant, and *v* is the frequency of the radiation.

TABLE 7·3   Colors and their complements

Wavelength, mμ	Color	Complementary color
400–435	violet	yellow-green
435–480	blue	yellow
480–490	green-blue	orange
490–500	blue-green	red
500–560	green	purple
560–580	yellow-green	violet
580–595	yellow	blue
595–610	orange	green-blue
610–750	red	blue-green

The mathematical expressions denoting wave motion of electromagnetic radiation and the energy of a photon may be combined to give

$$E = \frac{hc}{\lambda}$$

7·3 The Origin of Absorption Spectra

If a beam of white light impinges upon a medium, it may be reflected, absorbed, or transmitted. If a portion of the original white light is absorbed by the medium (usually glass or solution, or a combination of the two) the reflected or transmitted light appears as the complementary color of the region of the visible spectrum that has been absorbed. In Table 7·3 are listed various colors and their complements.

When a beam of white light passes through a cuvette (glass container) filled with solution, the radiation that emerges is less energetic than the original beam. The decrease in energy may spread equally over the visible wavelength, or it may be exhibited within particular colors. The various differentiations in colors, as a result of absorption, are indicated in Table 7·3. The loss of energy in the emergent beam results from absorption of energy to produce certain changes in atoms, ions, or molecules within the solution. As a consequence, radiation of a particular wavelength may be absorbed, whereas energies of other wavelengths may be relatively unchanged.

According to the photon concept of radiant energy, it may be postulated that the absorption of energy occurs in discrete units; therefore, the selective absorption may be designated as

$$E_1 - E_2 = h\nu$$

According to the quantum theory, a particle does not possess a definite quantity of energy, but rather it exists within certain energy states (or levels). Consequently, in passing from one energy state, to a higher one, the particle must absorb a discrete unit (or units) of energy to shift its normal level to the excited state.

The energy changes produced upon dissolved atoms, ions, or molecules by light absorption, may be classified as translational, rotational, vibrational, and electronic. If it is assumed that a photon (or photons) of a particular "energized" bundle produces excitation of a certain particle (or particles) beyond the original state, the excitation may result in any of the following energy processes:

(1) *Translational energy* is usually caused by increased motion (kinetic energy) of the particles as a whole. The effect of such absorption is slight, and usually the effect upon the wavelength of the particle, or particles, is considered to be negligible.

(2) *Rotational energy* may result in a rotation of the chemical dipole of the molecules, that are undergoing radiation, and may possibly produce spectral changes (by absorption of photons) within the infrared and longer wavelength regions.

(3) *Vibrational energy* may undergo spectral changes (through absorption of photons). This type of energy is associated with the average separation of nuclei of the atoms that make up the molecules in the solution.

Any spectral change thereby produced is invariably within the infrared region.

(4) *Electronic energy* absorption is caused by the motion of the electrons in the dissolved molecules. Electronic transitions can produce changes in the visible and the ultraviolet regions of the emergent beams.

It should be stressed again that the absorption of radiation (photons) is a selective proccss. In analytical procedures, it is customary to irradiate the sample solution with a wavelength range of only a few millimicrons. The particular wavelength must be determined (or known) before an absorption spectrum can be attempted. For example, in the analysis of ferrous ions with orthophenanthroline the wavelength for maximum absorbance is approximately 508 mμ. This wavelength must be either determined beforehand, or obtained from the chemical literature.

THE LAWS OF ABSORBANCE

This discussion is concerned only with the absorption of radiation, frequently designated as Absorptimetry. However, a student in elementary quantitative chemistry should be informed that this topic is only one aspect of optical analysis. Other interesting facets are the physico-chemical methods in which radiation may be emitted, and other procedures in which the radiation is simply scattered. Conditions which produce the latter phenomena are not described in this chapter.

7·4 Bouguer's Law (Lambert?)

Two simple elementary laws make up the basic foundation to colorimetric analysis. The first of these was formulated by Bouguer in 1729, and restated by Lambert *31 years later*. Through some peculiar quirk in chemical history, Lambert is usually given credit for the discovery of the law. According to this law, each layer of absorbing medium of equal thickness absorbs an equal fraction of the radiation transversing through it. In mathematical terms, the absorption capacity varies directly as the logarithm of the thickness of the absorber.

When a monochromatic beam of intensity, I, traverses the thickness of the absorber, b, the reduction in intensity is given by the equation,

$$-\frac{dI}{db} = k_1 I$$

which may be integrated to yield,

$$\log \frac{I_0}{I} = k_1 b$$

where k_1 is a constant depending on the nature of the medium, the wavelength, and the concentration of the solution; whereas b is the thickness of the absorbing medium; I_0 is the intensity of the incident radiation, and I is the intensity of the emergent radiation.

Transmittance, T, is a ratio of the energy transmitted by the sample, I, to the incident radiation, I_0. Both I and I_0 must refer to the same wavelength; therefore,

$$T = \frac{I}{I_0}$$

Since Transmittance is generally evaluated in percentage values,

$$\frac{I}{I_0} \times 100 = \%T$$

7·5 Beer's Law

The extent of absorption of light by a solution containing a colored solute is a function of concentration. Beer's Law states that the intensity of a beam of monochromatic light decreases exponentially as the concentration of the absorbing medium increases. It is analogous to Bouguer's Law in ascribing an exponential decrease in light intensity, which is accompanied by an arithmetical increase in concentration of the colored solute. The mathematical relationship is

$$\log \frac{I_0}{I} = k_2 c$$

TABLE 7·4   Summary of terms used in absorptometry

Symbol	Name	Definition
A	absorbance	$\log I_0/I$
T	transmittance	I/I_0
a	absorptivity	A/bc
b	length of path	A/ac
c	concentration	A/ab

where k_2 is a constant depending on the wavelength, the nature of the absorbing solution, and the sample thickness; c is the concentration of the solution. Beer's Law states the fundamental aspects of absorptiometry, but it is not complete unless combined with Bouguer's Law. No deviations from Bouguer's Law are known; however, those from Beer's Law are quite numerous.

7·6 The Bouguer–Beer Law

The Bouguer Law is usually combined with Beer's Law to give the following mathematical relationships:

$$\log \frac{I_0}{I} = A = abc = -\log T$$

where

$I_0 =$ incident radiation
$I =$ emergent radiation
$a =$ absorptivity (a constant depending on wavelength of radiation and nature of solution)
$b =$ length of light path of standard
$c =$ concentration of solute in solution
$A =$ absorbance
$T =$ transmittance

7·7 Deviations from Beer's Law

Deviations from Beer's Law are numerous. For example, the dilution of an orange dichromate solution produces a yellow chromate solution of a different pH.

$$\underset{\text{orange}}{Cr_2O_7^{--}} + H_2O \rightleftarrows \underset{\text{yellow}}{2CrO_4^{--}} + 2H^+$$

Any colored substance does not conform with Beer's Law if a chemical change in its chemical structure occurs during the dilution process. For example, if it is assumed that the highly colored complex ion has a chemical composition of $Fe(SCN)_6^{3-}$, it undergoes progressive hydrolysis upon dilution, with corresponding color changes as a result of its change in chemical structure. Without complete proof, it may be postulated that the chemical changes may be

$$Fe(SCN)_6^{3-}$$
(color A)
$$+ HOH \rightleftarrows Fe(SCN)_5(H_2O)^{--} + SCN^-$$
(color B)

$$Fe(SCN)_5(H_2O)^{--}$$
(color B)
$$+ HOH \rightleftarrows Fe(SCN)_4(H_2O)_2^{-} + SCN^-$$
(color C)

$$Fe(SCN)_4(H_2O)_2^{-}$$
(color C)
$$+ HOH \rightleftarrows Fe(SCN)_3(H_2O)_3 + SCN^-$$
(color D)

Consequently, the ferri-thiocyanate complex ion will never conform with Beer's Law in any experiments involving dilution with water.

THE INSTRUMENTS OF ABSORPTIOMETRY

Textbooks on optical analysis have traditionally included various devices that have been used in visual analysis during the past two centuries. Among these devices and instruments are Nessler tubes, Hehner cylinders, and the Duboscq colorimeter. However, many such pieces of equipment are inaccurate, and the best that can be said for them is that they illustrate the principles of colorimetry in an elementary fashion. At present, they are rarely used, or even found in a chemistry laboratory. Yet, it should be noted that items such as Nessler tubes are still widely used in field work, especially in biology.

7·8 Spectrophotometers

Many types of spectrophotometers are commercially available. They range in price from a few hundred dollars each to a sum of several thousand dollars for a highly sophisticated instrument. In general, the inexpensive spectrophotometers are designed and used for rough instructional purposes, whereas the expensive pieces of equipment are intended for research purposes. It is obvious that the less expensive instruments are of value in the laboratory training in elementary photometry, while the use of the expensive instruments is largely restricted to advanced undergraduate experimentation and graduate research. In the laboratory procedures in colorimetric methods described in this manual (Chapter 14), the less expensive instruments are specified.

Practically all electric absorption instruments are photoelectrometers. By definition, a spectrophotometer is an instrument capable of making fairly accurate measurements of radiant energy. The simple instruments are usually adapted to the visible spectrum. Research spectrophotometers may cover extended absorption ranges. For example, the UV spectrophotometer may absorb radiation within the visual spectrum as well as within the ultraviolet region. On the other hand, an IR spectrophotometer may extend its absorption region from the visual portion of the spectrum into the infrared area.

THE SPECTROPHOTOMETER AS AN INSTRUMENT FOR QUANTITATIVE ANALYSIS

Over a period of several years I have used inexpensive colorimeters as quantitative instruments, and have found the analytical errors resulting from their use to be approximately 2%. The samples used for analysis were of the "Thorn Smith" type, eliminating alloys and trace metals. The magnitude of this error may seem high when compared with standard analytical procedures, in which the error may be less than 1%. However, considering the limitations of optical absorption methods, the error in such analyses is not large. Furthermore, the absorption method of analysis has been applied to standard samples in order that a student may compare the accuracy of the optical procedure with that of another type of analytical procedure applied to an equivalent sample. For example, we might determine copper both by colorimetric methods and by the iodometric method, comparing the two.

7·9 Colorimetric Determination of the $Cu(NH_3)_4^{++}$ Ion

Many individuals have an unbounded confidence in various chemical laws. This belief in the impregnability of chemical laws may result from an inadequate knowledge concerning the distinction between "ideal behavior" and "actual results" relating to a chemical law. In absorptiometry very few dissolved substances follow Beer's Law with a high degree of exactitude. In fact, Körtum has stated that the validity of Beer's Law for ions must be considered as an exception rather than a rule.* However, the overall accuracy of results obtained through the Bouguer–Beer's Laws is surprisingly high, considering the deviations from Beer's Law.

In the experimental portion of this manual, copper ore is dissolved in nitric acid, and to the resulting solution is added a large excess of ammonia. The highly colored copper-ammonia complex is analyzed through the use of a Spectronic 20. This complex ion does not follow Beer's Law except in the presence of excess ammonia within the diluted samples. In the absence of an excess of ammonia the complex apparently undergoes a change in chemical composition similar to that for the $Fe(SCN)_6^{3-}$ ion as indicated in Section 7·7. One of the possible

---

* Körtum, G. B., *Physik Chem.* **B34**, 225 (1936).

chemical changes can be postulated as

$$Cu(NH_3)_4{}^{++}$$
(color A)

$$+ HOH \rightleftharpoons Cu(NH_3)_3(H_2O)^{++} + NH_3$$
(color B)

The details of the $Cu(NH_3)_4{}^{++}$ determination are given in Chapter 14; however, the fundamental processes in performing the experiment are outlined in this section. The essential steps are as follows:

(1) A series of solutions (in precise volumes) are prepared by weighing a calculated amount of reagent-grade copper, dissolved in nitric acid, and then diluted to exact volumes in the presence of excess NH_3.

(2) One of the known copper solutions, prepared above, is used for determining the most suitable wavelength for maximum absorption in the colorimetric analysis. This is obtained by plotting percent transmittance versus wavelength (mμ). In the experiment, at least 15 different wavelengths, varying from 475 mμ through 675 mμ, should be used in getting the absorption curve. Plot percent transmittance against wavelength mμ. A typical curve is shown in Figure 7·2.

(3) The transmittance values obtained for the series of solutions containing different concentrations of the diluted cupric-ammonia complex ion are plotted on semi-log paper, using a wavelength setting of 600 mμ. The plot should be a straight line. The solution of the unknown (or unknowns) in terms of mg of Cu/ml is plotted on the same straight line. Directions for calculating the percentage of copper in the sample are given in Section 14·12. A typical plot is shown in Figure 7·3.

THE SPECTROPHOTOMETER AS AN INSTRUMENT FOR QUALITATIVE ANALYSIS

Any spectrophotometer is a qualitative instrument, and the more accurately constructed instruments usually produce better qualitative results. Spectrophotometers are used for qualitative purposes in three regions of the spectrum— the ultraviolet, the visible, and the infrared.

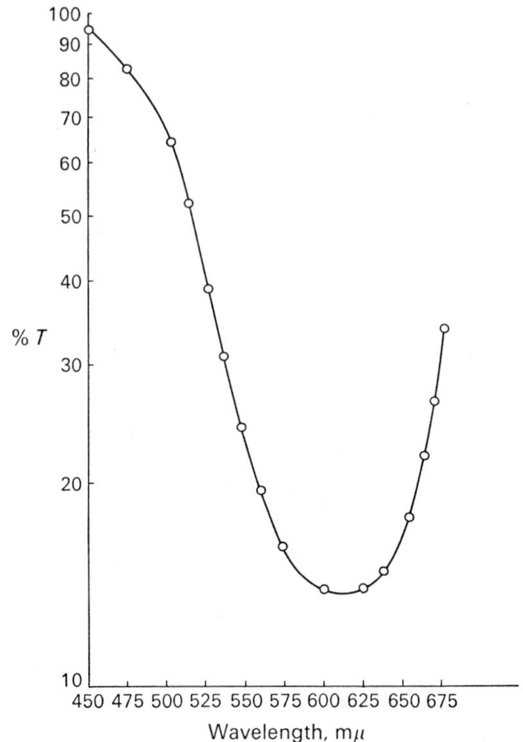

FIGURE 7·2 Plot of % transmittance against wavelength for cupric-ammonia complex

There is, of course, some overlapping between any two regions. The quantitative use of the inexpensive but excellent Spectronic 20 has been described in Section 7·9. A more expensive ultraviolet, or visible-range, spectrophotometer can be used in the same manner, and may be presumed to produce more accurate results. The infrared spectrophotometer is essentially an instrument of qualitative analysis, although it can be adapted to certain quantitative procedures.

7·10 The Use of the Ultraviolet and Visible Spectrophotometers for Qualitative Analysis

Although these instruments are most widely used in quantitative analysis, they are frequently used for identification of organic substances, or for the identification of functional groups that may have been inserted into the organic materials. This use is outside the field of quantitative analysis, but is mentioned as a

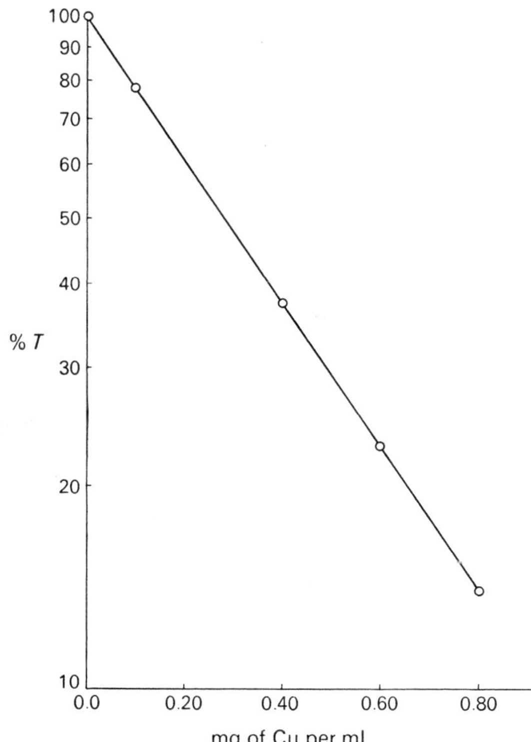

FIGURE 7·3 Transmittance for a series of cupric-ammonia complex ion solutions. Each solution is indicated by a circled dot. Concentration is in milligrams of Cu per milliliter. Wavelength setting is 600 mμ

TABLE 7·5 Approximate absorption band positions of certain chromophoric groups as appearing in the ultraviolet and visible spectral regions

Group	λ_{max} (mμ)
C=C	190
(C=C)$_2$	220
(C=C)$_3$	260
(C=C)$_4$	290
—COOH	210
C≡C	180
—CONH$_2$	210
C=N	190
C≡N	170
C=O	280
C=S	330
—N=N—	370
N=O	660
—NO$_2$	270
—NO$_3$	300
SH	230
—COCl	240
=C=S	330

usually referred to as chromophores. The most frequently occurring chromophores are listed in Table 7·5, with their characteristic wavelengths (λ) in millimicrons (mμ).

matter of interest in connection with the use of spectrophotometers.

In essence, the techniques involved in the identification of functional organic groups closely resemble the matching of fingerprints. If the spectroscopist has an extensive library of absorbance curves (percent transmittance versus wavelength) for known compounds, the identification of an unknown may be established rapidly from its qualitative spectrum. Instead of containing an index of fingerprint whorls, the spectral library includes a multitude of absorbance charts, which are usually too complicated for description. However, such a chart usually contains a complex series of peaks, valleys, and intervening bands arranged in a pattern so that each organic compound may be said to have a characteristic absorbance curve.

The electronic transitions, which may be responsible for absorption in the ultraviolet region, are produced by color-absorbing groups

7·11 The Use of the Infrared Spectrophotometer for Qualitative Analysis

Without going into the theory of vibrational energy levels of molecules, we may say that infrared absorption bands result from the excitation of atoms (within molecules) to higher levels. However, the vibrational spectra of a molecule usually result from a small group of atoms within the molecule. The absorption band, or bands, serves as a basis for the identification of the molecule, qualitatively, and may give some indication as to the structure of the molecule. Since the complexity of infrared spectra is so great, it is quite unlikely that any two compounds will have identical absorption curves. For this reason, an infrared spectrum of a pure compound provides the analyst with an

TABLE 7·6 Approximate absorption band positions for certain groups in the infrared region of the spectrum

Group	Wavelength, μ
C—H (aliphatic)	3.3 – 3.7
C—H (aromatic)	3.2 – 3.3
—NH$_2$ (primary amine)	2.9 – 3.0
N—H	2.96– 3.33
O—H (phenolic)	2.7 – 3.3
C≡N	4.2 – 4.6
C—O	9.55–10.0
C=O (aldehyde)	5.8 – 6.0
C=O (acid)	5.8 – 6.0
C=C	5.98– 6.17
C≡C	4.44– 4.76

almost foolproof method of identification, provided he has the original "fingerprint" spectrum of the compound. A few positions and their characteristic maximums, obtained by infrared absorptions are listed in Table 7·6.

INSTRUMENTS FOR SPECTROPHOTOMETRIC ANALYSIS

There are three regions in the electromagnetic spectrum that lend themselves to quantitative absorptiometry. These are the ultraviolet, visible, and infrared areas. As might be expected three general types of instruments have been designed for absorption in these regions. However, frequently one instrument may combine both the ultraviolet and visible regions.

7·12 Some Essential Components of a Spectrophotometer

The designs of spectrophotometers are changing more rapidly than the textbooks that describe their use, can be written. Usually any textbook containing an extensive discussion of spectrophotometry is out-of-date by the time it is published. However, there are several components of such instruments that are usually present, even though they are undergoing constant improvements. Components found most frequently in various spectrophotometers are as follows:

Radiation source. This may be an incandescent tungsten or hydrogen lamp, or a black body radiator. The tungsten lamp is used for the visible region, and the hydrogen lamp for the ultraviolet. The source of radiation in an infrared spectrophotometer is usually a thermopile, frequently called a *globar*.

Lens. This is for converging and focusing the beam of radiation.

Entrance slit. This is for selected wavelengths.

Color filter. A combination of filters can be used to give the desired ranges of transmission for narrow wavelength beams.

Prism, or diffraction grating. A source of monochromatic light.

Exit slit. This slit matches the entrance slit.

Mirrors. These are for focusing the light beam.

Absorption cell, or cuvette. This contains the sample for analysis.

Radiation detector. A phototube, or bolometer, is used to detect the magnitude of radiation energy, which has escaped through the sample under analysis.

Amplifier. This is used to increase output voltage from detector.

Recorder. A galvanometer, pen recorder, or potentiometer to record results.

7·13 A Spectrophotometer for the Visible Spectrum

The Spectronic 20 operates through a range of 340 to 625 mμ, and by the addition of a red filter and a red-sensitive photocell, the range may be

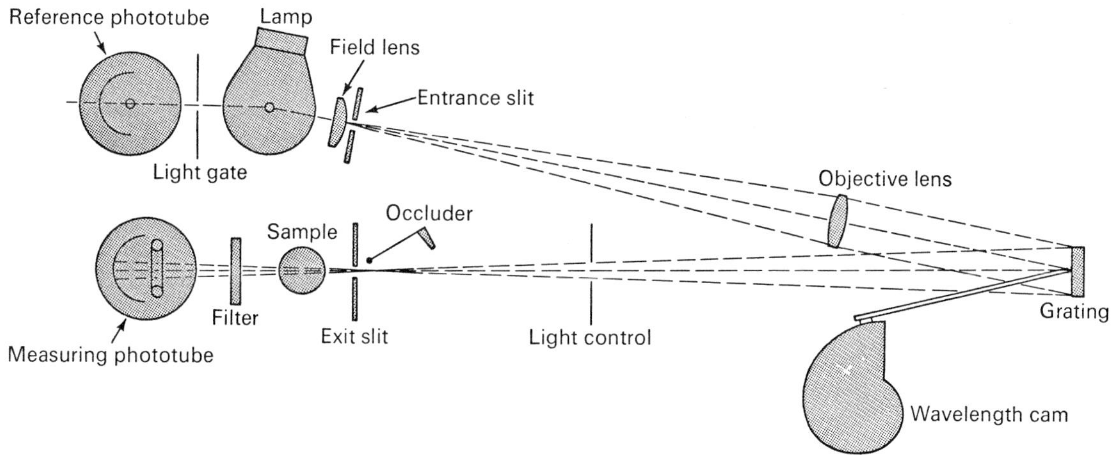

FIGURE 7·4 Schematic diagram of B. and L. Spectronic 20. [Courtesy Bausch and Lomb, Inc., Rochester, New York]

extended to 950 mμ. The nominal bandwidth is 20 mμ and is constant over the entire wavelength range. White light emanating from a tungsten lamp passes through the entrance slit (see Figure 7·4) and is focused by the field lens on the objective lens. The objective lens produces an image of the entrance slit at the exit slit after the light has been dispersed by the diffraction grating. To obtain various wavelengths the grating is rotated by a wavelength cam. The desired wavelength passes through the exit slit, and the monochromatic light that is transmitted passes through the cuvette placed in the path of the light beam. A light control is provided to set 100% transmittance or zero absorbance, on the meter, with a reference or standard solution in the sample compartment.

Unlike many colorimeters, the Spectronic 20 may be used as a spectrophotometer when choosing the proper wavelength for a substance. When no information is available for choosing the proper wavelength, it may be determined experimentally (see Figure 7·2). As indicated by the figure, it is necessary to plot spectrometric graphs (on semi-log paper) of percent transmittance versus wavelength for several known concentrations, which fall within the desired concentration range. Usually the desired wavelength for transmittance is the minimum (about 600 mμ according to the Figure 7·2 for the cupric-ammonia complex ion). The electrical output of the phototube is amplified and the resulting amplification is indicated on the calibrated scale of the meter of the instrument. Although the instrument employs a single phototube and one absorption cell, it is inexpensive and produces good results. It is particularly adaptable to routine analytical work. Results from the meter are interpreted according to the following equation:

$$\log \frac{I_0}{I} = \text{absorbance} = -\log T$$

Final results are obtained by plotting known concentrations, and the unknown concentration, of the substance under analysis on graph paper (see Figure 7·3).

7·14 A Spectrophotometer for the Ultraviolet Region

Some ultraviolet spectrophotometers can be used also in the visible spectrum. One is the Beckman DU-2 Spectrophotometer, which gives transmittance results in both the ultraviolet and visible spectrums. The spectrophotometer is designed for measurements in the 190 to 1000 mμ range. A hydrogen lamp is used as a source of radiation between 190 and 360 mμ, and an incandescent tungsten lamp extends the range of the visible spectrum up to 1000 mμ. It is a quality instrument, and under optimum conditions

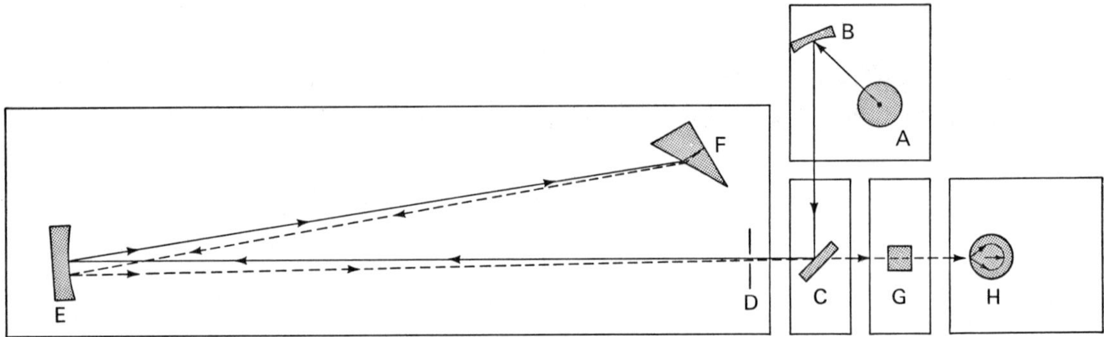

FIGURE 7·5   Schematic diagram of Beckman Spectrophotometer, Model DU-2. [Courtesy Beckman Instruments, Inc., Fullerton, California]

uncertainties in transmittance can be reduced to a few tenths of a percent.

The DU-2 Spectrophotometer consists of five major compartments: monochromator, cell compartment, mounting block, lamp house, and phototube house. The optical system is indicated in Figure 7·5. As shown in this figure, radiant energy is focused by means of the source-condensing mirror, and then directed to a monochromator-diagonal mirror. The latter reflects the beam to the collimating mirror, and from there to the prism. The prism is composed of fused silica, and serves as the component for dispersion. The backside of the prism is aluminized; therefore, it also serves as a mirror. As a result of this construction, the beam passes through the prism twice, undergoing refraction and reflection each time. The desired wavelength of light is selected by rotating the Wavelength Selector on top of the monochromator case; this control also adjusts the prism to the desired angle. The spectrum from the prism is directed back to the collimating mirror, which centers the chosen wavelength of light on the slit and the sample (G). Light passing through the sample strikes the phototube (H) causing a voltage to appear across the load resistor. The voltage is amplified and registered on the instrument's meter.

7·15 A Spectrophotometer in the Infrared Region

Infrared radiation (Section 7·1) is an electro-magnetic radiation of much longer wavelength than the visible spectrum. The infrared radiation most frequently used in qualitative chemistry falls between 2.5 and 15 μ. However, the near infrared, which is in the area between 0.8 and 2.0 μ, is also widely used. The principles of infrared spectrophotometry are the same as those of spectrophotometry in the visible and ultraviolet regions. However, the component parts of the infrared spectrophotometer are somewhat dissimilar, and different materials are required for their construction. Usually infrared radiation is detected by measuring the temperature rise of a "blackbody" material placed within the focus of a special mirror. A blackbody radiator is termed a *globar*, which may be a rod made of silicon carbide, or a bonded mixture of rare earth metals. Either type of globar is electrically heated to a temperature range between 1200 and 2000°K. The energy so produced changes with wavelength throughout the scanning range.

The infrared spectrophotometer utilizes front-surface mirrors instead of lens. Such a construction eliminates quartz and glass parts, which absorb in the infrared region. Most absorption cells are made of sodium chloride, which removes quartz or glass materials. Sodium chloride is almost transparent to infrared radiation, but it has the disadvantage of being soluble in water. Liquid samples are frequently used in infrared analysis, but water must be avoided as the solvent. Water is also an absorber of infrared radiation.

The prism, too, is made of sodium chloride, and dispersion is increased by passing a beam through the prism several times.

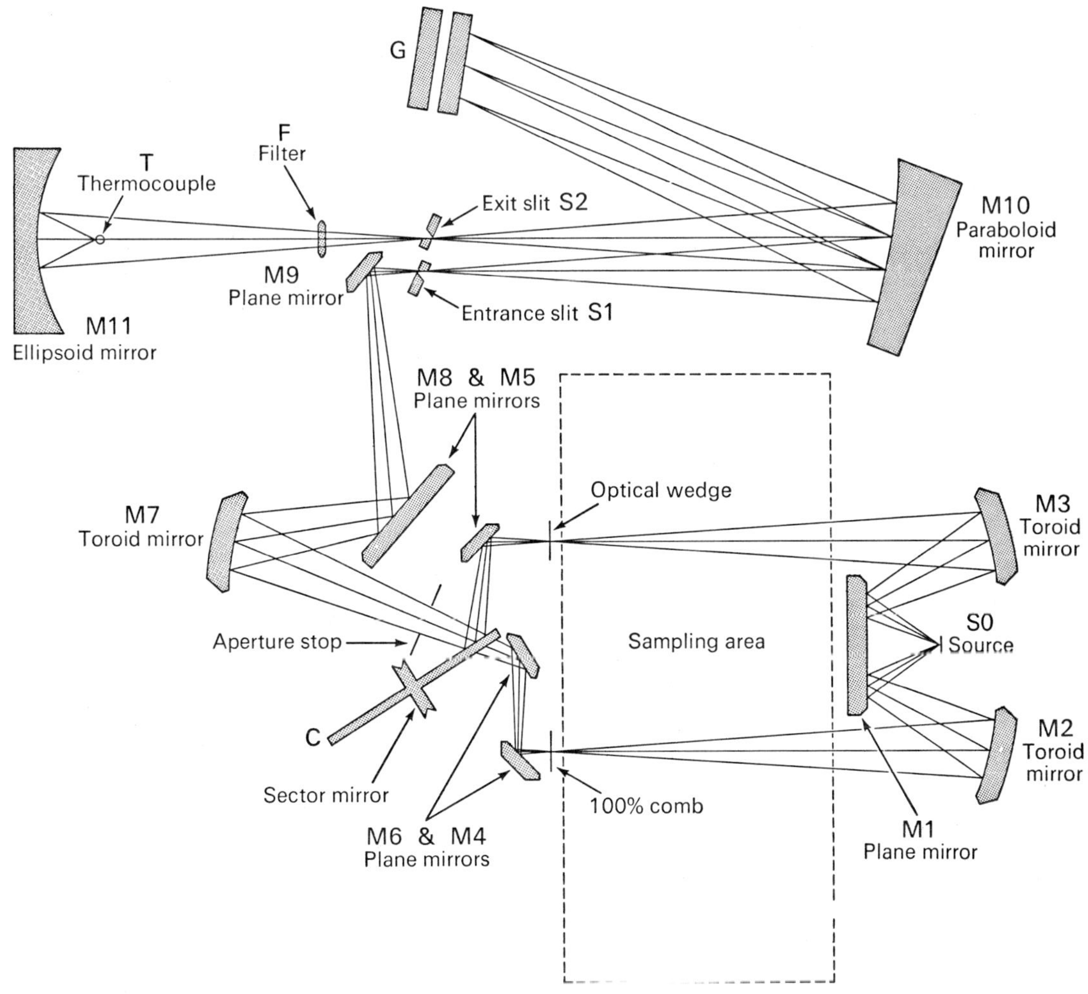

FIGURE 7·6   Optical schematic of Perkin-Elmer Infrared Spectrophotometer, Model 337. [Courtesy Perkin-Elmer Corporation, Norwalk, Connecticut]

The function of the detector is to convert infrared radiation into electrical energy, which may actuate the pen recorder. The detector is actually a unit composed of an ellipsoidal mirror, which focuses energy on a thermocouple, the latter being located within the focus of the range. The thermocouple is composed of two wires of different metals (or alloys) joined together for the electrical properties involved. Since the electrical signal is small, it must be amplified before a response can be produced on the pen recorder.

The infrared spectrophotometer is so widely different from other spectrophotometers, and so complex in its structure, that any extensive discussion is beyond the scope of an elementary course in quantitative analysis. The interested student is referred to textbooks on instrumental methods.

EXERCISES

1. The electromagnetic spectrum is described as being continuous. List shorter and longer wavelength regions beyond the extremes given in Table 7·1. Is it possible to use these regions for absorptiometry? Explain your answer.

2. List the reasons, and explain the apparent failures of some materials to follow Bouguer–Beer Laws. Give examples of systems that illustrate the errors producing such failures.

3. A certain spectrophotometer operates with a fairly constant error of 0.5% in transmittance. Compute the percentage uncertainty in the concentration, when this concentration has been adjusted to give 37.7% transmittance.

4. The molecule MA has an absorption peak of 550 mμ in water. The molecule dissociates somewhat into M^+ and A^- ions. If M^+ and A^- absorb only in the UV region, will a plot of absorbance at 530 mμ show a positive or negative deviation from Beer's Law? Explain your answer.

5. Predict the color of the following solutions; (a) one that absorbs all colors above 500 mμ; (b) a solution with a transmittance peak of 575 mμ; (c) a solution having a strong absorption band from 500 to 530 mμ?

6. When should copper be determined colorimetrically, and when should it be determined volumetrically?

7. Is a colorimetric method accurate when there is more than one colored substance present?

8. Of what particular (quantitative or qualitative) usefulness is; ultraviolet analysis by spectrophotometry?

9. Explain the practical uses of infrared analyses.

10. An unknown solution containing the cupric-ammonia complex ion gave a transmittance of 16.1%, when plotted on Figure 7·3. What is the percentage of copper in the original sample, which was weighed as 2.500 g?

CHAPTER 8

Precipitation and Complex Formation Titrations

INTRODUCTION

The majority of volumetric (titrimetric) procedures encountered by students in elementary quantitative chemistry involve soluble products, and the end points are usually detected by organic indicators. However, the types of reactions that involve volumetric precipitation and complex formation frequently deviate from the foregoing description. A *precipitation method* of titration, as the name implies, is based upon the formation of a slightly soluble substance. On the other hand, a *complexometric titration* produces a complex ion or molecule, which is usually soluble.

8·1 Prerequisites of Precipitation and Complexation Titrations

For any volumetric method of analysis, certain essential conditions are necessary:

(1) The reaction must proceed rapidly to completion.

(2) The titrimetric reaction should not be complicated by additional side-reactions.

(3) There should be some detectable change in the solution undergoing titration at the equivalence (stoichiometric) point.

(4) A suitable indicator should be available. (Occasionally a potentiometer is used in the titration to obtain a titration break in the potential versus titrant readings.)

Many precipitation and complex formations are not rapid, whereas others are associated with interfering secondary reactions. More pertinent is the fact that very few reliable indicators are applicable to such titrations. Those which have been adapted to the titration of the halide ions with silver, and of the sulfate ion with the barium ion are the most important. As mentioned above, it is also possible to titrate the halides with the use of a sensitive potentiometer.

The functioning of a titrimetric indicator usually falls into one of the following categories:

(1) Formation of a secondary precipitate (of different color) as the indicator.

(2) Formation of a colored complex ion at the equivalence point.

(3) Adsorption indicators.

(4) A potentiometric titration, which produces a sharp change in volume of titrant versus potential at the equivalence point.

(5) Appearance of a turbidity when complexation is barely complete.

(6) Special colored indicators for the titration (involving complexation) of certain metal ions with ethylenediaminetetraacetic acid. (The latter is usually abbreviated to EDTA.)

STOICHIOMETRY

8·2 Calculations Involved in Precipitations and Complexation Reactions

The computations are similar in principle to those in other branches of quantitative chemistry, with the exception that the materials involved are ions and compounds rather than hydrogen ions or electrons.

As in other calculations,

$$meq_1 = meq_2$$

and

$$meq = N \times ml$$

or

$$meq = \frac{grams}{meq\ w}$$

Consequently, the equivalent weight of an ion or compound, reacting in a titrimetric precipitation or a complexometric titration, is defined as the weight of the substance that will furnish or react with one gram atomic weight of a univalent cation, or $\frac{1}{2}$ gram atomic weight of a divalent cation, and so on. Thus, the equivalent weight of silver nitrate equals the formula weight of the substance, since the silver ion is an ion bearing one positive charge. Therefore, if exactly 16.988 grams of primary standard silver nitrate are dissolved in water, and diluted to the graduated mark on a one-liter flask, the normality may be computed as follows:

meq of $AgNO_3$ weighed

$$= meq\ of\ AgNO_3\ in\ solution$$

and

$$\frac{g\ of\ AgNO_3}{meq\ w} = N \times ml$$

$$N = 0.1000$$

Another example, involving the relationship between formula weight and equivalent weight, denotes that two Ag^+ ions react with one CrO_4^{--} ion in the following manner:

$$2Ag^+ + CrO_4^{--} \rightleftharpoons Ag_2CrO_4 \downarrow$$

In the ionic reaction, the equivalent weight of the CrO_4^{--} (a divalent ion) is fw/2, since one chromate ion reacts with two silver ions.

Another more complicated reaction, showing precipitation stoichiometry, is that of the divalent zinc ion reacting with potassium ferrocyanide according to the equation:

$$3Zn^{++} + 2K_4Fe(CN)_6 \rightleftharpoons$$
$$K_2Zn_3[Fe(CN)_6]_2 + 6K^+ \downarrow$$

In the reaction, the equivalent weight of the divalent zinc ion is fw/2. On the other hand, two ferrocyanide ions react with three zinc ions (a ratio of 2:3); consequently, the equivalent weight of potassium ferrocyanide is fw/3. As in other quantitative relationships, the ratio of $meq_1 = meq_2$ does not change.

PRECIPITATION TITRATIONS AND INDICATORS

8·3 Formation of a Colored Secondary Precipitate: The Mohr Method

Silver nitrate may be obtained in a primary-standard purity, which is readily soluble in water. However, it is customary to standardize a solution, if possible, against the substance to be analyzed. Procedures for standardizing silver nitrate against pure sodium chloride are given in the experimental portion of the text.

In the Mohr Method, the chloride ion is precipitated by titration with standard silver nitrate in the presence of a small amount of chromate ion. The end point is detected by the appearance of a colored precipitate of Ag_2CrO_4, which forms immediately after the precipitation of white silver chloride. The primary precipitation reaction, and the secondary indicator reaction are indicated as follows:

$$Ag^+ + Cl^- \rightleftharpoons AgCl\ (white) \downarrow$$

and

$$2Ag^+ + CrO_4^{--} \rightleftharpoons Ag_2CrO_4 \downarrow$$

(lemon-yellow colored solution to orange-colored precipitate)

The foregoing procedure is termed fractional precipitation, since two different precipitates (with different solubilities) are formed in the titration. In a saturated solution of these precipitates, the solubility products are as follows:

$$[Ag^+][Cl^-] = 10^{-10}$$

and

$$[Ag^+]^2[CrO_4^{--}] = 10^{-12}$$

The Ag_2CrO_4 is more soluble because two silver ions occur in the solubility product expression. For this reason, the relative solubilities of the two materials may appear misleading in terms of the solubility products. A saturated solution of AgCl is $10^{-5}F$ in respect to the silver ion, whereas the concentration of the silver ion in saturated Ag_2CrO_4 is $10^{-4}F$. The difference between the two values is much too small to produce a clean-cut separation by fractional precipitation. However, the apparent difference is magnified by using a small concentration of CrO_4^{--} ions in comparison to the Cl^- ion concentration, in the original solution. Even with this precaution there is an inherent error of approximately 0.1% in the Mohr Method. Since this error occurs in both the standardization of the silver nitrate solution and the analysis of an unknown chloride, the two errors should theoretically cancel each other.

In actual practice, the concentration of the chromate ion indicator is approximately $0.001F$ which causes the solution to have a slight lemon-yellow color. If the concentration of the indicator is much larger, its yellow color will prevent the proper detection of the formation of the orange-colored precipitate (Ag_2CrO_4).

It is emphasized that all silver precipitates are photosensitive, and darken fairly rapidly in the presence of light. This sensitivity can interfere with the distinction between the formation of white AgCl and orange-colored Ag_2CrO_4.

The Mohr Method has been a time-honored procedure for many years. The beginning student may have trouble in detecting the end point, but for an experienced analyst this difficulty does not exist.

8·4 Formation of a Colored Complex Ion at the Equivalence Point: The Volhard Method

The Volhard Method was originally developed as a determination for the silver ion, but is now largely used as an indirect method for the analyses of the halide ions. It is one of the few titrimetric precipitations conducted in an acid solution (HNO_3). Consequently, it has a decided advantage over most other precipitations, inasmuch as a regulation of pH is not necessary.

The equation for the primary reaction of the titration is

$$Ag^+ + SCN^- \rightleftarrows AgSCN\text{(a white ppt)}\downarrow$$

and the indicator reaction, utilizing ferric alum, is

$$Fe^{3+} + 6SCN^- \rightleftarrows Fe(SCN)_6^{3-}$$
$$\text{(a soluble reddish-brown complex ion)}$$

Nitric acid in excess produces no interference, but it is necessary to rid the solution of any nitrous acid that may be present, since the latter has a tendency to destroy the indicator (SCN^-) ion. The removal of HNO_2 may be accomplished by boiling the solution vigorously.

The Volhard Method is an indirect determination of certain halides $(Cl^-, Br^-, \text{ and } I^-)$. The essential equation for the reactions is

$$\underset{\text{(excess)}}{Ag^+} + X^- \rightleftarrows AgX\downarrow$$

where X^- may be the chloride, bromide or iodide ions. In theory, after the precipitation of the halides, the excess Ag^+ ions remaining in the solution are back-titrated with SCN^- until the reddish-brown color of the complex indicator ion is obtained. Therefore, the computation of a particular halide should appear as

$$[(\text{ml of } Ag^+ \times N \text{ of } Ag^+) -$$
$$(\text{ml of } SCN^- \times N \text{ of } SCN^-)] \times$$
$$\frac{\text{meq wt of halide} \times 100}{\text{wt of sample}} \times \% \text{ halide}$$

In practice, the above computation may not apply to the determination of the chloride, inasmuch as AgCl is more soluble than AgSCN,

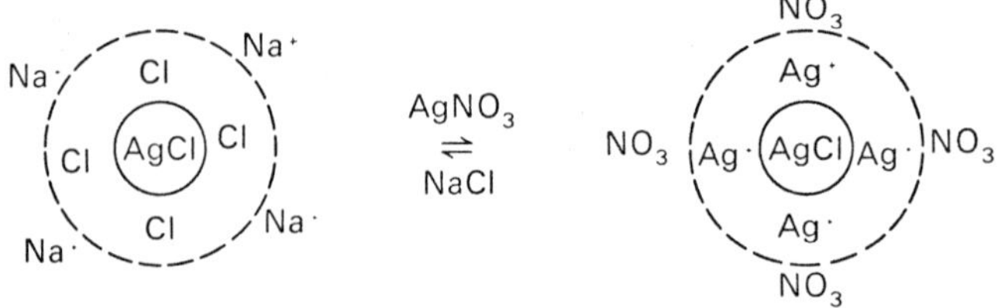

FIGURE 8·1   Adsorption of ions by colloidal AgCl particles

as indicated by the ratio

$$\frac{[Ag^+][Cl^-]}{[Ag^+][SCN^-]} = \frac{K_{sp} \text{ of AgCl} = 10^{-10}}{K_{sp} \text{ of AgSCN} = 10^{-12}}$$

$$= 10^2$$

Consequently, when SCN^- ions are added in sufficient excess to give an end point, a considerable amount of the Cl^- (from the precipitated AgCl) goes back into solution, resulting in an appreciable error.

The above error can be eliminated by filtering and washing the AgCl before the addition of the SCN^- ion. Then titrate the filtrate and washings (containing excess Ag^+ ions) with SCN^- in a separate back-titration. A simpler procedure is now available. If nitrobenzene is added, with stirring, after the precipitation of AgCl, the organic liquid forms an oily waterproof coating to the precipitate, thereby preventing it from going back into solution in the presence of SCN^- ions.

Silver bromide and silver iodide are so insoluble that the nitrobenzene treatment can be eliminated in their determinations. The approximate ratios of the solubility products involved are listed as

$$\frac{[Ag^+][Br^-]}{[Ag^+][SCN^-]} = \frac{K_{sp} \text{ of AgBr}}{K_{sp} \text{ of AgSCN}}$$

$$= \frac{10^{-13}}{10^{-12}} = 10^{-1}$$

and

$$\frac{[Ag^+][I^-]}{[Ag^+][SCN^-]} = \frac{K_{sp} \text{ of AgI}}{K_{sp} \text{ of AgSCN}}$$

$$= \frac{10^{-16}}{10^{-12}} = 10^{-4}$$

8·5 Adsorption Indicators: The Fajan Method

The adsorption-indicator method for the chloride ion employs a direct titration with standard silver nitrate solution, using dichlorofluorescein as the indicator. The standard silver nitrate is the only solution necessary for preliminary preparation. As a result of its composition, silver chloride first precipitates in a colloidal condition. There is a tendency for the colloidal formation, as is true of all inorganic colloids, to adsorb its own ions. Consequently, the silver chloride colloid adsorbs either chloride ions or silver ions (whichever are in excess) on the surface of the finely divided precipitated particles. It is generally assumed that an inorganic colloid is surrounded by two adsorption layers: a primary adsorption layer in which adsorbed ions are strongly held, and are relatively immovable; a secondary adsorption layer (slightly removed from the primary layer) in which ions are less strongly held and are somewhat mobile.

The condition of colloidal adsorption of ions is loosely depicted in Figure 8·1. When **excess** chloride ions are present, the colloid is negatively charged, since this ion appears in the primary layer. On the other hand, an excess of silver ions produces a positively charged particle, in which the primary layer contains silver ions. In Figure 8·1, the dotted circles indicate the secondary adsorption layers of the particles just described. In the reaction between silver nitrate and sodium chloride, with no other ions present, the ionic equation is

$$Ag^+NO_3^- + Na^-Cl^- \rightleftharpoons AgCl + Na^+NO_3^-$$
$$\downarrow$$
colloidal particle

The colloid particle produced by the reaction will adsorb either Cl^- or Ag^+ in the primary

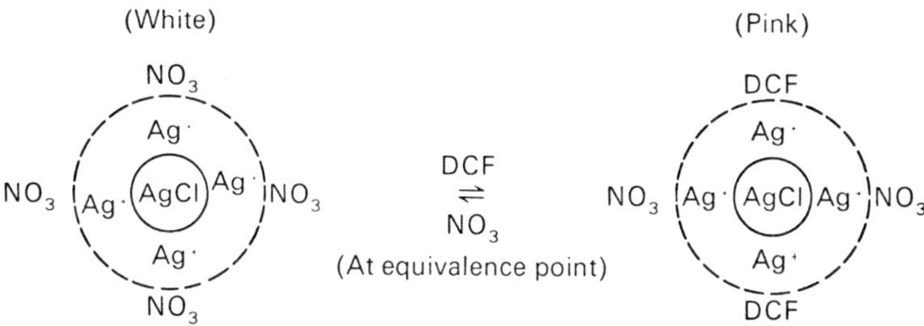

FIGURE 8·2   Color change due to adsorption indicator at equivalence point

layer, and either Na^+ or NO_3^- in the secondary layer, depending upon which original reactant is in excess. The electrical charge is thereby reversible as indicated in Figure 8·1.

Dichlorofluorescein (and any related dye) is an organic substance, which ionizes slightly in a solution undergoing titration, to give hydrogen ions and highly colored anions (designated as DCF^-). When unadsorbed, the dye has a yellowish-green color in solution. Immediately after the equivalence point is reached in the titration, the DCF^- ions replace some of the NO_3^- ions in the secondary layer. Under electrical stress (a supposition only), the yellowish-green color of the DCF^- changes to pink, and these ions impart their color to the colloidal AgCl particle. The color change, in the presence of DCF^- ions, is hypothetically indicated in Figure 8·2. Although the titration with its resulting color changes is theoretically reversible, in practice the reversibility is sometimes disappointing. As stated many times before, any silver precipitate is sensitive to light (darkens), and the changes in color at the equivalence points may not be sharp. As the pink colloidal particles appear, the solution should be whirled or stirred constantly, inasmuch as there is a tendency for the pink particles to settle to the bottom of the titration flask. The presence of a small amount of dextrin aids in prolonging the colloidal condition of the AgCl precipitate.

8·6 Potentiometric Precipitation Titrations

Potentiometric methods can be applied to a number of precipitation titrations, particularly to those determinations involving the chloride, bromide, and iodide ions, using silver nitrate as the titrant. The ordinary pH meter can be converted into a potentiometer by switching (by means of a control knob) to the scale calibrated in millivolts (usually ±700 mv). A silver electrode and a regular calomel electrode (Figure 4·2) are used to establish contacts with the solution undergoing titration. For the silver electrode, use a simple silver wire attached to the meter by a special connector, or a commercial silver electrode. The potential, E, of the silver electrode in volts is given by the equation

$$E = 0.80 - 0.059 \log \frac{1}{[Ag^+]} \quad \text{(at 25°C)}$$

Consider the titration of a chloride with silver nitrate as the titrant. During the course of the titration, there is eventually a rapid increase in the potential indicated on the meter, as well as a sudden decrease in volume of $AgNO_3$ necessary to produce the abrupt change in potential. The midpoint for these changes corresponds to the equivalence point (see Figure 15·2).

Potentiometric precipitation titrations are time-consuming, inasmuch as several minutes of mechanical stirring are required before equilibrium is attained, especially near the stoichiometric point. Furthermore, such a titration must be continued well beyond the hypothetical end point in order that the graph of the titration may attain a symmetrical curve (see Figure 15·2). Silver iodide is more insoluble than silver chloride and, as might be expected, a more satisfactory titration (as evidenced by the graph of results) can be obtained with a soluble iodide than with a chloride.

Because a detailed procedure for the potentiometric titration of the chloride ion is given in Chapter 15, it will not be described in this section. In conclusion, it should be noted that a titration of this nature is tedious, and requires considerable care and patience in finding the desired inflection in the potential curve. Otherwise, the equivalence point may be inaccurately placed on the graph of results. Also, the fact that the silver precipitate is photosensitive always produces a *methodic* error, which increases with the time necessary to complete the titration.

COMPLEX FORMATION TITRATIONS

The subject of complex ions is sufficient to fill more than one volume alone; yet the application of such ions to quantitative chemistry is not as extensive as might be expected. The accuracy demands of analytical chemistry are seldom encountered in complex-ion titrations. However, the potential use of complex ions and complex compounds in titrimetric analysis is quite varied, and should not be omitted from any discussion of quantitative chemistry. Their versatility is indicated by the fact that various complexes occur as ionic or molecular compounds, which may be soluble or insoluble in water. The subjects of coordination numbers (showing roughly the number of components added to the central ion) and of the possible geometric structures of complex substances provide a broad discussion area. However, in elementary quantitative chemistry the subject of complexation formations is limited to a few outstanding examples, which usually lend themselves to titrimetric analyses (although gravimetric methods are sometimes used). As in other titrimetric reactions, complex-formation reactions must be rapid, stoichiometric, and quantitative with a suitable indicator for the equivalence point.

8·7 Definition of a Complex Substance

The terms *complex ion* and *coordination compound* are frequently used very loosely. Particularly is this true of the first term, which is widely employed as a catchall for many meanings. Yet most of the loose usages cannot be labeled as erroneous, because of the large number of connotations that have become attached to the two terms. In a broad sense, any ion in which there is more than one atom may be called a complex ion. Also, any symmetrical arrangement of atoms about a central atom (or ion) in the construction of a molecule or ion, may be designated a coordination compound.

An incomplete definition of the two terms (complex ion and coordination compound), grouped together, states that such a compound contains a central atom or ion (usually metallic) surrounded by a cluster of ions or molecules. The clustering ions or molecules are termed *ligands*, and the structure of the total compound frequently assumes a definite geometric symmetry. Furthermore, if the central atom and its ligands produce a ring structure, the resulting substance is termed a *chelate*. Under certain conditions, the central atom forms more than one ring with its ligands, and such a structure is loosely defined as a polydentate compound, the forms being bidentate, tridentate, quadridentate, and so on. The polydentate nomenclature refers to the number of points of attachment for the ligands, and may be loosely described as two-toothed, three-toothed, four-toothed, and so on.

A simple form of complex ion involving a central ion with two ligands, but possessing linear symmetry, is exemplified by the silver–ammonia complex:

$$Ag^+ + 2NH_3 \rightleftharpoons Ag(NH_3)_2{}^+$$

(a soluble complex ion)

A compound involving a ring closure (a chelate) is typified by the union of the ethylene diamine molecule with a platinic ion, which may be symbolized by the formula:

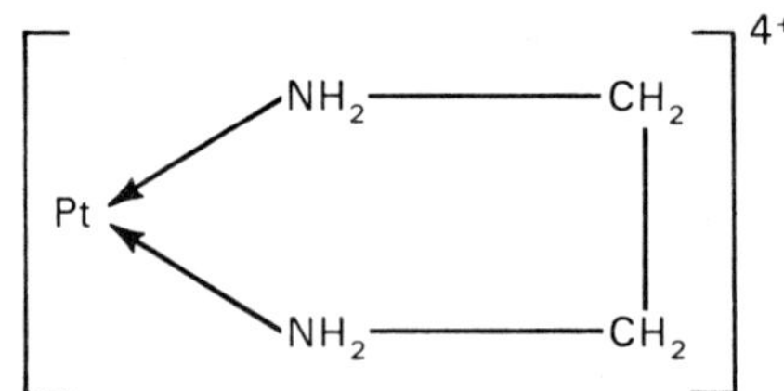

A complex involving geometric symmetry (but no ring formation) is illustrated by the union of six NH_3 molecules with one cobaltic ion to form an octahedral structure. The octahedron is not easily depicted on a planar surface, in which the cobaltic ion occurs at the center of the structure with one molecule of NH_3 on each corner of the octahedron. A somewhat inexact method of indicating the structure is given as follows:

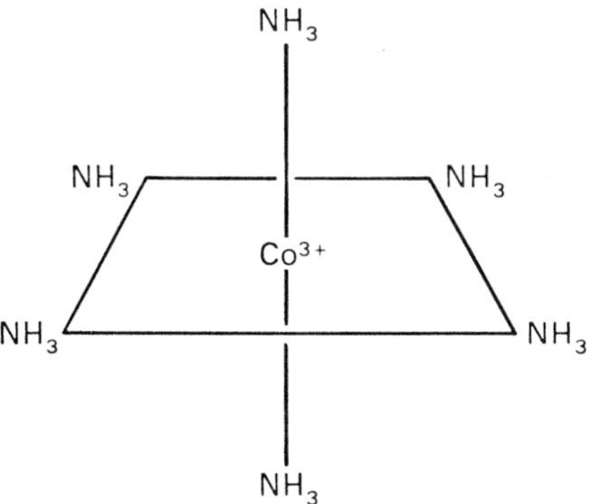

None of the three typical examples listed for complex-ion formations are suitable for quantitative titrations; however, they do indicate three different structures, which may be involved in quantitative titrimetric structures.

8·8 Dissociation of a Complex Ion

Monodentate ligands (coordinating agents) that occupy a single coordination position are invariably added or dissociated in single steps. For the purpose of simplification, the stepwise procedure will be ignored, and the dissociation of a simple complex with two ligands may be indicated as a single equilibrium, such as

$$Ag(NH_3)_2{}^+ \rightleftharpoons Ag^+ + 2NH_3$$

and

$$K_d = \frac{[Ag^+][NH_3]^2}{[Ag(NH_3)_2{}^+]} = 6.8 \times 10^{-8}$$

The dissociation of the silver–amino ion is not quantitative for several reasons; for example, the lack of a suitable indicator. However, it illustrates the fact that complexes do have dissociation constants, which might be quantitative, if all concentration values can be established.

8·9 Formation or Disappearance of a Solid Phase as an Indicator

A number of complexation titrations involving the CN^- ion are possible. Two typical methods are outlined below.

Formation of a Turbidity after Complexation: The Liebig Method. The Liebig titration is widely used for the determination of the cyanide ion; it also constitutes another procedure for the titrimetric analysis of the silver ion. Two chemical reactions are involved, as indicated by the equations,

$$Ag^+ + 2CN^- \rightleftharpoons Ag(CN)_2{}^-$$
(analytical reaction)

$$Ag^+ + Ag(CN)_2{}^- \rightleftharpoons Ag[Ag(CN)_2] \text{ (indicator}$$
reaction)
(white precipitate in the form of a turbidity)

If a solution of standard $AgNO_3$ solution is added, dropwise, to a solution of cyanide ions, the first reaction (complex formation) is the analytical one. At the equivalence point, a slight turbidity is caused by the formation of a very small amount of fairly insoluble $Ag[Ag(CN)_2]$ precipitate, which has a K_{sp} of 10^{-12}. The turbidity serves as an excellent indicator of the stoichiometric point.

The Disappearance of a Turbidity at the End Point. The procedure for such an analysis is a modification of the Liebig Method; an example is the titrimetric analysis of the complex nickel ion. The particular titration is performed in an ammoniacal medium, containing sufficient silver iodide to impart a slight turbidity to the original solution. The end point is detected by the disappearance of the dispersed AgI resulting from the dropwise addition of a standard cyanide ion solution. The equations for the reactions are,

$$Ni(NH_3)_4{}^{++} + 4CN^- \rightleftharpoons$$

$$Ni(CN)_4{}^{--} + 4NH_3$$
(analytical reaction)

$$AgI + 2CN^- \rightleftharpoons Ag(CN)_2{}^- + I^-$$
(indicator reaction)

In the indicator reaction, the equivalence point is signaled by the disappearance of the turbidity.

ETHYLENEDIAMINETETRAACETIC ACID AS A TITRIMETRIC COMPLEXATION AGENT

A family of tertiary amines, containing carboxylic acid groups, forms stable complexes with a large number of metal ions. The pioneer work of Schwarzenbach* on the volumetric determination of the hardness of water has apparently opened "a field within a field" involving titrimetric complex formations. Adaptations of his procedure have been extended to embrace titrimetric determinations of more than fifty metallic ions. The new complexometric field is sometimes called *chelometric titrations*, and the aminocarboxylic acids are referred to as *complexons*. The most important complexon is ethylenediaminetetraacetic acid. The present discussion will be confined to the chemistry of this unique compound.

8·10 The Structure of Ethylenediaminetetraacetic Acid (EDTA)

This versatile complexing agent, usually abbreviated to EDTA, has the structure

$$\text{HOOC-CH}_2 \diagdown \qquad \diagup \text{CH}_2\text{-COOH}$$
$$\text{N-CH}_2\text{-CH}_2\text{-N}$$
$$\text{HOOC-CH}_2 \diagup \qquad \diagdown \text{CH}_2\text{-COOH}$$

Only the four hydrogen atoms on the —COOH groups are ionizable into hydrogen ions. It is usually described as a fairly weak acid. The titrimetric methods based on its use arise from the formation of a stable 1:1 chelate complex between the complexon and a metallic ion, regardless of the valence of the ion.

The fairly high molecular weight of the amino acid predicts a limited solubility in water, which

* Schwarzenbach, G., W. Biedermann, and F. Bangerter, *Helv. Chim. Acta*, **29** (1946).

is supported by experimentation. Because of its low solubility, it is usually marketed as the soluble disodium salt with a structure, in solution, as follows:

$$2\text{Na}^+ +$$
$$\text{HOOC-CH}_2 \diagdown \qquad \diagup \text{CH}_2\text{-COO}^-$$
$$\text{N-CH}_2\text{-CH}_2\text{-N}$$
$$^-\text{OOC-CH}_2 \diagup \qquad \diagdown \text{CH}_2\text{-COOH}$$

8·11 The Ionization of EDTA

The chelon bears out the theory that predicts that when two protons are derived from opposite ends of a fairly long acid molecule, the ionization constants are fairly close together (Section 3·10).

The ionizations, in stages, give the following values, which are expressed in an abbreviated form:

$$H_4Y \rightleftharpoons H^+ + H_3Y^-$$
$$K_1 = 1.0 \times 10^{-2}$$

$$H_2Y^- \rightleftharpoons H^+ + H_2Y^{--}$$
$$K_2 = 2.0 \times 10^{-3}$$

$$H_2Y^{--} \rightleftharpoons H^+ + H_3Y^{3-}$$
$$K_3 = 6.3 \times 10^{-7}$$

$$H_3Y^{3-} \rightleftharpoons H^+ + Y^{4+}$$
$$K_4 = 5.0 \times 10^{-11}$$

Again, it is interesting to note the relative ionization constants with the positions of the ionizable hydrogens within the original molecule.

8·12 Complexes of Metals with EDTA

As stated previously, EDTA is unique in that it generally forms a 1:1 complex with metals regardless of their valencies; hence, it is more convenient to use formal concentrations of the solution, as a reagent, rather than normality. Thus, 1 ml of a 0.1 F solution is always equivalent to 0.1 mfw of the particular ion concerned. Again, there is the relationship of

$$\text{meq}_1 = \text{meq}_2$$

The unusual 1:1 complexes formed between the chelon and various metallic ions can be exemplified by the following equations, in which

M represents dissimilar metal ions in different valencies:

$$M^{++} + H_2Y^{--} \rightleftharpoons MY^{--} + 2H^+$$

$$M^{3+} + H_2Y^{--} \rightleftharpoons MY^- + 2H^+$$

$$M^{4+} + H_2Y^{--} \rightleftharpoons MY + 2H^+$$

As will be discussed in Section 8·13, it is obvious that the extent to which these complexes are formed is dependent upon the pH of the reacting solution.

With the exception of the alkali metals, all metallic ions react, to some extent, with EDTA in a ratio of 1:1 to form a complex ion. Many of the complexes are sufficiently stable for titrimetric analysis. The necessary details concerning the conditions for chelate structures are omitted (see advanced inorganic textbooks); nevertheless, certain functional groups must occur in the chelating molecule (or ion), in definite positions, before a chelate structure is possible. Chelate rings are usually five- or six-membered; however, the five-membered ring is the most stable.

EDTA forms octahedral complexes, and some of these may be optical isomers. The compound is potentially a sexidentate ligand although quinquedentate and quadridentate complexes are common. A simplified structure of a sexidentate complex, in combination with a divalent ion, is presented as follows:

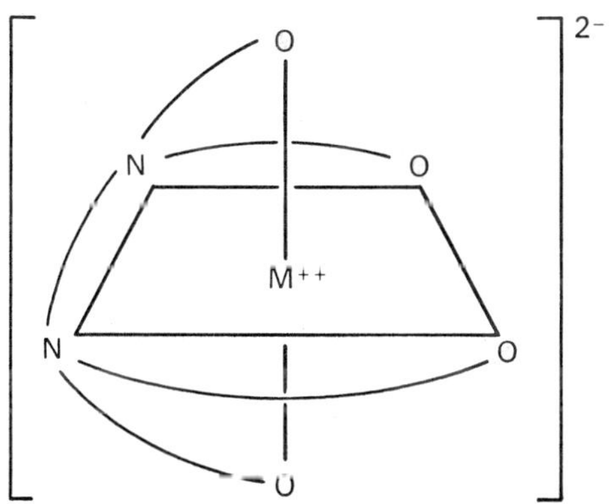

8·13 Solutions of EDTA

Because of the problem of solubility, the chelon is usually in the form of the soluble disodium salt of EDTA. Some hydrogen ions are liberated during the titration of metal ions with the ligand. In the absence of a complexing cation, EDTA reacts with either sodium hydroxide or potassium hydroxide as a weak acid, but in the presence of a complexing cation it acts as a strong acid, and may be titrated directly, with the use of ordinary acid–base indicators. Thus, the presence of a complexing cation has a profound effect upon the pH of the solution, as well as the extent of complex formation.

The fractions of EDTA present in various ionic forms have been calculated fairly exactly,* as a function of pH. Since these calculations are not used in this course, it is sufficient to indicate approximately the predominant ions as a function of pH, given as follows:

pH	Predominant Ionic Form
0–3	H_3Y^-
3–6	H_2Y^{--}
6–10	HY^{3-}
10–14	Y^{4-}

An approximate $0.1\,F$ solution of disodium EDTA (dihydrate) for titrimetric purposes is prepared by dissolving 37.22 g in distilled water, and diluting to mark in a one-liter volumetric flask. Several methods of standardization have been recommended, but it seems most desirable to standardize against a solution of the metal to be determined. If this is not possible, use a given weight of pure zinc dissolved in dilute HCl. The zinc solution is titrated in the presence of a buffer ($NH_4Cl + NH_3$), using a few drops of Eriochrome Black T as the indicator. The end point is a clear blue color.

8·14 Indicators for EDTA Titrations

When no complexing cation is present, EDTA reacts with KOH as a weak acid. However, in the presence of a complexing cation it acts as a strong acid, and may be titrated with ordinary acid–base indicators. Consequently, the ml of KOH consumed in the titration gives a method for computing the amount of cation. Many divalent cations can be analyzed by such a

* Latinen, H. A., *Chemical Analysis*, New York, McGraw-Hill, 1960, page 228.

procedure. For example, 4 gfw of KOH are equivalent to 1 fw of EDTA, and also 1 fw of the cation.

The foregoing procedure constitutes a direct titration; however, many EDTA analyses are not so simple (see Section 8·15), and may require special indicators. Such indicators are organic compounds that form colored compounds with the metal ions involved in the titrations. These are usually dyes of the o,o'-dihydroxy azo type. Probably the most widely used of these indicators are Eriochrome Black T and Calmagite, which are used in the titration of magnesium, as well as in the determinations of other ions. Eriochrome Black T (frequently abbreviated to Erio T) is a tribasic acid, which at a pH of 10 exists largely as the blue HIn^{--} ion.

$$H_3In \rightleftharpoons 2H^+ + HIn^{--}$$
$$\text{(blue)}$$

The blue form reacts with Mg^{++} ions to give a red complex

$$Mg^{++} + HIn^{--} \rightleftharpoons H^+ + MgIn^-$$
$$\text{(wine red)}$$

Although many indicators have been proposed and used for complexometric titrations, the majority of EDTA titrations are carried out with one of the following dyes: Eriochrome Black T, Calmagite, Naphthyl Axonine, Metalphthalein, or Xylenol Orange (all trade names). However, many other indicators, and indicator procedures, are available. Welcher* has given an excellent summary of such indicators and their uses with EDTA.

8·15 Titration Methods Using EDTA

A variety of procedures are used in the applications of EDTA to titrimetric analyses. The most common methods are outlined briefly:

Direction Titration. This method was mentioned in Section 8·14, as an example of the direct titration of the magnesium ion with the chelon,

---

* Welcher, F. J., *The Analytical Uses of Ethylenediamine-tetraacetic Acid*, Princeton, Van Nostrand, 1958.

using Eriochrome Black T as the indicator. The solution is initially red because of the large concentration of $MgIn^-$. At the equivalence point, the $MgIn^-$ is converted to Mg^{++}, which is shifted to the right in the reaction; therefore, the blue color at the end point in HIn^{--}, as shown by the equation:

$$MgIn^- + H^+ \rightleftharpoons Mg^{++} + HIn^{--}$$
$$\text{(blue)}$$

It may be necessary to add an auxiliary complexing agent, such as the citrate or tartrate ions, to prevent the precipitation of the hydroxides. Also, a buffer $(NH_3 + NH_4Cl)$ is frequently used to supply a complexing agent and to control the pH of the solution. However, a few ions such as iron(III) and bismuth can be titrated directly in acid solution.

At a pH of 10 (as provided by an ammonia buffer) some metals, including zinc, nickel, cadmium, and copper, can be held in solution in the form of ammine complexes, and titrated directly with EDTA. Yet in a few cases, the slow reaction of the metal ion with EDTA presents an obstacle to direct titration. Chromium(III) is an outstanding example of a reaction that is too slow for a feasible titration, and the same is true for the aluminum ion.

The calculations of EDTA titrations are quite simple, and follow the same stoichiometry, which holds true for titrimetric processes; therefore,

$$\text{meq of unknown cation} = \text{meq of standard EDTA}$$

in chelometric chemistry an even better relationship is indicated as

$$\text{mfw of unknown cation} = \text{mfw of standard EDTA}$$

However, in most cases, the values are so small that the final result is best expressed in parts per thousand, or parts per million.

Back-Titrations Involving EDTA. Various metal ions cannot be determined by the direct titration method because the high pH would result in the precipitation of the metal ion as a hydroxide. Nevertheless, many of the difficulties in direct EDTA titrations can be avoided by back-titrations. Such a titration requires two standard solutions other than the unknown metal ion undergoing analysis. The stoichiometric calculation may be indicated as

mfw of unknown cation =

mfw of standard EDTA

− mfw of standard metal ion

where the standard metal ion is that cation used in the standardization of EDTA.

An excess of standard EDTA is added to the unknown cation, and the excess back-titrated with a standard metal solution (used in the standardization) until the end point is reached. In such a procedure the unknown cation is already complexed, and the precipitation of the hydroxide is prevented. Usually trace metal impurities are also complexed by EDTA, but do not interfere with the analysis.

Inasmuch as two burets are employed, and since the reaction is reversible, it is possible to pass back and forth across the end point, until a satisfactory reading is obtained. Yet, it should be recalled that dual readings on two different burets constitute a *methodic* error in any titrimetric analysis.

The back-titration method is applicable to the analysis of cobalt, nickel, copper, aluminum, iron, titanium, as well as a number of other metals. Back-titrations are particularly useful for the analysis of metal ions that form stable complexes with EDTA, and for which a satisfactory indicator is not available. In such cases the excess EDTA may be back-titrated with standard magnesium solution, using Erio T as the indicator. However, the metal–EDTA complex must be more stable than the MgY^{--} complex, or the Mg^{++} will decompose the metal complex under analysis.

Displacement Titrations. The procedure avoids the use of two burets and two standard solutions. In displacement titrations, an excess of a solution containing EDTA, complexed with zinc or magnesium, is added. If the metal, undergoing titration, forms a more stable complex than zinc (or magnesium) the latter is displaced.

For example, the hypothetical M^{++} ion may replace Zn^{++} as follows:

$$M^{++} + ZnY^{--} \rightleftharpoons MY^{--} + Zn^{++}$$

The liberated zinc ion is then titrated with standard EDTA solution. This type of reaction is particularly useful when no satisfactory indicator is available for the metal being analyzed.

8·16 Errors Involved in EDTA Titrations

Complex formation with EDTA takes place to such an extent that various titrations serve for the determination of over fifty metal ions. Thus, EDTA is almost universal in its chelating action on the various cations. Yet, this convenience can also be an inconvenience. Stability constants of metal–EDTA complexes are listed in many textbooks, but such constants are affected by many conditions. Hence, a stability constant can be meaningless unless the conditions for the formation of the complex are also given.

According to Welcher,* there is an approximate correlation between the charge on the cation and the stability of the resulting complex. Moreover, both the charge and stability are functions of pH conditions. These relationships are listed as follows:

(1) The complexes of EDTA with divalent cations are very stable in basic and slightly acid solution.

(2) Complexes of trivalent cations are stable in a pH range of 1 to 2.

(3) Complexes of quadrivalent cations are frequently more stable at a pH of less than one.

The versatility of EDTA may create a *methodic* error when two stability (formation) constants are too close to be properly differentiated.

In addition to approximate errors resulting from the lack of selectivity between stability constants, certain specific errors may occur in EDTA titrations. Some of these are listed as follows:

(1) Impurities in the reagents used in the analysis.

* *Ibid.*

(2) Impurities in the water used as a solvent in the determination.

(3) Impurities from storage containers of the solutions used in EDTA titrations. Glass containers are unsuitable (even borosilicate glass), because of cations liberated from the surface of the glass. Plastic containers are commonly used as storage vessels, and are usually satisfactory provided the plastic material is unaffected by the chemicals in solution.

(4) Many of the indicators used in EDTA titrations are fairly unstable, and deteriorate over a period of time.

(5) The amount of buffer solution, necessary for some titrations, may be critical inasmuch as an excess of buffer removes certain metals from their chelates.

8·17 Some Questions Concerning the Future of EDTA Titrations

The reagent, EDTA, was first recommended as a titrant by Schwarzenbach in 1946, and since that date a large number of research papers on its uses have been published. It will be years before the reliability of the vast number of publications can be properly assessed. Similarly, the use of various indicators that have been used and suggested will have to be carefully adjudged. The very large number of complexes formed with EDTA raises a doubt concerning the specificities of so many titrations. Only time and experimentation can provide answers to settle this question.

EDTA was first suggested as a reagent for the estimation of hardness in water. Its introduction and expansion have replaced many time-consuming gravimetric procedures. It is widely asserted that, except for the alkali metals, the majority of cations can be determined by some method of EDTA titration. The statement concerning the versatility of EDTA is true, but how accurate are these many possibilities? Titrimetric uses of EDTA are largely for trace amounts of the various cations. Although the results may be relatively good, the accuracy in terms of quantitative standards is yet to be established.

TYPES OF EXERCISES

Type 1. Formality in Volumetric Precipitations

Problem: Calculate the formality of a silver nitrate solution, if 35.00 ml of the solution are required to titrate 0.1200 g of pure NaCl (dissolved in water).

$$\text{formality} = \frac{\text{wt of NaCl in g}}{\text{ml of AgNO}_3 \times \text{mfw of NaCl}}$$

$$\text{formality} = \frac{0.1200}{35.00 \times 0.05844} = 0.0586$$

Type 2. Titration Using the Mohr Method

Problem: What is the inherent error in the Mohr Method involving 35.00 ml of 0.1000 F NaCl versus 0.1000 F AgNO$_3$ solution? The end point occurs (including added water) in a volume of 100 ml. The K$_2$CrO$_4$ indicator in this volume is 0.001 F.

Solution:

$$[Ag^+][Cl^-] = 1.8 \times 10^{-10} \, (K_{sp})$$

$$[Ag^+] = 1.3 \times 10^{-5} \, F \text{ in a saturated solution of AgCl}$$

and

$$[Ag^+]^2[CrO_4^{--}] = 1.3 \times 10^{-12} \, (K_{sp})$$

$$[Ag^+]^2 = \frac{1.3 \times 10^{-12}}{1 \times 10^{-3}} \text{ (conc of indicator)}$$

$$[Ag^+] = 3.6 \times 10^{-5} F \text{ in saturated solution of } Ag_2CrO_4$$

$3.6 \times 10^{-5} \times 100 \text{ ml} = 3.6 \times 10^{-3} \text{ mfw of } Ag^+ \text{ from } Ag_2CrO_4$
$1.3 \times 10^{-5} \times 100 \text{ ml} = 1.3 \times 10^{-3} \text{ mfw of } Ag^+ \text{ from } AgNO_3$

$2.3 \times 10^{-3} \text{ mfw of } Ag^+ \text{ in excess to reach end point}$

$$\frac{2.3 \times 10^{-3}}{0.1 \text{ F}} = 2.3 \times 10^{-2} \text{ ml of excess } AgNO_3 \text{ for end point}$$

$$\frac{2.3 \times 10^{-2}}{35 \text{ ml}} \times 100 = 0.066\% \text{ error}$$

Type 3. Titration Using the Volhard Method

Problem: What is the percentage of iodide in an impure 0.4000 g sample to which has been added 37.50 ml of 0.1000 F AgNO$_3$, and which requires 7.50 ml of 0.1000 F KSCN solution for back-titration?

Solution:

$$\frac{[(\text{ml of AgNO}_3 \times F) - (\text{ml of KSCN} \times F)] \times \text{meq wt of } I^- \times 100}{\text{wt of sample}} = \% \text{ iodide}$$

$$\frac{[(37.50 \times 0.1000\ F) - (7.50 \times 0.1000\ F)] \times 0.1269 \times 100}{0.4000} = \% \text{ iodide}$$

$$\frac{(3.750 - 0.750) \times 0.1269 \times 100}{0.4000} = 9.52\% \text{ iodide}$$

Type 4. Titration Using the Fajan Method

Problem: Excess silver nitrate (0.1000 F) is used to titrate a sample of MgBr$_2$, using dichloro-fluorescein as an indicator. Predict the primary and secondary adsorbed ions at the end point.

Solution:

Primary adsorbed ion is Ag^+.
Secondary adsorbed ions are NO_3^- and DCF^-.

Type 5. Titration Using the Liebig Method

Problem: An impure 0.4000 g sample of NaCN, when analyzed by the Liebig Method required 35.00 ml of 0.1000 F AgNO$_3$ solution to titrate to an end point indicated by a turbidity. Calculate the percent of NaCN in the sample.

$$Ag^+ + 2CN^- \rightleftharpoons Ag(CN)_2^- \text{ (soluble complex)}$$

Therefore,

$$1 \text{ mfw of } Ag^+ \doteq 2 \text{ mfw of } CN^-$$

Thus,

$$\frac{(\text{ml of AgNO}_3 \times F) \times (2 \times \text{mfw of NaCN}) \times 100}{\text{wt of sample}} = \% \text{ NaCN}$$

and

$$\frac{(35.00 \times 0.1000) \times (2 \times 0.4901) \times 100}{0.4000} = 85.75\% \text{ NaCN}$$

Type 6. Titration with EDTA

Problem: A sample of pure zinc weighing 5.000 g is dissolved in dilute HCl and diluted exactly to mark in a 500 ml volumetric flask. The titration of 25 ml of the solution with EDTA requires 58.00 ml of the chelon. Calculate the formality of the EDTA solution.

Solution:

$$1 \text{ mfw of Zn}^{++} = 1 \text{ mfw of EDTA}$$

$$\frac{5.000}{65.37} \times 2 = 1.530 \text{ mfw of Zn}^{++} \text{ per liter}$$

$$\text{formality of Zn}^{++} \text{ solution} = \frac{1.530}{1000} = 0.00153 \ F$$

Therefore,

$$0.00153 \ F \times 25/58 = 0.000431 \ F \text{ (for EDTA solution)}$$

EXERCISES

1. Establish the relative formality of the analate and titrant in the reactions indicated by the following equations:

$$2\text{Ag}^+ + \text{CrO}_4^{--} \rightleftharpoons \text{Ag}_2\text{CrO}_4 \downarrow$$

$$3\text{Ag}^+ + \text{PO}_4^{3-} \rightleftharpoons \text{Ag}_3\text{PO}_4 \downarrow$$

$$3\text{Ca}^{++} + 2\text{PO}_4^{3-} \rightleftharpoons \text{Ca}_3(\text{PO}_4)_2 \downarrow$$

2. Calculate the formality of a silver nitrate solution, 40 ml of which are required to titrate 0.1500 g of primary standard NaCl.
3. Assume that a 0.4000 g sample contains 50.56% NaCl, and calculate the volume of $0.1000 \ F$ AgNO_3 that would be required for its titration.
4. How many g of AgNO_3 are necessary to titrate 30.00 ml of $0.1000 \ F \ \text{SCN}^-$?
5. What should be the concentration of the chromate ion so that there will be no *methodic* error in the determination of Cl^- by the Mohr Method? What would be the disadvantage of such a concentration?
6. Calculate the volume of $0.1000 \ F \ \text{AgNO}_3$ required to saturate 35 ml of $0.0010 \ F \ \text{K}_2\text{CrO}_4$ with Ag_2CrO_4?
7. In the titration of $0.1000 \ F$ KBr with $0.1000 \ F \ \text{AgNO}_3$, what formal concentration of chromate ion must be present to form Ag_2CrO_4 at the stoichiometric point?
8. Explain why the titration of the Cl^- ion in the Mohr Method must be within a pH range between 7 and 10.
9. In the Volhard Method, why must the AgCl precipitate be filtered or treated with nitrobenzene?
10. What weight of sample containing 40.00% NaCl should be taken for analysis by the Volhard Method so that 15.00 ml of $0.1000 \ F$ KSCN is used in back-titration? The total volume of $0.1000 \ F$ AgNO_3 used in the titration is 40.00 ml.
11. Why will the Volhard Method work only in an acid solution? What is the maximum permitted pH?
12. How many ml of $0.1000 \ F$ KSCN are needed to back-titrate 35.00 ml of $0.1000 \ F \ \text{AgNO}_3$ if added to 0.4000 g of KI in water solution?
13. In the Volhard procedure for the chloride ion, 35.00 ml of $0.1000 \ F \ \text{AgNO}_3$ was used, and 10.00 ml

of 0.1000 F KSCN was required for back-titration. What is the percentage of chloride ion in the sample?

14. A sample of primary standard sodium chloride weighing 0.3000 g is dissolved in water and 40.00 ml of $AgNO_3$ solution added. After inactivating the AgCl with nitrobenzene, the excess of $AgNO_3$ required 5.00 ml of KSCN for back-titration. If 1.000 ml KSCN $\approx$ 1.200 ml of $AgNO_3$, compute the normality of the $AgNO_3$ and KSCN solutions.

15. 40.00 ml of 0.1000 KSCN are added to 250 ml of 0.1000 F $AgNO_3$. What is the formality of the silver ion in the resulting solution?

16. Indicate a probable error introduced by failure to add dextrin in the Fajan Method for the analysis of the chloride ion.

17. What affect would the presence of a large amount of $Al(NO_3)_3$ have upon the Fajan titration of the chloride ion with dichlorofluorescein as the indicator?

18. Explain the tendency for coprecipitation of the ions of a reagent in a volumetric precipitation process.

19. Describe the relationship between the charge on an adsorption indicator ion and the charge of Ag^+, if the latter is used as the titrating agent.

20. A sample containing 0.3000 g of NaCN requires 20.00 ml of 0.1000 F $AgNO_3$ to obtain a faint turbidity (Liebig Method). What is the percentage of NaCN in the sample.

21. A sample of KCN weighing 10.000 g is dissolved in water and diluted to mark in a one-liter volumetric flask. What is the approximate formality of the KCN, as determined by the Liebig Method?

22. Calculate the solubility of silver chloride in 1.000 F KCN.

23. How many milliliters of 0.1000 F $AgNO_3$ are necessary to titrate 0.4000 g of KCN in water solution?

24. A 0.5000 g sample of pure $CaCO_3$ is dissolved in 6 F HCl, and diluted to mark in a 500 ml volumetric flask. A 25 ml aliquot required 24.50 ml of EDTA for titration. What is the formality of the EDTA solution?

25. What weight of disodium EDTA (dihydrate) is necessary to make exactly one liter of 0.0100 F solution?

26. A 200 ml sample of water was titrated with 0.0100 F EDTA solution, and 35.00 ml of the chelon was required. Calculate the hardness of the water in parts per million.

27. Why cannot iron(III) be titrated with EDTA in a basic solution?

28. Titration of a 100 ml water sample for total hardness required 10.00 ml of 0.0100 F EDTA. Compute the hardness of the water in parts per million.

29. Why is a buffer added with EDTA solution in the titration of certain cations?

CHAPTER 9

Equilibria in Precipitation Reactions

All ionic substances are soluble in water to some extent; consequently, the most insoluble electrolytes are slightly dispersed in this medium. Such a heterogeneous system is one of the quantitative applications of the equilibrium constant (Section 3·2).

THE PROBLEM OF SOLUBILITY

9·1 The Nature of Solubility

The three states of matter existing in nature can theoretically produce nine types of solutions or dispersions. However, the present discussion will be limited to the examination of heterogeneous systems of ionic materials in water or, more specifically, to equilibria existing in saturated solutions of difficultly soluble electrolytes. The application of equilibrium constants to such systems is known as the solubility product principle.

The nature of solubility is too complex for any single simplification, but as a general rule salts dissolve more readily in a solvent composed of molecules containing electrostatic *dipoles*. Such a molecule is electrically unsymmetrical with a separation of positive and negative charges within its structure. For example, in the water molecule, which has an angular shape, both hydrogen atoms are on the same side of the oxygen atom, so that a permanent dipole moment results. The electrical dissymmetry of

the water molecule may be sufficiently strong to remove ions from the surface of a fairly insoluble crystal. Solid electrolytes of the ionic type are composed of ions held in a crystal lattice by electrostatic charges upon the ions. In an ionic crystal, each ion tends to surround itself with as many oppositely charged ions as possible. Figure 9·1 shows a diagrammatic sketch of the possible ionic arrangement on a planar surface of such a crystal. It is assumed that the coulombic forces operating between ions are strong in a crystal of a fairly insoluble electrolyte; consequently, relatively few ions are pulled into solution as hydrated particles.

9·2 General Factors Affecting Solubility

Two factors that affect the solubility of solids in liquids are known: they are (a) temperature and (b) the physical and chemical structure of the solvent and solute particles.

Temperature. The solution of an ionic solid is analogous to a melting process. As stated in Section 9·1, the ions in crystal are geometrically interspaced so that each ion has the same pattern with respect to neighboring ions. The forces that hold the ions in their relative positions are strong within a fairly insoluble electrolyte. However, when a crystal is heated, the kinetic energy of the particles making up its composition is increased.

The increased energy is manifested, at first, by a greater vibration of each particle. When the amplitude of vibration becomes large enough, the coulombic forces holding the ions together are overcome, and the particles break away from their relatively fixed positions. In this manner a crystal melts, and the ions of which it was originally composed are free to move throughout the molten substance. The same conditions prevail when a solid enters into solution as a result of an elevation of temperature; consequently, as a general rule, a rise in temperature favors increased solubility.

Usually, it is desirable to form and wash precipitates for gravimetric analysis at elevated temperatures, provided such substances are relatively insoluble. It may be assumed that the impurities that contaminate a precipitate are usually small in amount, and are more soluble; hence, the hot-washing procedure does not result in serious errors. Although the foregoing statement is a general rule, it does not apply to fairly soluble gravimetric precipitates, such as magnesium ammonium phosphate and lead sulfate.

Physical and Chemical Structure of Solvent and Solute Particles. The solubility of most inorganic compounds is much smaller in a mixture of an organic solvent and water, than in pure water. Occasionally precipitation in such a mixture is necessary to reduce solubility losses; for example, the gravimetric precipitation of $CaSO_4$ is quantitative in a mixture of water and ethanol, but not in pure water.

It is generally assumed that the components of inorganic compounds are held together, to some extent, by ionic (coulombic) forces; however, pure ionic bonds are rare, and many bonds are partly ionic and partly covalent. It can be predicted that a compound such as BiI_3 is largely covalent; consequently, its slight solubility in water cannot be attributed to permanent dipole forces.

The crystal, or lattice, energy of an ionic compound may be defined as the energy required to separate a formula weight of its ions by an infinite distance; however, in terms of solubility the definition will be less exactly defined as the energy to remove the ions from the crystal into availability for hydration. The building up of electrostatic forces between ions in a crystal is an exothermic process, as is evidenced by the energy necessary to convert a crystal into a molten condition by heat.

The interaction that takes place when an ionic substance is converted into hydrated ions is called *hydration*, and the energy change involved in the process may be denoted as hydration energy.

The nature of the packing, or arrangement, of ions in the lattice is determined largely by the diameters of the ions. Furthermore, the ions are held in their relative positions by strongly electrostatic forces resulting from the charges on the ions. To break down the architecture of the crystal, the electrostatic forces must be overcome. For a substance to be soluble, more energy must be available to break down the crystal than was liberated in building up the ionic lattice. In other words, the energy of hydration must be greater than the lattice energy.

To produce ions in solution, the dipole attraction of the water molecules must be sufficient to pull ions away from the lattice to form hydrated ions. It may be deduced that the solvation of ions is an endothermic process. Consequently, the solubility of a salt is roughly proportional to the magnitude of, and the difference between, two energy factors: namely, the lattice energy and the energy of hydration.

Electrolytes in which the lattice energies are relatively large are usually not very soluble in water. The reaction of a solid electrolyte with the solvent may be indicated by the equation

$$MA_{(solid)} + xH_2O \underset{\text{crystallization}}{\overset{\text{solution}}{\rightleftharpoons}}$$

$$M(H_2O)_y{}^+ + A(H_2O)_z{}^-$$

The extent to which a solid will dissolve depends upon the amount of energy released in the hydration of the ions.

The foregoing explanation of lattice energy versus hydration energy is an oversimplification of the true solution process. Actually, other energy factors are involved, but a complete explanation is beyond the scope of a course in elementary quantitative chemistry.

THE SOLUBILITY PRODUCT PRINCIPLE

9·3 Derivation of Solubility Product Expression

In a saturated solution of an electrolyte, the equilibrium between the solid and its ions is in a dynamic state of balance between two opposing reactions. The dissolving reaction is the one in which ions are pulled away from the crystal to form soluble hydrated ions, whereas the precipitation reaction is the reverse, in which ions in solution lose their water of hydration and precipitate upon the surface of the crystal. The solubility product principle is only valid for fairly insoluble electrolytes. A good illustration of a reversible equilibrium of this type is a saturated solution of silver chloride. Most ions are assumed to be hydrated in water solution, and certainly the silver and chloride ions should be no exceptions. However, to simplify the derivation of the solubility product principle, the hydration of these particular ions is ignored.

Solid silver chloride forms a cubic lattice, similar to NaCl, in which there is an alternate spacing of silver and chloride ions. A diagrammatic sketch of the possible ionic arrangement on the planar surface of the cubic crystal is indicated in Figure 9·1. A positive silver ion must

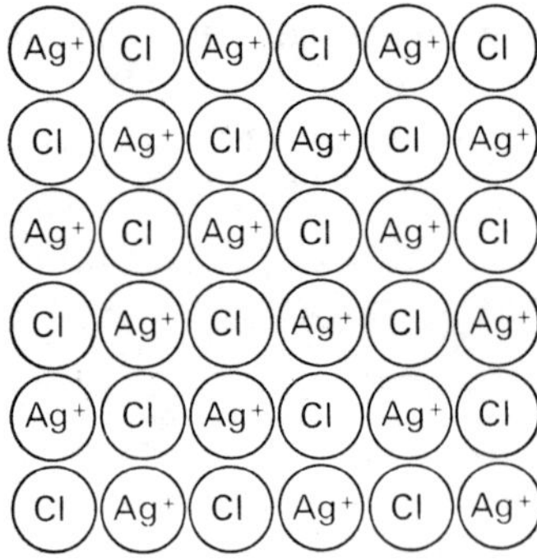

FIGURE 9·1   A diagrammatic sketch of the possible ionic arrangement on the planar surface of a crystal

be deposited above a negative chloride ion and vice versa, in order to continue the crystal lattice. The rate at which an ion leaves the surface is proportional to the number in the surface layer, whereas the rate of deposition of an ion is proportional to its concentration in solution and also to the number of places at the surface to which it can be attached. Equilibrium is attained when equal numbers of silver and chloride ions are dissolved and deposited in the same time.

The total surface of the solid phase, exposed within the saturated solution, may be represented as unity. If X is the fraction of the surface covered by the silver ions, then $1 - X$ is the fraction available to the chloride ions. The rate V_1 of silver ions leaving the surface may be expressed as $K_1 X$, in which K_1 is a proportionality constant, or

$$V_1 = K_1 X$$

The rate V_2 of silver ions deposited on the surface is proportional to the concentration of the silver ions in the solution, and to the number of available spaces at the surface for attachment. The concentration of silver ions is $[Ag^+]$, and $1 - X$ denotes the number of places to which silver ions may be attached. Therefore,

$$V_2 = K_2(1 - X)[Ag^+]$$

At equilibrium, the rates of solution and deposition are equal, or $V_1 = V_2$, and

$$K_1 X = K_2(1 - X)[Ag^+]$$

The above equation may be rearranged as follows:

$$[Ag^+] = \frac{K_1 X}{K_2(1 - X)}$$

In a similar fashion, the solution of chloride ions may be expressed as $K_3(1 - X)$, and the deposition of chloride ions at the surface as $K_4 X[Cl^-]$. At equilibrium, these expressions are equal to each other, or

$$K_3(1 - X) = K_4 X[Cl^-]$$

and upon rearrangement, the chloride concentration is

$$[Cl^-] = \frac{K_3(1 - X)}{K_4 X}$$

If the equations denoting $[Ag^+]$ and $[Cl^-]$ are multiplied together, the resultant is

$$[Ag^+][Cl^-] = \frac{K_1 X}{K_2(1 - X)} \times \frac{K_3(1 - X)}{K_4 X}$$

and the unknown factors of X can be eliminated to give

$$[Ag^+][Cl^-] = \frac{K_1 K_3}{K_2 K_4}$$

Since all of the K values are constant, they may be equated to single constant, which is the *solubility product constant.*

$$[Ag^+][Cl^-] = K_{sp}$$

In the foregoing derivation, the difficultly soluble salt chosen for an illustration forms only two ions in solution; therefore, the product of the concentration of the ions in solution is the solubility product constant. However, if the fairly insoluble salt produces more than two ions, as, for example, $Pb_3(PO_4)_2$ releases five ions in solution when it dissolves, or

$$Pb_3(PO_4)_2(s) \rightleftarrows 3Pb^{++} + 2PO_4{}^{3-}$$

then the solubility product expression for this compound is

$$[Pb^{++}]^3[PO_4{}^{3-}]^2 = K_{sp}$$

Therefore, the solubility product principle may be generally stated as follows: *In a saturated solution of a difficultly soluble electrolyte, the product of the formal concentrations of the ions, each concentration raised to the power equal to the number of times the ion occurs in the formula, is a constant at a given temperature.*

9·4 Relationship of Solubility Product to Solubility

In Sections 8·3 and 8·4, solubility product values were used for illustrative purposes to explain certain chemical procedures. The values were rounded to whole numbers for the purpose of simplification, consequently, they were somewhat inexact.

The student may be puzzled by divergencies in assigned values for solubility products when obtained from a variety of reference sources. These differences are the results of various methods of experimental determinations, and they may reflect activity corrections as well as other factors, such as hydration, crystal size, and so on. Henceforth, all solubility products used in various problems, as well as those listed in Appendix 6, will be those values which have attained wide acceptance and are apparently reliable.

The solubility product principle is an application of the law of chemical equilibrium (Section 3·1), and the solubility product constant is a special form of an equilibrium constant. In spite of its name, the solubility product is not in itself a quantitative value in terms of weight and concentration. For example, $AgCl$ and Ag_3PO_4 are soluble in water to the extent of 1.8×10^{-3} and 5.7×10^{-3} g/liter, respectively (as obtained from chemical handbooks), signifying that silver phosphate is more than three times as soluble as silver chloride in terms of weight. On the other hand, the solubility product constants of $AgCl$ and Ag_3PO_4 are 1.8×10^{-10} and 1.25×10^{-20}, respectively, or, in relative numerical terms, the K_{sp} of $AgCl$ is several billion times larger than the K_{sp} of Ag_3PO_4. From these examples, it is obvious that the numerical values of the solubility products cannot be compared in relation to the solubility of the two salts. The reason for the wide difference in the size of these solubility products is indicated by the mathematical expressions for them:

$$[Ag^+]^3[PO_4{}^{3-}] = K_{sp} \text{ of } Ag_3PO_4$$

$$[Ag^+][Cl^-] = K_{sp} \text{ of } AgCl$$

The concentration of the silver ion is cubed in the expression for the K_{sp} for Ag_3PO_4, whereas in the expression of the K_{sp} of $AgCl$, the silver-ion concentration is not taken to a power.

To indicate more precisely the relative solubility of these different types of salts, the concentration of the silver ion in a saturated solution of each salt is calculated. First of all, assume that the formal solubility of an ionic substance is x. When such a compound dissolves in water, each of its ions will also be x values, except that each x will be multiplied by the integer, and also raised to the power of the number of times that the particular ion occurs in the formula.

The calculations are as follows:

$$[Ag^+][Cl^-] = 1.8 \times 10^{-10}$$

$$(x)\quad(x) = 1.8 \times 10^{-10}$$

$$x = \sqrt{1.8 \times 10^{-10}}$$

$$= 1.3 \times 10^{-5}\,F$$

$$= \text{formality of AgCl}$$

Concentration of Ag^+ in saturated AgCl solution $= x = 1.3 \times 10^{-5}\,F$

$$[Ag^+]^3[PO_4{}^{3-}] = 1.25 \times 10^{-20}$$

$$(3x)^3\quad(x) = 1.25 \times 10^{-20}$$

$$27x^4 = 1.25 \times 10^{-20}$$

$$x = 1.4 \times 10^{-5}\,F$$

$$= \text{formality of Ag}_3\text{PO}_4$$

Concentration of Ag^+ in saturated $Ag_3PO_4 = 3x = 4.2 \times 10^{-5}\,F$.

Therefore, the solubilities of AgCl and Ag_3PO_4 in terms of g/liter, as obtained from their solubility product constants, are:

solubility of AgCl

$$= 1.3 \times 10^{-5}\,F \times 143.32\ \text{(fw)}$$

$$= 1.85 \times 10^{-3}\ \text{g/liter}$$

solubility of Ag_3PO_4

$$= 1.4 \times 10^{-5}\,F \times 418.58\ \text{(fw)}$$

$$= 5.7 \times 10^{-3}\ \text{g/liter}$$

Since the numerical value of the solubility product constant depends upon the concentration of ions produced by the relatively insoluble salt, this value is a rough qualitative indication of solubility. Consequently, a very small solubility product value indicates a very insoluble compound; and a large value, a more soluble substance. To illustrate this qualitative significance with two extremes, the solubility product constants of Ag_2S and $AgC_2H_3O_2$ are 6.3×10^{-50} and 1.8×10^{-3}, respectively, and from these values it may be deduced that silver sulfide is a very insoluble substance and that silver acetate is comparatively soluble. Only when two insoluble salts produce the same number of ions may numerical values of their solubility products be directly compared; even under these conditions, the comparison is somewhat relative because of the difference in the formula weights of the two compounds. For example, AgBr and AgI have solubility product constants of 5×10^{-13} and 4.5×10^{-17}, respectively, indicating that silver iodide is less soluble than silver bromide, but the relative values are not quantitative because the formula weights of the two compounds are different.

9·5 Calculations of Solubility Product Constant from Solubility

To illustrate the calculation of the solubility product constant from a known value of solubility, *consider the compound,* $BaSO_4$, *which has a solubility of 0.0024 g/liter.* Find the value of K_{sp} for $BaSO_4$.

Solubility dissociation:

$$Ba^{++}SO_4{}^{--}\ _{(solid)} \rightleftarrows Ba^{++} + SO_4{}^{--}$$

$$xF \qquad\qquad\qquad xF \qquad xF$$

Gram-formula weight of $BaSO_4 = 233.40$. The mathematical expression for K_{sp} is

$$[Ba^{++}][SO_4{}^{--}] = K_{sp}$$

Formal solubility of $BaSO_4$

$$= \frac{240.0 \times 10^{-5}}{233.4} = 1.02 \times 10^{-5}\ \text{gfw/liter}$$

Therefore:

$$(1.02 \times 10^{-5})(1.02 \times 10^{-5}) = K_{sp}$$

$$= 1.05 \times 10^{-10}$$

The calculations involving solubility product constants become more complex for a relatively insoluble salt which produces several ions in solution. The following problem illustrates this type: *Calculate the solubility product constant of lead phosphate, if the solubility of this salt is 1.27×10^{-6} g/liter.*

Gram-formula weight of $Pb_3(PO_4)_2 = 811.5$.

TABLE 9·1   Fractional Precipitation of $BaSO_4$ and $PbSO_4$

$10^{-3}\,F\,Ba^{++} \times 10^{-9}\,F\,SO_4^{--} = 1 \times 10^{-12}$	no pptn of $BaSO_4$, less than K_{sp}
$10^{-3}\,F\,Pb^{++} \times 10^{-9}\,F\,SO_4^{--} = 1 \times 10^{-12}$	no pptn of $PbSO_4$, less than K_{sp}
$10^{-3}\,F\,Ba^{++} \times 10^{-8}\,F\,SO_4^{--} = 1 \times 10^{-11}$	no pptn of $BaSO_4$, less than K_{sp}
$10^{-3}\,F\,Pb^{++} \times 10^{-8}\,F\,SO_4^{--} = 1 \times 10^{-11}$	no pptn of $PbSO_4$, less than K_{sp}
$10^{-3}\,F\,Ba^{++} \times 10^{-7}\,F\,SO_4^{--} = 1 \times 10^{-10}$	pptn of $BaSO_4$ begins, K_{sp} exceeded
$10^{-3}\,F\,Pb^{++} \times 10^{-7}\,F\,SO_4^{--} = 1 \times 10^{-10}$	no pptn of $PbSO_4$, less than K_{sp}
$10^{-4}\,F\,Ba^{++} \times 10^{-6}\,F\,SO_4^{--} = 1 \times 10^{-10}$	pptn of $BaSO_4$, K_{sp} exceeded
$10^{-3}\,F\,Pb^{++} \times 10^{-6}\,F\,SO_4^{-} = 1 \times 10^{-9}$	no pptn of $PbSO_4$, less than K_{sp}
$10^{-5}\,F\,Ba^{++} \times 10^{-5}\,F\,SO_4^{-} = 1 \times 10^{-10}$	pptn of $BaSO_4$, K_{sp} exceeded
$10^{-3}\,F\,Pb^{++} \times 10^{-5}\,F\,SO_4^{--} = 1 \times 10^{-8}$	pptn of $PbSO_4$ begins, K_{sp} exceeded
$10^{-6}\,F\,Ba^{++} \times 10^{-4}\,F\,SO_4^{--} = 1 \times 10^{-10}$	pptn of $BaSO_4$, K_{sp} exceeded
$10^{-4}\,F\,Pb^{++} \times 10^{-4}\,F\,SO_4^{-} = 1 \times 10^{-8}$	pptn of $PbSO_4$, K_{sp} exceeded

Formal solubility of $Pb_3(PO_4)_2$

$$= \frac{1.27 \times 10^{-6}}{811.5} = 1.56 \times 10^{-9}\ \text{gfw/liter}$$

$$Pb_3(PO_4)_{2\,(solid)} \rightleftharpoons$$

$$(1.56 \times 10^{-9}\,F)$$

$$3Pb^{++} \qquad + \qquad 2PO_4^{3-}$$

$$(3 \times 1.56 \times 10^{-9}\,F) \quad (2 \times 1.56 \times 10^{-9}\,F)$$

The expression for K_{sp} is

$$(Pb^{++})^3(PO_4^{3-})^2 = K_{sp}$$

Substituting formal-concentration values into this expression,

$$(3 \times 1.56 \times 10^{-9})^3(2 \times 1.56 \times 10^{-9})^2 = K_{sp}$$

$$K_{sp} = 1 \times 10^{-42}$$

9·6 Fractional Precipitation

If a difficultly soluble salt is to be precipitated from solution, the product of the concentration of the ions of the salt must exceed the solubility product constant, at a given temperature. Conversely, if the precipitation of a sparingly soluble salt is to be prevented, the concentrations of the ions must be kept below the values that exceed the solubility product of the salt. There are occasions in analytical chemistry in which it is necessary to precipitate one or more ions from a mixture, leaving other ions in solution that are precipitated later by the same reagent in a different concentration. The possibility of fractional precipitation brings up certain questions regarding the conditions necessary for separating two ions by means of the same precipitating agent. To consider these questions and their answers, suppose that a solution contains a mixture of barium ions and lead ions, and that the concentration of each of these ions is 0.001 F. If a solution of sulfate ions is added to this mixture in successive small portions, what will be the composition of the precipitate first formed? Will the compound having the smallest solubility product precipitate first, and, if so, will it precipitate completely before the other compound begins to precipitate? Under what conditions will simultaneous precipitation of both compounds take place? To answer these questions, examine Table 9·1, which indicates what happens in a solution containing such a mixture of barium and lead ions when the sulfate ion is added in a concentration of $10^{-9}\,F$ and then increased in successive steps to a concentration of $10^{-4}\,F$. The solubility product constants of $BaSO_4$ and $PbSO_4$ are considered to be 1×10^{-10} and 1×10^{-8}, respectively.

Since the concentration of barium and lead ions are equal in the original mixture, the successive addition of sulfate ions causes $BaSO_4$ to precipitate first, because its solubility product constant is exceeded first. Continued addition of sulfate ions precipitates $BaSO_4$ until the solubility product constant of $PbSO_4$ is exceeded, and then both compounds precipitate together. However, the concentration of Ba^{++} ions is reduced from $10^{-3}\,F$ to $10^{-5}\,F$ before Pb^{++} ions begin to precipitate as $PbSO_4$. Therefore, when Ba^{++} and Pb^{++} are in equal concentrations, there is a definite range of concentration for the two ions in which $BaSO_4$ can be precipitated alone. In practice, this range is difficult to attain for these two ions. The concentrations of $BaSO_4$ and $PbSO_4$ in the solid phase, as well as the concentrations of Ba^{++} and Pb^{++} ions in the solution, are not equal after the solubility products are exceeded. If a ratio is made of the two solubility product expressions for these compounds, it becomes

$$\frac{[Pb^{++}][SO_4^{--}]}{[Ba^{++}][SO_4^{--}]} = \frac{10^{-8}}{10^{-10}} = \frac{[Pb^{++}]}{[Ba^{++}]} = 10^2$$

The ratio shows that the solid phase contains $BaSO_4$ and $PbSO_4$ in a concentration ratio of $100:1$, whereas in the solution the ratio of the concentrations of Ba^{++} to Pb^{++} ions is $1:100$.

Separation of two ions from a solution by a common precipitating reagent is practical only when the equilibrium ratio between the two substances in the solid phase (sufficiently aged) is quite large. To make a fairly clean separation by fractional precipitation (under ideal conditions) requires a ratio (for a relatively insoluble two-ion salt) as high as one million to one. By way of illustration, the iodide ion may be separated from the chloride ion satisfactorily when the silver ion is used as the precipitating agent. The solubility product expressions, in the form of a ratio, for the two silver halides are

$$\frac{[Ag^+][Cl^-]}{[Ag^+][I^-]} = \frac{1.8 \times 10^{-10}}{4.5 \times 10^{-17}}$$

or

$$\frac{[Cl^-]}{[I^-]} = 4 \times 10^6$$

The equilibrium ratio of the two precipitates, AgCl and AgI, in the solid phase is $4 \times 10^6:1$, and under ideal conditions for precipitation such a separation may be regarded as complete. Unfortunately, most of the calculations and conclusions therefrom are necessarily approximate because such calculations are derived from conditions which may not strictly apply. Explanations for this statement are contained in some of the sections which follow.

SPECIFIC FACTORS AFFECTING SOLUBILITY OF IONIC COMPOUNDS

9·7 Solubility and Particle Size

It is a well-known fact that rapid precipitations favor the formation of unstable states such as gels and nonstable colloids. Also, considerable time may elapse before the unstable forms undergo necessary changes to attain a stable equilibrium state. Explanations of these transformations are fairly complex. Some of them are given in Chapter 10, but in this section only a few pertinent facts are presented.

The solubility of very small crystals of a sparingly soluble salt is greater than that of larger crystals of the same salt. The difference is more pronounced for substances that form hard crystals than for those composed of soft crystals. The effect of particle size upon the solubility of hard crystals may be illustrated by crystals of barium sulfate. Particles of this salt with an average diameter of 0.0018 mm have a solubility of 0.00229 g/liter, whereas particles with an average diameter of 0.0011 mm are soluble to the extent of 0.00415 g/liter. In other words, for the particular dimensions, the smaller crystals of $BaSO_4$ are approximately 80% more soluble than the larger crystals. From experimental data* it has been calculated that barium sulfate particles with an average diameter of 0.00004 mm are about 1000 times more soluble than large crystals. On the other hand, substances that form soft crystals show little difference in solubility with decrease in particle size. For instance, particles of PbI_2 with an average diameter of 0.00004 mm are only about 1.4 times as soluble as large crystals. Other examples of soft crystals with solubilities independent of crystal size are the halides of silver.

For substances that form hard crystals, the decrease in solubility with crystal growth is explained in terms of reduced surface energy. The surface tension that exists at the boundary surfaces of the solid and liquid phases tends to decrease the total surface to the smallest possible

* Dundon, M. L., and E. Mack, *J. Am. Chem. Soc.*, **45**, 2479 (1923).

area. The reduction in surface area is accomplished by small crystals going into solution and then precipitating upon the surface of larger crystals. Usually, when a precipitate is first formed, the solid phase consists of particles of varying sizes. Also, the first precipitated particles are more imperfect in terms of lattice structure. If the precipitate is of the type which may form hard crystals with closely-packed ions, the imperfect or irregular crystals will dissolve more readily than perfect crystals. Again, this may be attributed to the larger surface area of the imperfect crystals.

A finely divided precipitate is difficult to filter; consequently, a precipitate of this nature is usually allowed to age before such an operation is attempted. The aging of a precipitate is customarily called *digestion*, which amounts to letting the precipitate stand until the average particle size has increased enough for easy filtering. Heating increases the rate of digestion because it speeds up the rate of solution of solid phase of most substances. Stirring also favors crystal growth by keeping the concentration of the solution uniform throughout.

Since the solubility of small crystals is greater than that of large crystals, it follows that the solubility product of a difficultly soluble substance, existing as very small crystals, is a larger value than if the solid phases were composed of particles of larger dimensions. However, solubility product constants are not calculated from the solubility of very small crystals, but from ionic concentrations of solutions which have stood in contact with the solid phase during an interval sufficient for complete digestion, or, in other words, until a stable equilibrium is attained.

9·8 Common Ion Effects

The effect of a common ion upon a precipitate in equilibrium with its ions is analogous to the effect of a common ion upon a weak electrolyte in equilibrium with its ions, which was discussed in Chapter 3.

Accordingly, if a solution contains a sparingly soluble salt in equilibrium with its ions, it is to be expected that an increase in the concentration of one of the ions will cause a corresponding decrease in the concentration of the other ion to satisfy the value of the solubility product constant, or, in other words, to maintain the constant value of the solubility product. For instance, in a saturated solution of silver chloride, at 25°C, the concentration of the chloride is $1.3 \times 10^{-5} F$. This value is obtained from the solubility product expression as

$$[Ag^+][Cl^-] = 1.8 \times 10^{-10}$$

$$x = [Ag^+] = [Cl^-]$$

$$x^2 = 1.8 \times 10^{-10}$$

$$x = 1.3 \times 10^{-5} F$$

If the concentration of the silver ion is increased, by adding excess silver ions, to $10^{-4} F$, this increase causes a corresponding decrease in the chloride-ion concentration to $1.8 \times 10^{-6} F$ to satisfy the solubility product constant.

$$[Ag^+][Cl^-] = 1.8 \times 10^{-10}$$

$$(1 \times 10^{-4})[Cl^-] = 1.8 \times 10^{-10}$$

$$[Cl^-] = 1.8 \times 10^{-6} F$$

A decrease in the chloride-ion concentration can be accomplished only by the precipitation of more silver chloride; consequently, the addition of a small amount of a common ion produces a decrease in the solubility of the solid phase.

The decrease in the solubility of a precipitate by common-ion effect is utilized in quantitative analysis to lower the concentration of an ion that is under analysis and that remains in solution. It is common practice to add a slight excess of a precipitating reagent to produce a more nearly complete precipitation of the ion undergoing analysis. On the other hand, a large excess of the common ion is to be avoided, because such an excess may cause increased solubility either by neutral salt effect (Section 9·9) or by the formation of a soluble complex ion (Section 9·10). Experimental evidence indicates that the solubility of AgCl cannot be reduced by common ion effect below the value of $3.6 \times 10^{-7} F$.

The common-ion effect is illustrated by the following problem: *What is the silver-ion concentration in equilibrium with a precipitate of silver chromate, when the solution contains a*

TABLE 9·2   Approximate values of activity coefficients for ions in aqueous solution at 25°C

Ionic charge	Ionic strength, μ				
	0.0001	0.0005	0.001	0.005	0.01
± 1	0.988	0.974	0.964	0.921	0.889
± 2	0.959	0.900	0.862	0.717	0.625
± 3	0.900	0.790	0.716	0.473	0.348

chromate-ion concentration of 0.01 F? Substituting known values into the solubility-product expression:

$$[Ag^+]^2(0.01) = 1.3 \times 10^{-12}$$

$$[Ag^+] = 1.14 \times 10^{-5} \, F$$

9·9 Solubility Product Principle and Salt (Diverse Ion) Effect

From solubility data obtained in pure water, the calculation of the solubility product constant is fairly accurate for a sparingly soluble salt. But such a condition as a difficultly soluble salt existing in pure water is rarely encountered in ordinary laboratory procedures. Precipitations of most compounds usually occur in solutions containing other ions. Also, the presence of other ions that are not common to a precipitate increases its solubility and, in consequence, the value of its solubility product. The effect of an "inert" electrolyte on the solubility of a slightly soluble precipitate is referred to by various names, such as diverse ion, neutral salt, uncommon ion, or activity effect. As an illustration, the K_{sp} of $BaSO_4$ in pure water is 1.07×10^{-10} at 25°C, whereas in a solution of 0.01 F KCl the K_{sp} is increased to 2.78×10^{-10}. In other terms, $BaSO_4$, which has a solubility of 2.3 mg/liter in pure water, dissolves to the extent of 3.9 mg/liter in 0.01 F KCl. The increased solubility change is explained by taking into consideration the activity effects.

In Section 3·3, the relation between actual concentration, effective concentration, and activity coefficient was indicated mathematically as

$$a = fC$$

In 1923, P. Debye and E. Hückel derived a theoretical expression for the calculation of activity coefficients of ions. According to their derivation, the activity coefficient of an ion i is given by the following equation:

$$-\log f_i = \frac{AZ_i^2\sqrt{\mu}}{1 + aB\sqrt{\mu}}$$

where a, A, and B are constants that depend on ionic size, temperature, and dielectric constant of the solvent. The symbol Z_i denotes the number of positive or negative charges on the ion, and μ is the ionic strength of the solution.

In dilute aqueous solutions, the value of aB is close to unity, and $\sqrt{\mu}$ is negligible: consequently, the original equation may be simplified to

$$-\log f_i = AZ_i^2\sqrt{\mu}$$

The resulting equation is designated as the DEBYE–HÜCKEL LIMITING LAW, in which the constant A has a value of 0.51 for aqueous solutions at 25°C, regardless of the nature, size, or charge of the ion. However, this reduced equation only applies to the calculations of ionic strengths in dilute aqueous solutions at 25°C.

The Debye–Hückel Limiting Law is useful in obtaining activity coefficients for ions of various charges and at various ionic strengths. However, such values are only approximate, and the error in the approximation increases with increasing ionic charge and ionic strength of the solution. A limited number of activity coefficients may be obtained from Table 9·2.

The concept of activity explains the increased solubility, and the larger solubility product constant, of a sparingly soluble salt in the presence of an inert electrolyte. For example, *calculate the difference in the formal solubility of* $BaSO_4$ *in (1) pure water bh conventional calculations, and (2) 0.01 F KCl in terms of activities.*

(1) The equilibrium in pure water may be indicated as

$$BaSO_{4\,(solid)} \rightleftharpoons Ba^{++} + SO_4^{--}$$

Therefore,

$$[Ba^{++}][SO_4^{--}] = K_{sp} = 1.07 \times 10^{-10}$$

$$x = [Ba^{++}] = [SO_4^{--}]$$

$$x^2 = 1.07 \times 10^{-10}$$

$$x = 1.03 \times 10^{-5}\,F$$

(2) The equilibrium in $0.01\ F$ KCl, if we make use of activity coefficients (see Table 9·2), is as follows:

$$K_{sp} = [Ba^{++}][SO_4^{--}] \times f_{Ba^{++}} \times f_{SO_4^{--}}$$

$$= 1.07 \times 10^{-10}$$

$$\text{solubility} = \sqrt{\frac{1.07 \times 10^{-10}}{(0.625)^2}}$$

$$= 1.65 \times 10^{-5}\,F$$

Consequently, if we use activity coefficients, we calculate $BaSO_4$ as being 1.54 times more soluble in the presence of diverse ions (0.01 F KCl) than it is if we employ the usual fw concentration method in pure water; there is, then, an increase of 44.9%.

9·10 Complex Ion Formation and Solubility

A portion of Chapter 8 was devoted to complex ion formations, inasmuch as such complexations are widely used in analytical chemistry. The discussion was concerned with examples, which may be easily recognized as results of the velocity and completion of the reactions. For example, AgCl dissolves quickly and completely in excess ammonia to form a complex ion, indicated in the following equation:

$$AgCl_{(solid)} + 2NH_3 \rightleftharpoons Ag(NH_3)_2^+ + Cl^-$$

Another aspect of complex-ion formation, which is not so easily recognized, is the fact that the majority of anions enter into complex formations with many cations. Frequently, the amount of complexation between a sparingly soluble electrolyte with an excess of its anion is slight, but it may be sufficient to produce a decided error in the quantitative analysis of the fairly insoluble salt. The various possibilities of complexations of this type are too extensive to be examined in detail; however, the action of excess chloride ions on insoluble chlorides is representative. The product of such a reaction exhibits a slightly increased solubility as a result of the complexation. Typical equations for such reactions are

$$AgCl_{(solid)} + Cl^- \rightleftharpoons AgCl_2^-$$

$$AgCl_{(solid)} + 2Cl^- \rightleftharpoons AgCl_3^{--}$$

$$PbCl_{2\,(solid)} + 2Cl^- \rightleftharpoons PbCl_4^{--}$$

Although soluble complexes formed by cations with an excess of anions, in precipitation processes, are not very stable, their formations may constitute definite inherent errors. Such a complexation can nullify the common effect (Section 9·8), and give rise to an increased solubility of a precipitate in a high concentration of the common ion. In Section 9·8, it was stated that the common ion effect could not reduce the solubility below $3.6 \times 10^{-7}\,F$. However, beyond this reduced solubility, additional common ion concentration (the chloride ion) increases the solubility of AgCl by complex-ion formation. The unusual possibility of increased solubility by a common ion (the action of Cl^- on AgCl) is depicted in Figure 9·2, which clearly shows that a small excess of chloride-ion concentration decreases the solubility of AgCl, and that a large excess of chloride ion produces an increased solubility.

9·11 Effect of pH on Solubility of a Precipitate

Certain ionic precipitates may act as proton acceptors and, by acid–base reactions, are converted from insoluble substances to ions and

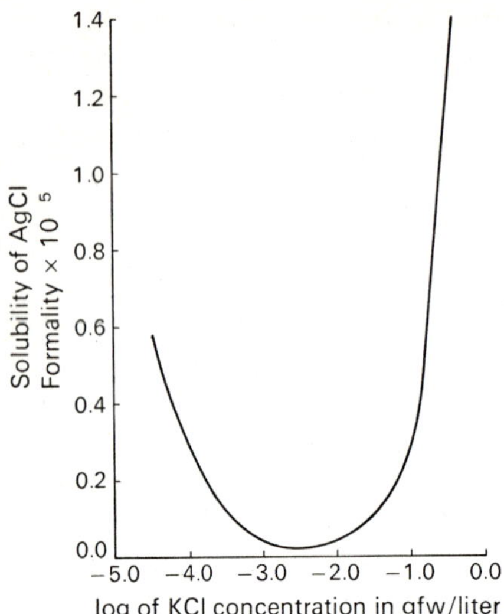

FIGURE 9·2   The solubility of AgCl as a function of the chloride concentration

soluble molecules. For convenience, these insoluble substances are classified as (1) insoluble hydroxides, and (2) salts of weak acids such as insoluble carbonates, sulfides, phosphates, oxalates, and so on.

Hydroxides.   The hydroxides of many cations are insoluble, and the solubility products vary over a wide range. Consequently, it is possible either to precipitate, or to prevent the precipitation of, such hydroxides from aqueous solution by carefully controlled pH. As an illustration of the effect of controlled pH on precipitations of hydroxides, consider the following problem: *Calculate the pH at which 0.1 F ferric ions and 0.1 F magnesium ions begin to precipitate as the hydroxides.**
Ferric hydroxide:

$$[Fe^{3+}][OH^-]^3 = 2.0 \times 10^{-39}$$

$$(0.1)[OH^-]^3 = 2.0 \times 10^{-39}$$

$$[OH^-]^3 = 20 \times 10^{-39}$$

$$[OH^-] = 2.7 \times 10^{-13}$$

$$pOH = 12.57$$

$$pH = 1.43$$

* Ferric hydroxide is a hydrated oxide with a probable formula of $Fe_2O_3 \cdot xH_2O$; consequently, the formula of $Fe(OH)_3$ is a simplification not entirely justified.

Magnesium hydroxide:

$$[Mg^{++}][OH^-]^2 = 1.1 \times .10^{-11}$$

$$(0.1)[OH^-]^2 = 1.1 \times 10^{-11}$$

$$[OH^-]^2 = 1.1 \times 10^{-10}$$

$$[OH^-] = 1.05 \times 10^{-5}$$

$$pOH = 4.98$$

$$pH = 9.02$$

The foregoing calculations indicate that an acid solution containing the two ions, Fe^{3+} and Mg^{++}, may be slowly neutralized with a base so that ferric hydroxide will precipitate first. The precipitate can be separated by filtration before the pH is increased sufficiently to precipitate magnesium hydroxide. In actual practice, the ferric hydroxide is precipitated with ammonia solution buffered with ammonium ions, and, after the removal of $Fe(OH)_3$, the magnesium ion is precipitated as $Mg(OH)_2$ with NaOH solution.

Insoluble Salts of Weak Acids.   Ions of most of the heavy metals form fairly insoluble substances with many weak acids, such as H_2S, $H_2C_2O_4$, and so on. Because of the large number of these compounds, an extensive discussion concerning the complexities involved in their dissolution by hydrogen ions is impossible. Among these salts, the solubilities of the sulfides have been studied in detail. The following problem is a typical example: *Calculate the theoretical formality of hydrogen ions necessary to dissolve 0.01 g fw of ZnS in a liter of solution. Perform the same calculations for 0.01 g fw of PbS.* (Assume that K_a for $H_2S = 6.8 \times 10^{-23}$.)

If the divalent sulfide is represented by the formula, MS, it may react with hydrogen ions according to the following equation:

$$MS + 2H^+ \rightleftharpoons M^{++} + H_2S$$

The over-all ionization of H_2S may be indicated as

$$H_2S \rightleftharpoons 2H^+ + S^{--}$$

The two expressions can be combined into the form of a ratio as,

$$\frac{\dfrac{[M^{++}][S^{--}]}{[H^+]^2[S^{--}]}}{[H_2S]} = \frac{K_{sp}}{K_a}$$

The ratio combination may be simplified to

$$\frac{[M^{++}][H_2S]}{[H^+]^2} = \frac{K_{sp}}{6.8 \times 10^{-23}}$$

The formality of hydrogen ions necessary to dissolve 0.01 gfw of ZnS in a liter of solution is indicated by the equation

$$ZnS \;+ H^+ \; \rightleftharpoons \; Zn^{++} \; + \; H_2S$$
$$\text{0.01 gfw} \qquad x \qquad \text{0.01 gfw} \quad \text{0.01 gfw}$$

Substituting these values into the above simplified equation gives

$$\frac{(0.01)^2}{[H^+]^2} = \frac{4.5 \times 10^{-24}}{6.8 \times 10^{-23}}$$

$$[H^+] = 3.9 \times 10^{-2} \, F$$

Similar calculations for 0.01 gfw of PbS are

$$\frac{(0.01)^2}{[H^+]^2} = \frac{1.25 \times 10^{-28}}{6.8 \times 10^{-23}}$$

$$[H^+] = 7.4 \, F$$

EXERCISES

1. Given the following solubility data, calculate the K_{sp} for each of the following compounds: (a) AgI, 1.6×10^{-6} g/liter; (b) $PbCl_2$, 4.2 mg per ml; (c) Hg_2Cl_2, 3.4×10^{-5} g/100 ml.

2. From the solubility products given in Appendix 6, calculate the following solubilities: (a) $CaCO_3$ in g/100 ml; (b) $Ba(IO_3)_2 \cdot 2H_2O$ in mg/100 ml; (c) Bi_2S_3 in g/liter.

3. Calculate the formal solubilities of the following compounds (neglect complex-ion formations or diverse-ion effects): (a) AgBr in a solution which is $0.0010 \, F$ in Br^-; (b) Ag_2CrO_4 in a solution which is $0.0010 \, F$ in Ag^+; (c) Ag_3PO_4 in a solution which is $0.0010 \, F$ in PO_4^{3-}.

4. Compute the formal solubilities of the following: (a) $Al(OH)_3$ at a pH of 4.0; (b) $Zn(OH)_2$ at a pH of 5.5.

5. Calculate the formal concentration of hydroxide ion necessary to start the precipitation of a $0.0010 \, F$ solution of each of the following cations: (a) Cr^{3+}; (b) Cu^{++}; (c) Fe^{++}; (d) Mn^{++}.

6. Calculate the formal solubilities of each of the following: (a) CdS at a pH of 1.0, in a solution of $0.1 \, F \, H_2S$; (b) MnS at a pH of 1.0, in a solution of $0.1 \, F \, H_2S$.

7. Calculate the formal concentration of iodide ion required to start precipitation from each of the following solutions: (a) $0.0010 \, F \, Ag^+$; (b) $0.0010 \, F \, Pb^{++}$; (c) $0.0010 \, F \, Cu^+$; (d) $0.0010 \, F \, Hg_2^{++}$.

8. What is the solubility of $MgNH_4PO_4$ in gram-formula weights per liter?

9. What is the sulfide-ion formality in a saturated solution of Bi_2S_3?

10. What weight of barium sulfate will remain in solution when 250 ml of $0.10 \, F$ barium nitrate solution is added to 600 ml of $0.10 \, F$ potassium sulfate?

11. Calculate the formal-concentration ratio $[Ba^{++}]/[Mg^{++}]$ in a solution which is saturated with $BaCO_3$ and $MgCO_3$.

12. What is the maximum concentration of Mg^{++}, in formula weights per liter, that can exist in a solution which is $0.5 \, F \, NH_4Cl$ and $0.1 \, F \, NH_3$? (K_b for $NH_3 = 1.8 \times 10^{-5}$.)

13. A $0.6 \, F$ solution of NH_3 is made $0.1 \, F \, Mg^{++}$ ions. What is the minimum concentration, in terms of formality, of NH_4^+ ions necessary to prevent the precipitation of magnesium? (K_b for $NH_3 = 1.8 \times 10^{-5}$.)

14. How many milliformula weights of $Mg(OH)_2$ remain unprecipitated in a liter of $0.2\ F$ NaOH?

15. If K_2CrO_4 is added to 10 g each of Ag^+ and Pb^{++} per liter, which ion will precipitate first?

16. If K_2CrO_4 is added to 10 g of Ag^+ and 1 g of Pb^{++} per liter, which ion will precipitate first? What is the formal-concentration ratio of $[Ag^+]/[Pb^{++}]$ when the solution is saturated with both chromates?

17. If the dissociation constant of $AgCl_2^-$ ($AgCl_2^- \rightleftharpoons AgCl + Cl^-$) is 9.1×10^{-6}, calculate the solubility of AgCl in $1.0\ F$ KCl.

18. Calculate the solubility of ZnS in $1\ F$ acetic acid. (K_a for acetic acid $= 1.75 \times 10^{-5}$, and K_a for $H_2S = 6.8 \times 10^{-23}$.)

19. Calculate the difference in the formal solubility of AgCl in (a) pure water by conventional calculations, and (b) $0.10\ F$ KNO_3 in terms of activities (see Table 9·2).

CHAPTER 10

Principles of Gravimetric Analysis

Gravimetric analysis is based entirely upon weight. In a gravimetric procedure the original substance is weighed, and from it the constituent to be determined is isolated and also weighed. The percentage of the component sought is calculated from the two weights.

A successful gravimetric procedure postulates two major conditions:

(1) The separation of the desired constituent must be sufficiently complete so that the amount left in solution cannot produce a detectable error. In macroanalysis the amount left in solution should not exceed one milligram.

(2) The final substance that is weighed should be relatively pure and have a definite composition, since the analysis and resultant calculations are based on these assumptions.

The ideal conditions, so described, are seldom obtained; consequently, the problem confronting the analyst is to approximate the stated conditions as nearly as possible. This chapter is devoted largely to a discussion of those factors which affect both the completeness of precipitation and the composition and purity of the precipitate.

STOICHIOMETRY

10·1 Calculation of a Desired Constituent

A gravimetric analysis is based upon the experimental measurements of the weights described above, namely, the weight of the original sample and the weight of the pure compound isolated in the procedure. The weight of the purified component is always placed in the numerator, and that of the original sample in the denominator. The ratio is usually expressed in terms of percentage composition. For example,

% of desired component

$$= \frac{\text{wt of isolated compound}}{\text{wt of sample}} \times 100$$

10·2 Gravimetric Factor

When the isolated compound is not the constituent sought, it becomes necessary to use a gravimetric factor to calculate the desired component. *The gravimetric factor is a number used to multiply a given weight of some compound in order to obtain the equivalent weight of some other constituent.* This number is usually obtained from a ratio of the formula (or atomic) weight sought, divided by the formula (or atomic) weight of the substance weighed in the analysis. As an illustration, the weight of chlorine in silver chloride is a ratio of the atomic weight of chlorine to the formula weight of silver chloride, or $\text{Cl}/\text{AgCl} = 0.2474$.

In a chloride analysis the calculation becomes

$$\frac{\text{Cl}}{\text{AgCl}} \times \frac{\text{wt of dried AgCl}}{\text{wt of sample}} \times 100 = \% \text{ Cl}$$

In the gravimetric factor, Cl/AgCl, the *constituent weighed* is placed in the *denominator*, and the *constituent sought* in the *numerator*.

Although gravimetric factors are valuable arithmetical aids, they cannot be used indiscriminately. It is necessary to know the chemical equations involved in the gravimetric procedures before a chemical factor can be used intelligently. For example, the magnesium ion may be precipitated by the phosphate ion in an ammoniacal water solution to give

$$Mg^{++} + NH_4^+ + PO_4^{3-} + 6H_2O \rightleftarrows$$
$$MgNH_4PO_4 \cdot 6H_2O \downarrow$$

and the precipitate may be ignited to constant weight as follows:

$$2MgNH_4PO_4 \cdot 6H_2O \rightleftarrows$$
$$Mg_2P_2O_7 + 2NH_3 + 13H_2O$$

Therefore, the gravimetric factor for the amount of magnesium in magnesium pyrophosphate (the substance produced by ignition) is $2Mg/Mg_2P_2O_7$. It is obvious that the number of atoms of magnesium appearing in the numerator must be the same as the number in the denominator. The complete gravimetric calculations are:

$$\frac{2Mg}{Mg_2P_2O_7} \times \frac{\text{wt of ignited precipitate}}{\text{wt of sample}} \times 100$$
$$= \% \, Mg$$

THE PRECIPITATION PROCESS

Precipitation is one of the oldest and most widely used techniques available to analytical chemists. In quantitative analysis, the formation of a precipitate may be the fundamental process of an analysis. If the precipitate is eventually weighed, it is the basis of a gravimetric method. On the other hand, if the precipitate is dissolved and titrated, the procedure becomes a volumetric method.

The chemical and physical changes that occur during a precipitation are quite complex. The explanations of some of these changes are frequently speculative; however, the chemist has learned the probable conditions which produce a precipitate of variable physical properties, such as gelatinous or crystalline consistency. He has also gained some insight into how various impurities are incorporated in a precipitate, and some knowledge of the precautions that may prevent or remove such contaminants.

10·3 Supersaturation and Nucleation

The concept of supersaturation and its effect upon particle size was developed by P. P. Von Weimarn during an extended period in the early part of the century. It was his belief that the extent of supersaturation determined the size of the particles of a precipitate.

Nucleation may be defined as the production of a more stable phase from the metastable phase denoted as supersaturation. A solution is said to be *supersaturated* when it contains a higher concentration of solute than may exist in a saturated solution. If solutions of two ions, which normally form a precipitate, are brought together without precipitation occurring, the solution is supersaturated. Nucleation eventually takes place, and the processes involved determine the nature and the purity of the resulting precipitate. If the precipitation is rapid, many nuclei will result and the individual microcrystals, so produced, will be characterized by colloidal dimensions and low purity. Such micelles are impossible to manipulate through conventional gravimetric procedures. On the other hand, if the precipitation produces few nuclei, the crystals will be larger and of higher purity, and they may be filtered and washed without difficulty.

The degree of supersaturation, which appears to determine the number of initial nuclei, is expressed by the Von Weimarn ratio as

$$\text{relative supersaturation} = \frac{Q - S}{S}$$

where Q is the concentration of the solute in solution at a given time, and S is the equilibrium solubility of the insoluble salt.

Any explanation of the first step in precipitation (nucleation) is speculative. Some authors postulate that adsorption of ions on dust particles in solution produces the first nuclei. Others believe that random factors result in clusters of ions which take on a definite geometric pattern. However, either speculation recognizes that the process of precipitation goes through several stages before large crystals can be produced.

10·4 Particle Size

Before we discuss crystal growth, it seems worthwhile to list the dimensions and dimensional units most frequently used in describing dispersions of particles in water on the basis of diameter. Also listed are the possible separation factors resulting from particle dimensions.

Simple ionic dispersions contain particles with diameters of less than 10^{-7} *cm* (1 angstrom unit $= 10^{-8}$ cm; one millimicron, $m\mu = 10^{-7}$ cm). Particles of this size are in true solution, and mechanical separation is virtually impossible.

Colloidal dispersions consist of particles with diameters of between 10^{-7} *and* 10^{-5} *cm* (1–100 $m\mu$). True colloids within these dimensions will not settle to the bottom of the container; however, they may be separated with an ultracentrifuge.

Fine-crystal dispersions are composed of particles having diameters of between 10^{-5} *and* 10^{-3} *cm* (100 $m\mu$ to 10 μ). Particles of these dimensions may be separated with a centrifuge, but a precipitate of such particles is undesirable for gravimetric manipulations. The particles will slowly settle out on standing.

Coarse dispersions contain particles larger than 10^{-3} *cm* (larger than 10 μ). Precipitates of these dimensions are filterable with filter paper or with filter crucibles.

10·5 Stages in the Precipitation Process, and Crystal Growth

From Von Weimarn's ratio, it is apparent that precipitation cannot occur until the solution is supersaturated with respect to a given com-

pound. Once this metastable phase is attained, primary nuclei of submicroscopic dimensions will eventually be formed. The formation and growth of crystals may result in a variety of end products. The final precipitation product may be in the form of coarse crystals, fine crystals, or colloidal aggregates.

The velocity of crystal growth is dependent on the relative supersaturation according to the following relation:

$$V = k\frac{(Q - S)}{S}$$

where V is the velocity and k is a proportionality constant. Therefore, to obtain a pure precipitate of large particles, the relative supersaturation should be low in order that the rate of precipitation may also be reduced. For the best possible precipitate, conditions should be adjusted to keep Q as low as possible.

A flow diagram of crystal growth is given in Figure 10·1, which also includes types of impurities that may be expected to contaminate the resulting precipitates.

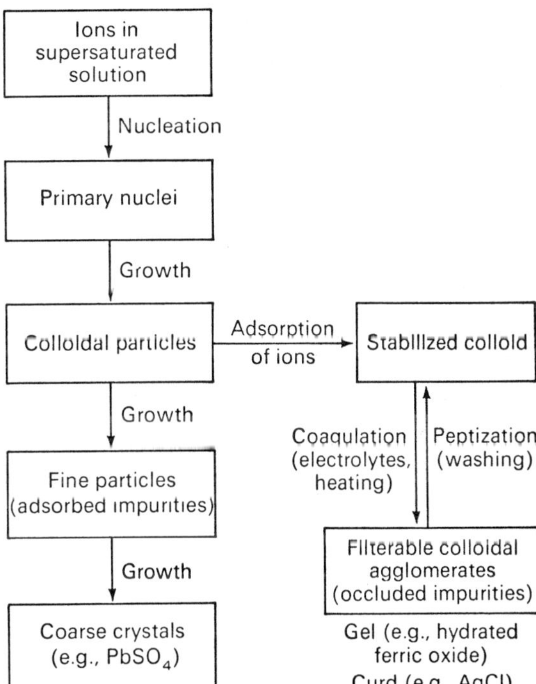

FIGURE 10·1   Flow chart depicting crystal growth

Nucleation (see Section 10·3) may be described as the process by which ions assemble into a cluster large enough for crystal lattice forces to become operative. To obtain coarse and pure particles, it is necessary that nucleation produce as few crystal nuclei as possible; if there are too many nuclei, crystal growth will be slow. Growth is merely the addition of ions to a stable crystal nucleus. The operation should be as rapid as is feasible; this is generally accomplished by decreasing the concentration of the solute. Crystal growth continues as long as the state of supersaturation exists. If crystals of different sizes are produced at the same time, the larger crystals should grow, and the smaller crystals may disappear as a result of their higher solubility.

10·6 Optimum Conditions for the Formation of Large Crystals

The particle size of a precipitate is increased by a reduction in the relative supersaturation of the precipitation solution. This reduction favors the formation of a minimum number of primary nuclei, so that subsequent growth of the precipitate occurs by deposition of ions on fewer crystals. Usually, optimum conditions for analytical precipitations may be attained through the following techniques:

Precipitation from Dilute Solution. Dilution slows the nucleation process, thereby aiding the growth of large crystals. However, there are practical limitations to dilution. Solubility losses increase with dilution, and the manipulation of a large volume expands the time required for analysis.

Slow Addition of Precipitating Reagent with Constant Stirring. The slow addition of a precipitating agent gives larger particles (and also keeps Q low), but unfortunately such a procedure is time consuming. Constant stirring prevents locally high concentration of the precipitating ion.

Precipitation at Maximum Solubility. Two techniques are exploited: (a) precipitation from hot solution, and (b) control of pH when precipitate is soluble in an acid solution. An example of these techniques is the precipitation of calcium oxalate from a hot acid solution by the slow addition of ammonia solution.

Digestion of the Precipitate. The process of digestion consists in allowing the precipitate to stand (that is, to age) in contact with the mother liquor (the solution remaining after precipitation). It is generally assumed that the digestion period enables small crystals to dissolve and reprecipitate on the surface of large crystals. In other words, large crystals grow at the expense of small crystals. However, for a curdy precipitate, such as silver chloride, the digestion process is not necessarily a growth in crystal size, but rather an agglomeration of aggregates, which renders the precipitate more filterable.

10·7 Precipitation from a Homogeneous Solution

In this method the precipitant is added indirectly to the solution as a result of internal hydrolysis (or another similar chemical reaction which proceeds slowly). Although many homogeneous precipitations are utilized, the most outstanding example is the hydrolysis of urea (in the precipitating solution), according to the following equation:

$$CO(NH_2)_2 + 3H_2O \rightleftarrows$$
$$CO_2 + 2NH_4^+ + 2OH^-$$

The hydrolysis ions that are provided can precipitate many cations. The hydrolysis is slow at room temperature, but more rapid at an elevated temperature. In the analysis of either Fe^{3+} or Al^{3+} ions, from acid solutions, the hydrous oxides are more easily filtered when formed by the slow generation of hydroxide ions from urea. Furthermore, such precipitates are more compact and less contaminated.

COLLOIDAL PARTICLES

The term *colloid* is derived from the Greek word for glue. It was first used by Thomas

TABLE 10·1   Increase of surface area with subdivision

Volume of cube, cm^3	Total surface produced, cm^2
1	6
1×10^{-1}	6×10^1
1×10^{-2}	6×10^2
1×10^{-3}	6×10^3
1×10^{-4}	6×10^4
1×10^{-5}	6×10^5
1×10^{-6}	6×10^6
1×10^{-7}	6×10^7

Graham in 1861 to distinguish noncrystalline from crystalline material. The original meaning has little significance at present; strictly speaking, colloids are not a class of materials but are more aptly described as materials in a state of restricted subdivision. Although many types of colloids are possible, this discussion is concerned only with solid particles dispersed within an aqueous medium. A colloidal state may exist between primary nuclei and a suspension of fine crystals, as is diagrammed in Figure 10·1. Such a colloid is sometimes called a *sol*.

10·8 Some Properties of Colloidal Suspensions

In a true solution, the freezing point, boiling point, vapor pressure, and osmotic pressure are changed by the presence of the dissolved particles. A colloidal suspension resembles a true solution inasmuch as these colligative properties are altered by the presence of the dispersed phase. Yet, the total effect is quite small, since the particles are larger, and less in number, than particles in a true solution.

A colloidal dispersion may appear completely dissolved under diffused light. However, if a beam of light is passed through the dispersion, its path becomes visible by the light reflected by the dispersed particles. Under similar conditions, a true solution shows only a faint beam, which is produced by dust particles that are present everywhere. This test for a colloidal dispersion is known as the *Tyndall effect*.

Colloidal particles will not pass through animal or vegetable membranes. In contrast, substances in a true solution do pass through such barriers. The separation of a dispersed phase by means of a membrane is called *dialysis*. Colloidal particles pass readily through ordinary filtering media used in quantitative chemistry.

The particles of a colloidal suspension are so small that they are unaffected by the pull of gravity. On the other hand, when pseudo-gravitational forces are added, in the form of a high-speed *ultracentrifuge*, colloidal particles are deposited as a colloidal precipitate. Such a precipitate is gelatinous in appearance, although X-ray analysis reveals that it has a lattice structure.

10·9 Surface Area of Colloids

The distinguishing properties of colloids, as compared with larger dispersed particles, are due almost entirely to the larger area per unit mass for the colloidal particles. If a given particle of matter with a known surface area is subdivided progressively, in steps, the total area increases with each subdivision. As a definite illustration, assume that a cube, with a volume of $1 \, cm^3$ and an area of $6 \, cm^2$, is subdivided in consecutive steps so that each subdivision reduces the particle size by one-tenth in volume, until the final particles are of colloidal dimension. The successive increases in area are given in Table 10·1. The total of the subdivisions represents an increase in area from $6 \, cm^2$ to 60 million cm^2 or from approximately 1 square inch to an area greater than that of a football gridiron.

Adsorption is a surface phenomenon which is dependent upon surface area. One theory of adsorption postulates that the electrostatic forces that bind ions into a crystal lattice extend in all directions, and that ions at the surface of the crystal have unsatisfied electrostatic forces that extend outwardly, and that may hold atoms, molecules, or ions upon the surface of the crystal.

10·10 Adsorption of Ions by Colloids

A surface composed of ions will absorb ions of opposite charge. As a general rule, electrovalent colloids tend to adsorb ions which are common to them; thus, colloidal silver chloride tends to adsorb either silver or chloride ions. This tendency is to be expected, since it simply illustrates the manner in which an electrovalent crystal may extend its space lattice. In the building up of a crystal, only ions of the proper size will fit into the lattice. It is probable that the same is true when an ion is adsorbed at the surface of the colloidal particle. If the ion is of suitable diameter to fit into the crystal lattice, it should be held more tightly upon the surface of the ionic colloid. Moreover, the presence of adsorbed ions imparts a charge to the colloidal particle. Since all the particles have the same charge, they repel each other and tend to prevent combination and coagulation of the colloidal material.

The adsorption of a monatomic layer of ions upon the surface of a particle results in an electrical unbalance in the solution. This unbalance is corrected by the formation of another layer of ions, of opposite charges, extending into the solution that surrounds the colloidal particle. The resulting double layer resembles a parallel-plate capacitor, except for the fact that the primary layer of charges on the particle is monatomic in depth and immovable, whereas the secondary layer is diffuse and mobile. The presence of the electrical double layer accounts for the stability and electrical properties of electrovalent colloids.

The formation of the electrical double layer at the surface of a colloidal particle may be illustrated by the consideration of a silver chloride sol. When a solution of sodium chloride is added dropwise to a solution of silver nitrate, colloidal particles of silver chloride result. Since

electrovalent sols tend to adsorb their own ions, the particles of silver chloride are positively charged in the presence of an excess of silver ions. The primary layer of ions adsorbed on the surface of the particle is composed of silver ions, and the secondary layer extending into the solution is composed of nitrate ions. These conditions are indicated diagrammatically in Figure 10·2. On the other hand, if a solution of silver nitrate is added dropwise to a solution of sodium chloride, the silver chloride particles will be formed in the presence of an excess of chloride ions. Consequently, the sol particles will adsorb chloride ions and have a negative charge. The secondary layer of ions contains the positively charged sodium to counterbalance the

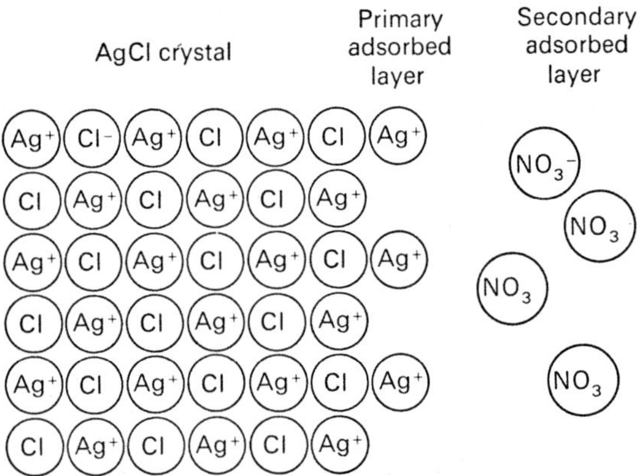

FIGURE 10·2 Diagrammatic representation of a colloidal silver chloride particle in the presence of excess silver nitrate solution

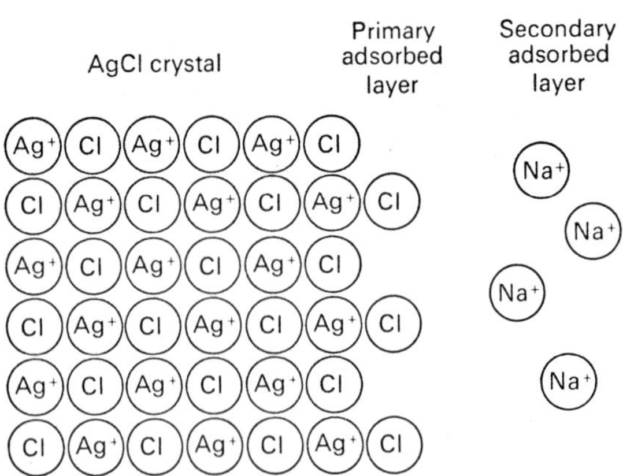

FIGURE 10·3 Diagrammatic representation of a colloidal silver chloride particle in the presence of excess sodium chloride solution

negatively charged $AgCl \cdot Cl^-$ particles. The diagrammatic representation of these conditions is given in Figure 10·3.

10·11 Coagulation of Colloids

To coagulate a colloid, the colloidal particles must coalesce into larger aggregates. In a stable sol, coalescence is prevented by the mutual repulsion of the micelles, all of which carry a similar electric charge. To coagulate a colloid, it is necessary to neutralize the charges carried by the dispersed particles. Although low concentrations of ions are essential for the stabilization of ionic sols, higher concentrations cause the micelles to aggregate until coagulation is accomplished. The ions which effect the coagulation of sols have a charge opposite to that of the ions which impart the charge to the particles. As a general rule, ions with multiple charges are more effective in producing coagulation than ions with single charges.

The increase in precipitating effect with increasing valence (charge) of an ion is not a quantitative increment. Trivalent ions are more effective in coagulating colloids of opposite charge than are divalent ions, and divalent ions are more effective than univalent ions. This is not to indicate that a divalent ion has twice the coagulating power of a monovalent ion. Svedberg found the efficiency of the ions K^+, Ba^{++}, and Al^{+++} as coagulants of an arsenious sol to stand in a ratio of approximately $1:20:1000$.

A less effective method of coagulating certain colloids is by raising the temperature of the dispersed phase. It may be assumed that the elevation of temperature reduces the number of ions adsorbed on the particles. Heating also increases the kinetic energy of the particles, and thereby increases the probability of collisions, which may lead to flocculation.

10·12 Peptization of Colloids

When a coagulated particle is washed with pure water, the stabilizing counter ions may be leached from the secondary layer. If a coagulated silver chloride particle (in an aggregate) is

denoted as $AgCl \cdot Ag^+NO_3^-$, the nitrate ion can be assumed to have been the coagulant. When the coagulating ion is removed, the colloidal aggregate will revert back to the colloidal state. This phenomenon is known as *peptization*. In some respects, peptization can be considered the opposite of coagulation (see Figure 10·1). To prevent peptization, colloidal aggregates are usually washed with solutions of electrolytes. It is important, however, that such electrolytes will not be a source of contamination in additional analytical steps.

CONTAMINATION OF PRECIPITATES

All ionic precipitates pass through the colloidal stage during their formations; consequently, all final precipitates will be contaminated with the surface-adsorbed ions which are characteristic of colloids. Contamination may also occur through other processes.

10·13 Types of Contamination

Some of the various sources of contamination are outlined as follows:

Incomplete Separation in Fractional Precipitation.   For example, silver chloride precipitated in the presence of small amounts of either bromide or silver ions will be contaminated with the silver salts of these ions.

Postprecipitation.   A commonly occurring illustration is the behavior of calcium and magnesium ions in the presence of excess oxalate ions. Ordinarily, calcium oxalate precipitates satisfactorily in the presence of Mg^{++} ions, but if the precipitate stands in contact with the mother liquor for an appreciable time, magnesium oxalate will precipitate on top of the original precipitate. This postprecipitation can be avoided by filtering the calcium oxalate within one or two hours after precipitation.

Coprecipitation.   In coprecipitation the main precipitate and the contaminating materials usually come down together. The mechanisms of coprecipitation may be divided into two types, namely, *adsorption* and *occlusion.*

Contamination by incomplete separation and by postprecipitation can be predicted by the analyst, and frequently avoided or counteracted. But because the two types of coprecipitation may not be foreseen or prevented, it seems pertinent to discuss these types more completely.

10·14 Contamination of Gravimetric Precipitates by Adsorption

Ions are adsorbed at the surface of the particles of all precipitates. However, the total contamination constitutes a significant error only in those precipitates which present a large surface area for adsorption. As described elsewhere, ions are adsorbed from the mother liquor on the surface of a precipitated particle to produce a tightly held primary layer. The counter ions of the secondary layer are held more loosely. Yet, in the process of precipitation both types of adsorbed ions are dragged down as a part of the main precipitate. For example, in the determination of the sulfate ion by precipitation with barium ions, the external surface of the crystal lattice is made up of particles, which may be designated as $BaSO_4 \cdot Ba^{++} \cdot 2Cl^-$. The barium ions are part of the initial primary layer, and the chloride ions were originally the counter ions. It may be assumed that the chloride ions were part of the coagulant. On the other hand, the internal lattice (after aging) contains only barium and sulfate ions, unless occluded impurities are present. The small amount of barium chloride on the surface constitutes the principal contamination, which may result from adsorption. For large crystals, the total contamination is slight.

10·15 Contamination of Gravimetric Precipitates by Occlusion

Occlusion is a form of coprecipitation in which impurities are enclosed, or entrapped, within the lattice structure of a crystalline precipitate. Three types of occlusion are recognized:

Mechanical Entrapment of Contaminating Ions and Molecules. When crystals grow rapidly, the counter ions on small crystals may become surrounded and enclosed into the structure of the resulting larger crystals. It is also possible for water molecules, with their dissolved impurities, to become entrapped in a similar manner.

Mechanical Entrapment of Colloidal Aggregates. As has been stated before, all precipitates pass through the colloidal state in their formations. Fast-growing crystals may entrap some of their colloidal aggregates before the aggregates have had time to be converted into crystals. Such an aggregate contains the usual contamination of colloidal adsorption.

Isomorphoric Inclusions. Occasionally two ions in a solution have dimensions that are relatively the same. For example, the radii of the K^+ and NH_4^+ ions are of the same magnitude. In the case of a fast-growing crystal such as those of $MgNH_4PO_4$, it is possible for K^+ ions, if present, to enter into the lattice structure. The potassium compound could be designated as $MgKPO_4$, although the K^+ ions are only part of a mixed crystal. Thus, the four ions may produce an isomorphoric-crystal formation, which does not possess a definite composition.

Contamination by occlusion in any of the foregoing three types can produce serious errors in gravimetric analysis. It has been said that the best way to contend with occlusion is to avoid it; however, this is not always possible. Contamination of the first two types may be reduced by digestion, but some impurities will always remain. There is no remedy for isomorphoric inclusions. Fortunately, isomorphoric replacements are rarely encountered in analytical precipitations.

EXERCISES

1. The chemical factor for determining sulfur in a weighted precipitate of barium sulfate is $S/BaSO_4$. In like manner, indicate the correct gravimetric factor for the following:

	Weighed	Sought	Factor
(a)	U_3O_8	U	
(b)	B_2O_3	$Na_2B_4O_7$	
(c)	Mn_3O_4	Mn_2O_3	
(d)	KBF_4	$Na_2B_4O_7$	
(e)	Bi_2O_3	$BiCl_3$	
(f)	$Mg_2P_2O_7$	P_2O_5	
(g)	MnO_2	Mn_3O_4	
(h)	$Mg_2P_2O_7$	MgO	
(i)	$BaSO_4$	$Al_2(SO_4)_3$	
(j)	MoS_3	MoO_3	
(k)	Sb_2O_4	Sb	
(l)	$Ca_3(PO_4)_2$	$CaSO_4$	
(m)	$BaSO_4$	Sb_2S_3	
(n)	SiO_2	$KAlSiO_3O_8$	
(o)	Fe_2O_3	$FeSO_4 \cdot (NH_4)_2SO_4 \cdot 6H_2O$	
(p)	$BaSO_4$	FeS_2	
(q)	$PbSO_4$	Pb_3O_4	
(r)	Fe_2O_3	Fe_2CO_3	
(s)	K_2PtCl_6	KNO_3	
(t)	Cu_2S	CuO	
(u)	CaO	$H_2C_2O_4 \cdot 2H_2O$	
(v)	Pb_3O_4	Pb	
(w)	Fe_2O_3	Fe	
(x)	$Mg_2P_2O_7$	$MgNH_4PO_4$	

2. Calculate the percentage of Zn in an ore, 1.000 g of which yielded 0.7500 g of $Zn_2P_2O_7$.
3. A 0.4000 g sample of impure NaCl yielded a silver chloride precipitate of 0.7000 g; calculate the percentage of chlorine in the sample.
4. A sample of iron ore weighing 0.6000 g gives an ignited precipitate of 0.1500 g of ferric oxide. What is the percentage of iron in the sample?
5. Distinguish between nucleation and crystal growth.
6. Explain why small particles of a substance are more soluble than large particles.
7. Calcium oxalate is digested before filtration, but this procedure is not followed for ferric hydroxide. Why?
8. Hydrated ferric oxide is gelatinous when precipitated from a homogeneous solution. Explain
9. What is the minimum diameter of dispersed particles in a water suspension, in order for them to be visible to the naked eye?
10. Describe the Tyndall effect.
11. A precipitate of ferric hydroxide is contaminated with magnesium hydroxide. How can the contamination be removed?
12. Explain why two colloids containing particles of opposite charges will precipitate when mixed.
13. Define or characterize the following terms:

(a) Colloid	(d) Peptization
(b) Supersaturation	(e) Occlusion
(c) Coagulation	(f) Counter ions

14. Explain why most precipitates from homogeneous precipitations contain larger and purer crystals.
15. Compare a colloid with a true solution.
16. When is postprecipitation likely to occur? Why?

17. Discuss proper procedures for minimizing coprecipitation.
18. Describe the various types of contaminations resulting from occlusion.
19. Distinguish between adsorption and occlusion.
20. Explain the role of digestion in the purification of a precipitate.
21. What determines whether the final form of a precipitate is a gel, a curd, or a crystalline precipitate?
22. How can stabilized colloids be coagulated?
23. Illustrate how surface area increases as particle size decreases.
24. How can the charge of a colloid be determined?
25. What conditions are necessary to produce colloidal dispersions?
26. What is the relation between the charge of an ion and its ability to coagulate a colloid?
27. How may the addition of an electrolyte to wash water prevent peptization of a coagulated colloid?
28. Discuss Von Weimarn's concept of precipitation. Why is a low Von Weimarn ratio desirable?
29. Explain isomorphic replacement, and give an example of it.
30. What happens when a coagulated colloid is washed with pure water?

PART 2

EXPERIMENTAL PROCEDURES

CHAPTER 11

General Laboratory Directions

BEGINNING THE LABORATORY WORK

11·1 Planning and Schedule

The laboratory work in analytical chemistry is usually planned so that students can complete the various assignments within the scheduled periods. It is essential that the individual student formulate his own schedule to conform with the group assignments. To make effective use of the laboratory period the student must be familiar with the analytical operations to be performed. The only way such familiarity can be attained is by an appropriate amount of homework before coming to the laboratory. The procedures for an assigned period should be studied carefully, and an outline may be made of the successive operations to be encountered. Laboratory work should be confined, as much as possible, to *manipulations* and *operations*. Use outside time for calculations and the filling in of routine forms in the notebook.

The student might consider a quantitative analysis as being divided into three steps: (1) the attainment of a summarized knowledge of the analysis; (2) the physical accomplishment of the experiment with attendant and pertinent observations; (3) the calculations of the results from the observations. Only the second step should utilize laboratory time.

11·2 Economy of Time

Within the laboratory hours the student may learn to accelerate his laboratory production by attention to various analytical techniques which are described elsewhere. These techniques are mastered only through practice, and are not acquired "overnight." It may not be easy to perform techniques properly the first time, but it should be remembered that one of the principal objectives of this course is to attain good laboratory skills.

One technique that the student should learn from the beginning is the efficient utilization of laboratory time. Part of this is proper organization of operations, and part of it may be sound judgment. A trained analyst learns to do many things at the same time. For example, certain evaporations, filtrations, and titrations may be performed simultaneously. The average student will probably not become a laboratory analyst, but he should recognize the advantage of doing many operations in the minimum amount of time.

It is common sense for a student to foresee each step in analytical manipulations, and to plan these steps so that no period of waiting occurs between them. If, for some reason, such a period should occur, the resourceful student will be prepared to begin the next analysis in the assigned schedule. For example, practically all samples for analysis must be dried beforehand, and before one analysis is complete, the student should have a dried sample ready for the next analysis.

11·3 Desk Equipment

In Appendix 1 will be found a list of the apparatus issued to each student. However, the list may be modified by the instructor to fit the particular needs of the course. Certain general procedures concerning assignment and care of equipment are widely accepted. Some of the directions for these are as follows:

(1) Copy desk number and lock combination in your laboratory manual, and also on a piece of paper to be carried in your billfold.

(2) Check desk equipment against the apparatus list; any items missing should be obtained from the main stockroom. Glassware which is cracked, chipped, or badly etched should also be replaced from the storeroom. It is usually necessary to get the instructor's approval for such replacement to avoid extra charge against laboratory deposit.

(3) All equipment will be checked by the instructor at the end of the term, and any defective equipment will be charged to the student.

(4) It is usually customary for the student to make a breakage deposit to take care of wear and tear on equipment. Part of this deposit may be returned.

(5) The equipment should be arranged in the desk to provide maximum availability of the most widely used items, and to insure minimum breakage of fragile items. If the desk has not been in constant use, it may be necessary to remove all equipment before rearrangement, and to remove any accumulation of dust.

(6) Clean the desiccator, and add fresh anhydrous calcium chloride, if necessary. Replacement is indicated when there is evidence of caking. The ground-glass rim of the desiccator should be lightly greased with petroleum jelly to maintain an air-tight seal. Additional instructions concerning the desiccator are to be found in Section 15·9.

11·4 Apparatus to be Constructed

In addition to the standard items of equipment listed in Appendix 1, it will be necessary for the student to construct a few pieces of apparatus which have special uses in some of the laboratory procedures.

Wash Bottle. The wash bottle is a familiar object in any analytical laboratory. There are many types, but the one produced from a flat-bottomed flask (500 or 1000 ml Florence flask)* is typical. The simplest form of this piece of equipment is shown in Figure 11·1. The fittings are usually made from three pieces of 6 mm glass tubing, cut to the approximate lengths desired. The bends, indicated in the diagram, should be made in the same plane and without kinks. Use a wing tip on the burner so that the tubing can be heated over a wide distance. The tip should have an opening that is about the same size as that found on either a standard burette or an ordinary pipette, and is approximately 1 mm in diameter. The tip is connected to the main delivery tube by a short piece of rubber tubing of the size to fit snugly. The ends of the pieces of glass tubing should be fire-polished before the wash bottle is assembled. The instructor may be consulted for additional direction in the techniques of glass working.

Safety Bottle. Ordinary suction filtrations are accomplished by means of a water aspirator connected to a suction flask. Inasmuch as there are usually wide variations in water pressure, such changes may cause water to back up into the flask from the water pump and contaminate the filtrate. A safety bottle inserted between suction flask and the aspirator will act as a safety valve to prevent such contamination. The safety bottle is constructed by inserting two pieces of glass tubing, bent as indicated in Figure 11·1, into a two-hole rubber stopper which fits an ordinary wide-mouth 250 ml bottle.

Stirring Rods. Cut six stirring rods, 5 mm in diameter, each to a length of approximately

---

* A polyethylene squeeze bottle may be substituted, but such a bottle has many disadvantages.

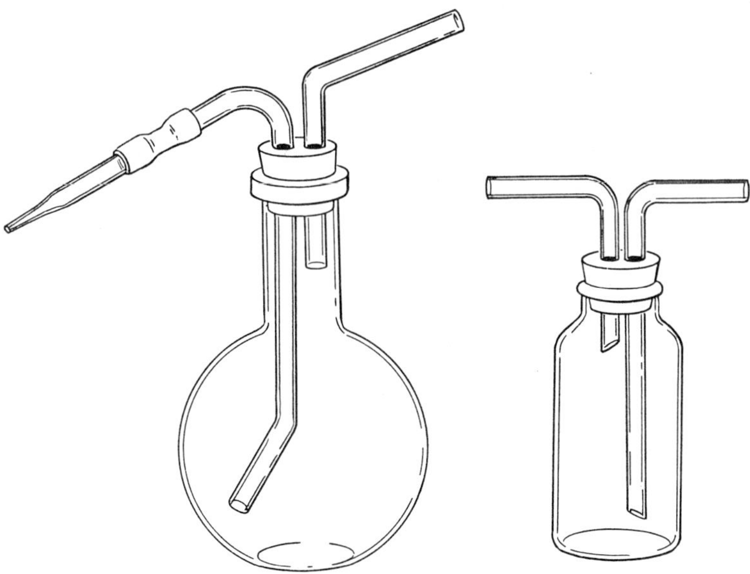

FIGURE 11·1   Wash bottle and safety bottle

150 mm. Fire-polish both ends of each rod to prevent scratching of beakers and test tubes. If rods are already available, you do not need to construct these.

CLEANLINESS AND SAFETY

11·5 Preparation and Use of Cleaning Solution

One type of reagent used for cleaning purposes is a mixture of concentrated sulfuric acid and sodium dichromate. This cleaning solution is prepared by adding 20 g of sodium dichromate to 400 ml of hot (not more than 100°C), concentrated H_2SO_4 contained in a large beaker. The mixture is stirred and allowed to cool. After cooling, the clear liquid is decanted into a 500 ml glass-stoppered bottle, leaving any undissolved crystals behind. Cleaning solution may be used over and over as long as it retains its red color. A green color indicates that it has lost its oxidizing power, and such a solution should be discarded. If kept at room temperature the solution usually retains its cleaning property over a period of months.

Cleaning solution is highly corrosive and must be used with extreme care. It will destroy clothing, and will produce severe burns on the skin. Any amount, even a drop, spilled on any surface should be neutralized with sodium bicarbonate, and then washed completely with water. If it is spilled on the skin, it should be removed with running water as quickly as possible.

Hot cleaning solution cleans within ten minutes, but is more dangerous to handle and may cause errors in the volume of calibrated glassware. Cold cleaning solution is equally effective, but slower in action, usually requiring an overnight period of contact.

11·6 Cleaning Glassware

Most glassware is best cleaned with a commercial detergent, applied with a brush, and then removed with several rinsings of water. Usually the detergent is supplied by the instructor or from the stockroom. Use tap water for all washing and rinsing purposes; under no conditions is distilled water to be used except in a *small* quantity for a final rinse.

The insides of volumetric glassware (burets, pipets, and volumetric flasks) should be clean enough that an even film remains after they have been filled and drained with distilled water. A number of recipes for satisfactory cleaning solutions are to be found in the various chemistry handbooks. One that is widely used, although hazardous and corrosive in nature, is the $H_2SO_4 \cdot Na_2Cr_2O_7$ mixture described in Section 11·5.

A buret may be cleaned rapidly by drawing warm cleaning solution from a beaker through the action of a water aspirator into the inverted tube of the burette. It is customary to insert a safety bottle in the suction line to prevent the cleaning solution from being drawn into the water pump. The apparatus is illustrated in Figure 11·2. Pipets may be cleaned by a similar procedure. It is more convenient to clean volumetric flasks by allowing cold cleaning solution to stand in the flasks overnight. The cleaned volumetric equipment should be rinsed thoroughly with distilled water. If any drops of water adhere to the walls, the item must be recleaned.

Clean burets should be filled with distilled water and stored in an upright position. If it is necessary to store them in a horizontal position they should be well corked. The new types of burets with Teflon stopcocks do not leak, and even those with glass stopcocks do not leak when used with a suitable spring clip. Burets thus stored will stay clean indefinitely.

11·7 Disposal of Waste Products

Waste solids such as match stems, used filter paper, used corks, broken glassware, and old towels must be discarded into stone crocks, which are distributed in convenient places in the laboratory. Under no conditions are solids to be allowed in the troughs or drains. A student who unthinkingly discards solids into a sink may cause extensive damage to the laboratory plumbing.

Waste liquids are to be discarded directly in the main sink, not in the laboratory trough, and flushed with large quantities of tap water. The disposal of corrosive acids and alkalies must be done in this manner to avoid damage to troughs and plumbing.

If a liquid contains particles of a solid, decant the liquid into the sink, retaining as much solid as possible in the glass container. Dump any remaining solid into a stone crock.

11·8 Laboratory Safety

Serious injuries in the laboratory are infrequent. Students have learned from high school and general chemistry that laboratory equipment and chemicals are to be respected and, consequently, they use certain precautions dictated by common sense. On the other hand, it sometimes happens that "familiarity breeds contempt", and students may not realize the hazardous and poisonous nature of certain common chemicals such as hydrogen sulfide, mercury, carbon tetrachloride, and so on. It is the

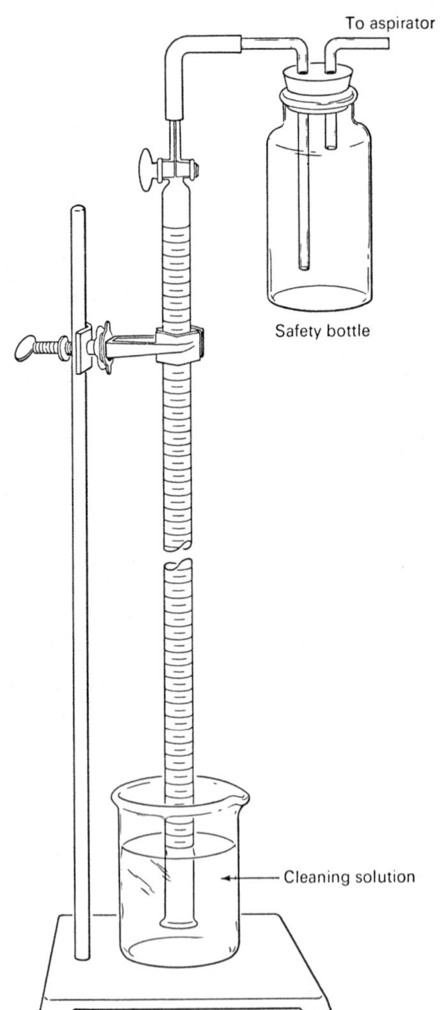

FIGURE 11·2   Cleaning a buret

responsibility of every instructor to emphasize the importance of safety in the laboratory, and it should be the duty of every student to observe laboratory safety rules. In every laboratory there should be an emergency chart to specify treatment of all types of possible accidents. One excellent chart of this type is available from the Fisher Scientific Company.

The best way to treat injuries is to avoid them; however, this is not entirely possible, and there will always be a few accidents in all laboratories. Injuries usually fall into one of the following categories: (1) poisoning, (2) burns and scalds, (3) cuts from broken glass, and, rarely, (4) explosions.

Poisoning. This can result from so many chemicals—in fact, from nearly all chemicals—that the cautious handling of laboratory reagents cannot be overemphasized. The treatment of various specific poisons varies so widely that no simple, over-all antidote is available. Treatments for some specific poisons are designated on Laboratory Emergency Charts. The labels on many chemical reagent bottles give directions for emergency treatment. As precautionary measures, all laboratories should be well ventilated, and they should have adequate hoods and, if possible, an emergency shower. As a general rule, rubber gloves should be worn while handling poisonous chemicals.

Burns and Scalds. All severe burns and scalds must be treated by a physician. A slight burn may be treated with butesin picrate ointment or sterile petroleum jelly. Do not use ointments on severe burns, but leave any treatment to a physician. Minor burns arising from spilled acids and alkalies should be washed thoroughly with water as quickly as possible, and then treated with neutralizing agents recommended by the instructor or specified on the emergency chart. Burns in the eyes should be washed with large quantities of water. If the burn is due to acid, wash with 5% solution of sodium bicarbonate; if due to an alkali, wash with a saturated solution of boric acid. Eye burns should always be sent to a physician after the preliminary emergency treatment. The use of safety glasses is highly recommended.

Cuts. Severe cuts are rare, but if they should occur, consult the emergency chart for instructions on the application of a tourniquet until a physician arrives. Minor cuts may be treated with tincture of merthiolate, and then covered with sterile gauze.

11·9 Terminating the Laboratory Period

The laboratory desk and its contents should be kept orderly, clean and neat at all times. Satisfactory laboratory techniques will permit the student to spend a minimum amount of time in terminating a laboratory period. The instructor will expect each student to store all equipment in his desk at the end of each laboratory period, and each desk should bear inspection if necessary. The working area should be sponged lightly with a small amount of cleaning powder, and then thoroughly rinsed with water to remove excess powder and dirt, before the student leaves the laboratory. In return the student should expect to find the top of his desk completely clean at the beginning of a work period.

REAGENTS

11·10 Care and Use of Reagents

A list of reagents used in the experiments is given in Appendix 2. Recipes for making special solutions are given in the same table. As a general rule, only reagents of highest purity are purchased for use in the quantitative analysis course. Because many of these reagents are expensive, they should be used sparingly and with reasonable care.

The following precautions should be generally observed in the use of reagents in a quantitative analysis laboratory:

Reagents in General. Reagents must not be contaminated by inserting dirty instruments into reagent containers. Reagent bottles should be kept tightly stoppered when not in use.

Liquid Reagents. A solution should be poured from its bottle, in the volume estimated as necessary, into a clean, dry beaker or graduate. Never insert a pipet into a reagent bottle. In order to avoid waste, remove only what you will probably need; but do not return any unused portion to the bottle. Stoppers of reagent bottles should be held in the hand while pouring and then replaced immediately.

Solid Chemicals, used in preparing solutions, are to be weighed on a general laboratory balance (*not* the analytical balance) unless otherwise directed. Remove directly from the bottle— by slightly tilting—to a piece of glazed paper on the balance pan. It is permissible to use a *clean* spatula to loosen the solid chemical within its container. Never place a chemical directly on the balance pan. If a small excess is poured out, discard into a stone waste jar. If any large quantity of a solid is spilled, consult the instructor at once. Keep the top of the general balance, and the desk top about it, free from chemicals and waste paper. Remove spillages at once with a sponge or towel.

11·11 Distilled or Deionized Water

Distilled (or deionized) water is the most important reagent in the analytical chemistry laboratory. Even distilled water may contain impurities. Soluble or insoluble impurities may occasionally be carried over in the distillation process, and gases from the atmosphere will contaminate distilled water that is standing. The procedures of elementary quantitative analysis rarely require ultrapure water, and usually ordinary distilled or deionized water is satisfactory. Dissolved gases from the air may be removed by boiling, after which proper storage is necessary.

Tap water is always contaminated with impurities which may affect chemical analyses, but tap water is usually pure enough for washing purposes. Distilled water is so expensive and difficult to produce and store, that it should never be used for washing purposes except under unusual conditions specified by the laboratory procedures or by the instructor. In other words,

the student should appreciate the presence of sufficient distilled water, and should not abuse this convenience by wasting the water.

Distilled water is used for all solutions. It may be used for rinsing, but only *from a wash bottle. Under no conditions is washing or rinsing permitted at the distilled water tap.*

USE OF THE ANALYTICAL BALANCE

Most courses in quantitative analysis make use of one-pan electric balances for analytical weighings. Such balances are made by a number of manufacturers, and the majority are excellent instruments. The balance described in these procedures is the Mettler H15; however, the use of any other analytical one-pan balance involves essentially the same manipulations.

11·12 Fundamental Rules for Maintaining Accuracy and Reproducibility

Each balance is assigned to a group of individual students; once having been assigned to a particular balance, a student should use it exclusively. Although modern analytical balances are quite accurate, there are always slight differences in the weights which are built into the balance. Thus if the same balance is used for a given analysis, errors caused by slight differences in weights usually cancel out. But such a cancellation of errors might not occur if more than one balance were used in a single analysis.

The following rules for the use of an electric one-pan balance should be observed:

(*1*) *The balance should be placed where it will be free from vibration.*

(*2*) *The balance should be leveled by means of the built-in spirit level before attempting a weighing operation.*

(*3*) *Objects to be weighed should be placed on the balance pan only when the balance is arrested. The same rule applies for the removal of objects.* The Mettler balance has a knob on the left side of the balance case (Figure 11·3), and it may be turned to three different positions. When the knob pointer is vertical (pointing upward) the

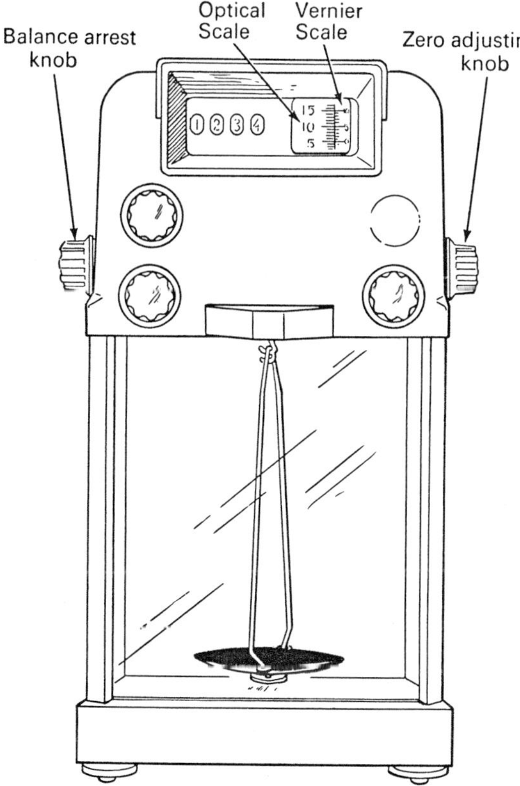

FIGURE 11·3 An electric one-pan balance

balance is arrested; if the knob is moved forward (toward the operator) the balance is semi-released; when the knob is moved backward (away from the operator) the balance is fully released.

(*4*) *Turn the weight-setting knobs only when the balance is semireleased.* As has been stated before, the semireleased condition is attained when the knob on the left side of the balance is turned toward the operator.

(*5*) *The side windows of the balance must be shut before the balance is fully released.* Air motion through the balance can cause the pan to fluctuate enough to cause inaccurate weighings.

(*6*) *If the balance is out of adjustment, consult your instructor.*

11·13 Procedures for Weighing

The following rules should be observed in a weighing operation:

(*1*) *The balance pan must be clean.* If necessary open the window and brush the pan with a camel's hair brush.

(*2*) *Set all weight knobs* (located on front of the balance, as shown in Figure 11·3) *on zero.*

(*3*) *The optical scale and vernier scale should coincide on the zero positions.* Release the balance completely, by moving the knob on the left side of the balance backward. When the optical scale comes to rest, bring the zero of this scale (using the adjusting knob on the right side of the balance case) to a position such that it will coincide with the zero line on the vernier scale. These two scales are illuminated from within the balance case: the movable scale on the left is designated as the optical scale; the smaller fixed-position scale on the right, as the vernier scale.

(*4*) *After the zero point has been checked and adjusted, return the balance to the arrested position.*

(*5*) *Place objects on the pan when the balance is in the arrested position.* All objects to be weighed must be dry and at room temperature. Use tweezers or a piece of folded paper to handle objects to be weighed. (Moisture from the fingers will affect weighings.) Close the balance window immediately after the object is placed on the pan.

(*6*) *Adjust weight-setting knobs with the balance in the semireleased position.* Starting with the knob controlling the heaviest weights, turn the knob clockwise until the optical scale moves upward; then turn it back one step. Adjust the other two knobs in the same way. The two lower knobs adjust the number of whole grams; the upper left knob adjusts the number of tenth-grams.

(*7*) *Make the final weight reading with balance fully released.* After the weight-setting knobs have been adjusted as described in Step (6), release the balance completely, and record the final weight reading from the knob indicators and the optical and vernier scales.

The optical scale is read by observing where the *zero line on the vernier* contacts the optical scale. For example, if the zero line falls between calibration lines 50 and 51 on the optical scale, the optical scale reading is 50– the smaller of the two values.

The vernier scale is read by observing which *vernier line* coincides exactly with a calibration line on the optical scale. If vernier line 3 happens to coincide, the vernier reading is 3. In the vernier reading, the mass of the object is recorded to the nearest tenth of a milligram. For example, 129.7503 g.

11·14 Weighing Exercise

Determine the weight of the bottom of a clean, dry weighing bottle. Next weigh the top of the same bottle, and record both weights. Now combine the top and bottom and weigh the complete weighing bottle. Compare the weight of the complete bottle with the sum of the weights of the top and bottom. If they differ more than 0.0003 g (0.3 mg) the weighing should be repeated. For example, the weighings might give the following results:

   14.6728 g = bottom of weighing bottle
    8.2416 g = top of weighing bottle
  22.9144 g = sum of above weights
  22.9146 g = weight of complete bottle
   0.0002 g = difference between weighings

The slight difference in weighings is generally unavoidable. The balance is so sensitive that weighings are affected by temperature variations, dust particles from the air, relative humidity, adsorption of traces of moisture by the weighed object, and uncertainties in reading the vernier scale. However, differences greater than 0.3 mg imply misreading of the scales or other experimental errors.

THE QUANTITATIVE LABORATORY NOTEBOOK

11·15 Suggestions for Keeping Notebook

Different teachers may have slightly different preferences as to the keeping of laboratory records. However, through many years of experience, the method of keeping a laboratory notebook has evolved along certain general lines, so that there is not much difference between the form used by a beginning student in quantitative analysis and the form used by a research chemist.

Laboratory records are usually kept in a hard-back bound notebook with pages numbered, measuring $7\frac{1}{2} \times 9\frac{1}{2}$ inches. The notebook is expected to be a complete, accurate, and original record of all experimental work and the related calculations. The following suggestions are offered, but they may be modified to the needs of the student and the instructor:

(1) *Never* tear a page from the notebook (see suggestion 7 below).

(2) Use a *medium, black* pencil instead of a pen.

(3) Reserve pages 1–4 for an *index* to be compiled as work progresses.

(4) Commence the record of each experiment at the top of a *right-hand* page, headed as follows:

Experiment No....   ....
                        (Object of   (Date)
                        experiment)

(5) Enter on this page, *at the time they are made, all observations* (weighings, buret readings, and so on) following the model forms given in this section. Continue with the necessary equations, calculations, and summary of results. Do not crowd your work; take all the space needed for clarity; if one page is not sufficient, pass to the next right-hand page.

(6) Use the *left-hand* pages for making incidental calculations and for any purpose that scratch paper is ordinarily used. The use of loose pieces of paper for any kind of laboratory records or calculations is expressly forbidden.

(7) Correct *minor* errors by neat erasures. Such errors are those which are recognized at the time they are recorded. When errors that necessitate extensive recalculations are discovered, do not try to erase but strike out the original work with a single diagonal line and refer, by number, to the page on which the new work is found.

(8) It is expected that all calculations will be based upon actual observations, and that observations or figures will not be altered for the purpose of forcing a better agreement of results.

11·16 Calculations

Calculations in quantitative analysis are customarily performed with the aid of five-place

TABLE 11·1   Specimen of a volumetric determination

Exp. _________   Percentage composition of impure   Date _________
potassium acid phthalate

	Sample A	Sample B	Sample C
Wt of weighing bottle + sample	31.6242 g	30.5177 g	29.6569 g
Wt of weighing bottle − sample	30.5177 g	29.6569 g	28.6662 g
Wt of sample	1.1065 g	0.8608 g	0.9907 g
Buret reading (initial)	0.30 ml	0.14 ml	0.20 ml
Buret reading (final)	36.30 ml	28.17 ml	32.44 ml
Ml of NaOH added	36.00 ml	28.03 ml	32.24 ml

General formula:
$$\frac{\text{ml NaOH} \times N \text{ NaOH} \times \text{meq wt KHP} \times 100}{\text{wt of sample}} = \% \text{ KHP}$$

Sample A
$$\frac{36.00 \times 0.1020 \times 0.2042 \times 100}{1.1065} = 67.77\% \text{ KHP}$$

Sample B
$$\frac{28.03 \times 0.1020 \times 0.2042 \times 100}{0.8608} = 67.82\% \text{ KHP}$$

Sample C
$$\frac{32.24 \times 0.1020 \times 0.2042 \times 100}{0.9907} = 67.78\% \text{ KHP}$$

		Deviation	Deviation parts/thousand
Sample A	67.77	0.02	0.3
Sample B	67.82	0.03	0.4
Sample C	67.78	0.01	0.1
	3)‾203.37	3)‾0.06	3)‾0.8
	67.79	0.02	0.3

% potassium acid phthalate = 67.79   Deviation = 0.3 parts/thousand

---

logarithm tables. Within recent years electric automatic calculators* have become available in some laboratories for student use. The operation and care of such calculators will be explained by the instructor, but the student is forewarned that calculators may be easily jammed,† so that expensive repairs are required. Two cardinal rules should be stressed:

(*1*) *Do not punch a button on a calculator while it is in operation.*

(*2*) *Do not punch two control buttons at the same time.*

As a general rule, round off all final results to four significant figures. The retention of more figures frequently assumes a degree of accuracy that is not warranted by the equipment and methods of the laboratory procedures.

* Also electronic calculators.
† Electronic calculators cannot be jammed, but they are quite expensive.

The student may make his calculations in the laboratory if time permits; see Section 11·1 concerning planning and scheduling of time. Unless calculations are performed immediately after a series of simple titrations a student may not know whether he has attained the required precision, as, for example, in the determination of the ratio between a base solution and an acid solution. Thus, time may be gained by doing simple calculations at the time the observations are made. On the other hand, if the calculations are complex enough to be time-consuming, they should be delayed until the laboratory work is completed.

Specimen forms illustrating hypothetical notebook records are given in Tables 11·1 and 11·2.

11·17 Number of Determinations

The actual number of samples that a student should carry through a given analysis may

TABLE 11·2   Specimen of a gravimetric determination

Exp. __________	Gravimetric determination of chloride in a soluble salt mixture	Date __________
	Sample *A*	Sample *B*
Wt of weighing bottle + sample	26.0297 g	25.5976 g
Wt of weighing bottle − sample	25.5976 g	25.0940 g
Wt of sample	0.4321 g	0.5036 g
Wt of empty Gooch crucible		
After 1st heating	16.7423 g	16.8926 g
After 2nd heating	16.7421 g	16.8923 g
Wt of Gooch crucible + AgCl	17.2826 g	17.5205 g
Wt of Gooch crucible − AgCl	16.7421 g	16.8923 g
Wt of AgCl	0.5405 g	0.6282 g

General formula: $\dfrac{Cl}{AgCl} \times \dfrac{wt\ AgCl}{wt\ of\ sample} \times 100 = \%\ Cl$

Sample *A* $\dfrac{35.45}{143.3} \times \dfrac{0.5405}{0.4321} \times 100 = 30.94$

Sample *B* $\dfrac{35.45}{143.3} \times \dfrac{0.6282}{0.5036} \times 100 = 30.85$

		Deviation	Deviation parts/thousand
Sample *A*	30.94	0.04	1
Sample *B*	30.85	0.05	2
	2) 61.79	2) 0.09	2) 3
Average	30.90	0.045	2

% chlorine = 30.90   Deviation = 2 parts/thousand

depend upon a variety of tangibles and intangibles. These factors are analyzed at the beginning of Chapter 1 under the topic. *The Evaluation of Analytical Data.* However, it frequently becomes necessary for a student to make a rough evaluation of his laboratory work before the subject is discussed elsewhere.

Two of the factors which may influence the number of determinations are *precision* and *accuracy.* These two terms have quite different meanings. The *precision* of a group of results refers to the agreement among the various numerical results obtained by the experiments. In other words, precision indicates the *degree of reproducibility.* The term *accuracy* refers to the agreement between the numerical result and a hypothetical true value. Absolute true values can never be attained but they may be approached. In actual practice, true values are defined within certain limits, which are dependent upon the accuracy of the instruments and the skill of the operators.

It is sometimes frustrating to a student when he obtains results which are close together (precise), but which are found to be far from the true value (inaccurate). The accuracy of student determinations is usually evaluated by the instructor, on the basis of results obtained from multiple analyses of professional analysts.

The precision (and also accuracy) of analytical results is customarily expressed as an average deviation in parts per thousand. The computation of this expression involves four essential steps:

(1) Calculate the arithmetical mean of the results.

(2) Compute the deviation of each result from average, ignoring its algebraic sign.

(3) Obtain the average deviation of all observations.

(4) Multiply the average deviation by 1000 and divide by the average result to get *average deviation in parts per thousand.*

These steps are illustrated by using the following hypothetical observations:

	Observations	Deviations from the average
	35.46	0.09
	35.62	0.07
	35.57	0.02
Average observation	3)106.65 35.55	Average deviation 3)0.18 0.06

$$\frac{0.06 \times 1000}{35.55} = 2 \text{ parts per thousand}$$

(approximately)

Duplicate determinations must be made. If they are not, a student has no criterion for calculating the precision of his work. Because a beginning student frequently ruins one sample of a determination, it is usually desirable to make triplicate determinations. The student will soon realize that the three determinations can be performed simultaneously almost as quickly as a single determination.

The statistics of the variations in accuracy and precision has been studied for more than 300 years, and has become progressively more complex. Such studies, including the rejection of results, are considered in Chapter 1. However, it can be said that an average result is more apt to be accurate when a large number of determinations are involved. But this accuracy is increased only in rough proportion to the *square root* of the number of determinations; thus, many determinations are not significantly more accurate than a few.

Agreement of duplicate or triplicate results, in carefully tested procedures, is a good indication that the error in the analysis is small, but does not constitute proof of such. In other words, close agreement indicates excellent precision, but not necessarily close accuracy. A student should strive to attain a precision—that is, agreement of results—of two parts per thousand. Most quantitative laboratories have equipment, for student use, that permits the attainment of this range of precision. Three parts in one thousand is usually satisfactory, but when the precision drops below five parts in one thousand, the accuracy is very likely to be poor. A precision as low as ten parts per thousand is not acceptable.

CHAPTER 12

Neutralization Methods

PREPARATION OF ACID AND BASE SOLUTIONS

12·1 Introductory Statements

The reaction of hydroxide and hydronium ions with each other, or with other acids or bases, forms the foundation for neutralization methods. For a quantitative determination the reaction should produce a rapid change in pH near the equivalence point. This change may be observed either with a pH meter or by color changes in suitable acid–base indicators.

The best solutions for acid–base titrations contain strong acids or bases. Such reagents must be fairly soluble and relatively inexpensive. The resulting solutions should be stable, and should not produce side reactions when used in acid–base determinations. Only a few acids and bases meet these criteria. The most satisfactory acids are hydrochloric and sulfuric acid; and the most frequently used bases are sodium hydroxide and potassium hydroxide. However, strong bases have a tendency to react with glass containers, and to absorb CO_2 when exposed to the atmosphere; consequently, certain precautionary measures are necessary in the preparation and use of base solutions.

12·2 Preparation of Approximately 0.1 _N_ HCl Solution

By means of a 25 ml graduated cylinder, measure 17.5 ml of reagent-grade concentrated HCl (specific gravity 1.185–1.192), and pour this amount into a one-liter volumetric flask. Rinse the graduate with two portions of distilled (or deionized) water, and add rinsings also to the flask. Fill the volumetric flask approximately to mark with distilled water, and then transfer the solution to a clean five-pint bottle. (Non-returnable acid bottles are excellent for this purpose.) Fill the flask once more to mark with distilled water, and pour the additional liter of water into the bottle. This gives a total volume of two liters approximately measured. Stopper the bottle tightly, and then thoroughly mix the contents by inverting repeatedly and shaking for several minutes. Label the bottle neatly as follows:

0.1 _N_ HCl

Your name   Date

Store the bottle of 0.1 _N_ HCl inside the desk cabinet until needed.

Concentrated HCl is very close to 12 normal; hence, when 17.5 ml is diluted to a total volume of 2 liters (2000 ml) the resulting solution will be approximately 0.1 _N_:

$$ml_1 \times N_1 = ml_2 \times N_2$$

or

$$17.5\ ml \times 12\ N = 2000\ ml \times N_2$$

$$N_2 = 0.105$$

Normality, molarity, and formality may be used interchangeably for HCl and NaOH; the distinction between these three terms is discussed in Section 2·3.

12·3 Preparation of Approximately 0.1 *N* NaOH Solution

Using a one-liter volumetric flask, measure approximately two liters of distilled water into clean beakers. Heat the water to boiling, cool until warm, and then transfer to a 2-liter borosilicate glass bottle. Stopper tightly with a solid rubber stopper.

By means of a small graduated cylinder, measure 10.5 ml of clear 50% reagent-grade sodium hydroxide solution, and add the measured quantity to the two liters of freshly boiled water. (*Warning.* Use care to prevent the concentrated NaOH from coming into contact with the skin, since this substance can produce severe burns. See Section 11·8 for treatment of such burns.) Mix the resulting solution *thoroughly* by inverting *repeatedly*, and shaking, for several minutes. Label the bottle 0.1 *N* NaOH, with your name and date, as indicated in the preceding section and store in the desk until needed. It is essential that the solution should be kept well-stoppered at all times to prevent absorption of carbon dioxide from the air.

Fifty percent sodium hydroxide is approximately 19 *N* and is fairly free of carbonate ions, since sodium carbonate is insoluble in this high concentration of hydroxide ions. When 10.5 ml is diluted to an approximate volume of two liters the resulting solution will be approximately 0.1 *N*.

COMPARISON OF STRENGTH OF ACID AND BASE SOLUTIONS BY TITRATIONS

12·4 Preliminary Comments on Titrations

A titration involves the addition of an equivalent amount of a known material to a measured amount of unknown substance. When a buret is used to measure out the added amount, the procedure is known as volumetric analysis. An *end point* occurs when there is a definite recognizable change in the solution being titrated. The end point may be a change in the color of a suitable visual indicator, or it may be some other sharp change in a physical or chemical property of the solution being titrated. The *equivalence point* in a titration occurs when the number of equivalents of material in the titration flask is just equal to the number of equivalents of titrant added. Some acids, possessing more than one removable proton, may show more than one equivalence point during a titration.

Titrations with visual indicators are stopped exactly at the end point and are selected so that the end point will occur *near* the equivalence point (for example, see the procedures given in Sections 12·6 and 12·7). However, where graphical methods of analysis are used (for example, Section 12·8), the equivalence point may be determined more precisely if you continue to take measurements beyond the end point.

When the pH of a solution is plotted versus the ml of titrant added, an S-shaped curve is obtained. The S-shape occurs as a result of the very rapid change in pH in the area of the equivalence point. The equivalence can be detected graphically by determining the point at which the rate of change of pH is greatest. This point is usually midway along the portion of the S-curve that is most nearly parallel to the pH axis. It is difficult to locate the point graphically unless the entire S-curve is plotted.

A number of highly colored organic dyes have the property of changing color when the pH of a solution is varied between certain limits. These dyes are called visual acid–base indicators. For a visual indicator to be satisfactory for a quantitative acid–base determination, it must change color abruptly at a point which coincides closely with the equivalence point. Since most visual indicators require a change of 2 pH units before their color change is distinct enough for satisfactory detection, there must be a sharp change in the pH of the solution in the vicinity of the equivalence point to permit the use of such an indicator. Phenolphthalein is useful for the titration of weak acids by a carbonate-free base; it undergoes a color change from colorless to pink at a pH of approximately 9.5. The most suitable indicator for carbonate titrations is a mixture

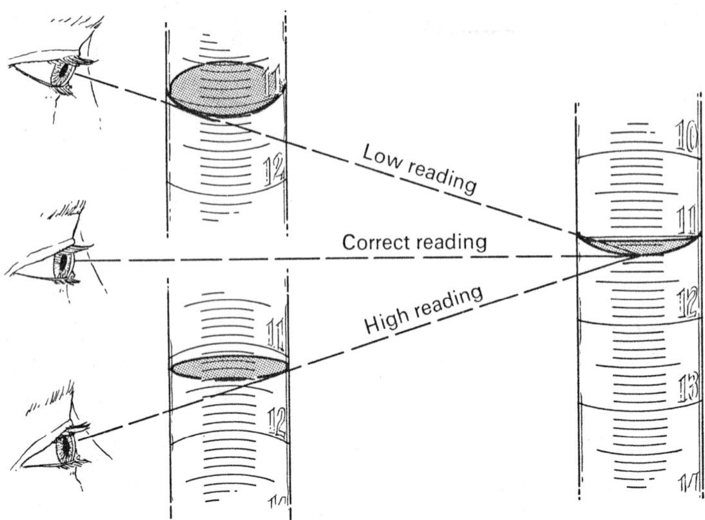

FIGURE 12·1 Correct method for reading a buret to avoid parallax

of two particular dyes: methyl red and brom-cresol green. This mixed indicator, when used after boiling to remove carbon dioxide, produces a satisfactory color change from pink to green at a pH of approximately 5.0. The two indicators jointly produce a more distinct and more abrupt color change than either indicator does separately.

12·5 Preparation of Burets for Titrations

With small paper labels, identify each of your two burets near their tops as HCl and NaOH. For convenience and to avoid error, place the burets in a given order; for example, the *acid* might be always to the left. Rinse the previously cleaned burets (see Section 11·6) at least four times with *small* portions of the respective solution to be placed in the particular buret; namely, HCl in the left buret, and NaOH in the right buret. Discharge rinsings through the tips of the burets.

Fill the burets with their respective solutions to a point considerably above the top graduation mark, and then open the stopcock completely on each buret until all air bubbles have been removed from the tip. The final solution level should be just below the 0.00 ml mark. Do not try to adjust the level so that it is exactly on the zero mark. Remove drops clinging to the tips of the burets by touching them against the side of the waste-solution beaker. Read the lower level of the meniscus formed by the solution in each buret in such a way that parallax is avoided (see Figure 12·1).* Record these readings in an orderly fashion on a right-hand sheet in the laboratory notebook (see Table 12·1). The apparatus is now ready for performing a titration.

12·6 Acid–Base Ratio with Phenolphthalein as the Indicator

Once you have recorded the original readings of both burets, place a clean 500 ml wide-mouth Erlenmeyer flask on the white tile of the buret stand under the acid buret. Run into the flask, down the side to avoid spattering, approximately 30–35 ml of the HCl solution. (It is not necessary to record the buret reading at this stage.) It is correct technique to use the left hand, with fingers curled around the buret, to control the buret cockstop. Use the right hand for rotating the titration flask to produce proper mixing.

Wash down the sides of the flask with distilled water from a wash bottle, and then add 2 or 3

---

* A white piece of paper held behind the graduations will aid in establishing a correct level for the meniscus.

TABLE 12·1 Volume-ratio of HCl and NaOH with phenolphthalein as the indicator

HCl	NaOH	HCl/NaOH	Deviation
36.46 ml	36.10 ml		
0.36	0.20		
36.10	35.90	1.016	0.002
37.21 ml	36.90 ml		
0.34	0.46		
36.87	36.44	1.012	0.002
38.41 ml	36.80 ml		
0.35	0.21		
38.06	37.59	1.013	0.001
		Average 1.014	Average 0.002

$$\text{Average deviation per thousand} = \frac{0.002 \times 1000}{1.014} = 2/M$$

Therefore the acid–base ratio is 1.014 ml HCl/ml NaOH with phenolphthalein indicator.

drops of phenolphthalein solution. Since phenolphthalein behaves as a weak acid in the acid–base reaction, more than 2 or 3 drops of the indicator will produce an error in the determination of the ratio. Place the titration flask under the NaOH buret, and run in this solution fairly rapidly, with constant slow rotation to secure proper mixing, until a red color pervades the entire solution. Discharge the color by the cautious addition of HCl. Finally, add NaOH, *drop by drop*, with constant rotation after each addition until one drop gives the end point. This is a *faint* pink color, just visible against a white background, but persisting at least 15 seconds. Immediately take and record final readings of both burets (see Table 12·1). Discard the contents of the Erlenmeyer flask, and rinse at once with distilled water.

Refill both burets (rinsing at this time is of course unnecessary), read, and record. Make a second acid–base ratio determination as before. Then complete a third in the same manner.

Before proceeding to Section 12·7, which involves the use of a mixed indicator, calculate the three ratios just determined and compute the mean deviation (see Table 12·1). Use electric calculator or five-place logarithms for all calculations; it is usually wise to double check all such computations.

If the mean deviation is less than 4 parts per thousand (a good analyst can secure 1 to 2 parts per thousand or less), take the average of the three as the volume of HCl neutralized by 1.000 ml of NaOH. If this agreement is not obtained, it will be necessary to determine additional ratios until satisfactory results are secured. Erratic results may signify that the HCl and NaOH stock solutions have not be shaken sufficiently to be homogeneous (see Sections 12·2 and 12·3).

12·7 Acid–Base Ratio with Mixture of Methyl Red and Bromcresol Green as the Indicator

Begin the record on a fresh right-hand page of the notebook. Having refilled, read, and recorded the buret readings, run approximately 35 ml of the NaOH solution into a clean wide-mouth Erlenmeyer flask, taking care to avoid spattering. Wash down the walls of the flask with distilled water from a wash bottle. Add one drop of methyl orange indicator and titrate with HCl until a definite color change is obtained, and then add one ml of HCl in excess. Do not take a buret reading at this point since more acid is required to complete the titration. The methyl orange indicator is used to make certain that the pH of the solution is definitely in the acid range.

Place the flask on a wire gauze and heat over a free flame until the solution comes to a boil; then continue to boil *gently* for about one minute. Cool to room temperature; cooling may be hastened by running cold tap water on the outside of the titration flask. Add 2 or 3 drops of the mixture of methyl red and bromcresol green, and titrate slowly with NaOH solution until a green color is obtained. Now back-titrate with acid to a pink color; then add the base a drop at a time until one drop gives a sharp change to green. It is advisable to approach the end point by the addition of base in each titration. Read both burets and record the results.

In a like manner, run two more determinations, using the mixed indicator solution. Calculate the ratios and deviations (see Table 12·1) and, if necessary, make additional determinations until a satisfactory agreement is obtained. A careful analyst can secure an average deviation of two parts in a thousand, or less.

Theoretically the difference between the ratio obtained with phenolphthalein and that obtained by the mixed indicator should be about 3 or 4 parts in a thousand. However, in practice the difference between the ratios obtained in Sections 12·6 and 12·7 may be five parts in a thousand or more, depending upon the analyst's judgment of the end points. Since the phenolphthalein end point is obtained in a basic solution, and the end point for the mixed indicators occurs in an acid solution, the acid–base ratios for the two different indicators cannot be the same. The reasons for a difference are explained in Section 4·4.

12·8 Titrations Involving a pH Meter

The purpose of this titration is to demonstrate the use of a pH meter in performing an acid–base titration, and to illustrate how the pH of a solution changes as the titration proceeds. Owing to the limited number of pH meters available for use by the class, each student will be assigned use of a pH meter for only *one* particular day. *All necessary experimental work with the pH meter must be performed on that one day.*

Principle. A pH meter is, in effect, a high-impedance voltmeter which measures the difference in potential between two electrodes, when they are immersed in a solution. One of the electrodes is a saturated calomel electrode, with internal parts composed of a mixture of Hg and Hg_2Cl_2 (calomel) in contact with a saturated KCl solution (Figure 4·2); it is known as a *reference* electrode, E_{ref}^0, because its potential remains constant. The other electrode is a glass electrode, also called a pH indicator electrode; it contains a special glass membrane (Figure 4·1), which is sensitive to hydronium-ion concentrations. The potential of the indicator electrode varies according to the Nernst equation (Section 6·6) as indicated in the following equation:

$$E_{ind} = E_{ind}^0 - 0.059 \log \frac{1}{[H^+]} \qquad \text{(at 25°C)}$$

Since pH $= -\log_{10}[H^+]$, the pH of the solution is directly related to the potential of the electrode and can be measured on a linear scale. The net difference in the potential between the two electrodes may be expressed as

$$(E_{ind} - E_{ref}^0) = (E_{ind}^0 - E_{ref}^0)$$
$$- 0.059\,(-\log[H^+])$$

or

$$\Delta E_{obs} = \Delta E^0 - \log 0.059 \text{ pH};$$

therefore

$$\text{pH} = \frac{\Delta E^0 - \Delta E_{obs}}{0.059}$$

Differences in electrode potential are customarily expressed in volts or millivolts.

Before a pH meter is used for measuring the pH of an unknown, it must be calibrated with a solution of constant known pH, a *buffer solution.* (Use one certified buffer tablet dissolved in 100 ml of water.) One dial is used to account for the temperature of the solution. In effect, it adjusts the proportionality factor, 0.059, to the proper value via electrical circuitry. The other dial is used to set the meter needle to read the pH value of the buffer; it compensates for the value of the constant, ΔE^0, so that the meter reads directly in pH units rather than in volts (ΔE). Once the pH meter has been calibrated, it may be used to measure the pH of an unknown solution or of a solution during titration.

More details about the theory and operation of pH meters can be found in numerous textbooks

TABLE 12·2   Data for a pH titration of NaOH versus HCl

HCl, buret readings	pH reading	Additional directions and observations to be made
0.00 ml	————	
2.00	————	
10.00	————	
20.00	————	Add 2–3 drops of phenolphthalein solution at this point in the titration.
22.00	————	
23.00	————	
24.00	————	
24.20	————	Indicate with an arrow the reading where the pink color of phenolphthalein disappears.*
24.40	————	
24.60	————	
24.80	————	
25.00	———— **	
25.20	————	
25.40	————	To colorless solution, add 2–3 drops of mixed indicator solution.*
25.60	————	
25.80	————	Indicate with an arrow the reading where the green color of the mixed indicator solution turns pink.*
26.00	————	
28.00	————	
30.00	————	
40.00	————	

* The indicators are added only for the purpose of showing that phenolphthalein changes in a basic solution, whereas the mixed indicator changes in an acid solution. If you missed color changes do not repeat the titration.

** This should be the *true equivalence point*, if the directions have been followed properly. The equivalence point occurs at the half-way point along the steep portion of the titration curve.

of advanced quantitative analysis, analytical chemistry, and instrumental analysis. See your instructor concerning operating instructions for the particular pH meter used in your pH titration.

Graphical titration requires no visual indicators, since the graph provides needed data; however, the directions for this experiment call for the addition of indicators so that you can observe the pH ranges over which some indicators change color, and can judge their applicability in various pH titrations.

Since the following pH titration is intended merely to be illustrative, it is not necessary to check for precision. Unless the student obtains a completely unsatisfactory set of pH readings, a highly unlikely possibility, he should make only *one* NaOH versus HCl titration. However, in the procedures described in Sections 12·6 and 12·7, in which accuracy was a primary concern, it was necessary to make several titrations to obtain a check on experimental precision.

Preliminary Operations. (1) There may be more than one type of pH meter in use in the laboratory. Carefully examine the one assigned to you, and consult with your instructor for directions for the proper operating procedure. *Caution:* At all times when they are not in use, the electrodes should be kept immersed in distilled water.

(2) Fill your burets with the approximately 0.1 N NaOH and 0.1 N HCl solutions, following the instructions given in Section 12·5 for the proper methods of rinsing and filling burets. If the calomel reference electrode is not at least one-third full of saturated KCl solution, see the instructor for directions on filling the electrode.

(3) Using the buffer tablet furnished by your instructor, standardize the pH meter according to the directions you have received. Record in your notebook the pH obtained; it should be approximately 7.0. You are now ready to continue to the titration.

Procedure for the pH Titration of NaOH *versus* HCl. (*Caution:* During this titration do not allow the electrodes to remain in the strongly basic solutions for prolonged periods of time; a strong base will etch the glass electrode and ultimately ruin it.)

(1) Set up data in your notebook as indicated in Table 12·2.

(2) Find in your notebook the HCl/NaOH volume-ratio with phenolphthalein indicator (the procedure given in Section 12·6). Divide your acid–base ratio, as obtained in that experiment, into the number 25.00. The result will be the number of milliliters of NaOH which are equivalent to 25.00 ml of your HCl solution. For example, if the HCl/NaOH ratio is 1.014, then:

$$25.00/1.014 = 24.65 \text{ ml of NaOH}$$

Transfer the above amount of NaOH (25.00 divided by your acid/base ratio) from your buret into a 250 ml beaker. Place the beaker on a magnetic stirrer and add enough distilled water so that the fragile electrodes can be immersed in the solution without the possibility of being struck by the stirring bar. Immerse the (previously rinsed) electrodes. Turn on the stirrer, and gradually increase its speed until the middle range is reached.

(3) Arrange your burets so that the one containing your HCl solution extends over the beaker. Run in the amounts of acid indicated in Table 12·2. *After each addition, having allowed time for temperature equilibration, read the pH meter and record your observation.*

(4) When you have finished, remove the beaker containing the titrated solution. Discard the solution and wash the beaker thoroughly. Rinse off the electrodes and immerse them in a beaker of fresh distilled water. Unplug the pH meter. Clean off the work space, and store the marking pencil, magnetic stirrer, and stirring bar where you found them.

Special Laboratory Report for the pH Meter Titration. On graph paper, prepare a curve of your pH meter titration, plotting pH vertically and ml of titrant horizontally. Plot each pH as a point of the graph, and enclose each point in a small neat circle. Using the data points as guides, draw a *smooth* curve for the complete titration. Label the following clearly on the graph:

(1) Your name, and the date of the experiment
(2) Solution being titrated
(3) Titrant.
(4) Location of the true equivalence point, at which the moles of added titrant exactly equals the moles of NaOH solution being titrated
(5) Point at which each indicator changed color
(6) What pH meter was used.

STANDARDIZATION OF ACID AND BASE SOLUTIONS

12·9 Preliminary Comments

A standard solution is one in which the concentration of the solute is known to a high degree of accuracy. In elementary quantitative chemistry, a student strives for an accuracy of 2 parts per thousand; therefore, the concentration of an approximately 0.1 N solution of base should be determined to within 2 parts in the fourth decimal place; for example, for a solution whose actual concentration is 0.1098 N NaOH, determined values may fall between 0.1096 N and 0.1100 N and still provide the desired accuracy.

The standardization of a solution of base is usually accomplished by determining experimentally the volume of the base, which is equivalent to a known weight of a pure acidic substance. Likewise, a solution of an acid may be standardized by measuring accurately the volume of the acid that is equivalent to a pure basic substance. The extremely pure chemicals which are used for standardization are known as *primary standards*. Frequently it is more convenient, in neutralization methods, to standardize either the base or acid solution against a primary standard, and then to determine the concentration of the other solution by a comparison of the acid to base volume–ratio (see

Sections 12·6 and 12·7). When a standard solution is used to determine the strength of another solution by means of a volume–ratio, the standard solution is referred to as a *secondary standard*. Theoretically, a secondary standard is not as accurate as a primary standard, but in practice the convenience may outweigh the inaccuracy of the method.

A primary standard is usually a solid substance about 99.95% pure (impurity of 0.5 parts per thousand). It should be relatively soluble in water, and it must react quantitatively with the solution to be standardized. The composition of a satisfactory primary standard is not affected by drying, and ideally its equivalent weight should be fairly high in order that a large sample may be weighed without requiring more than a buretful of solution in the standardization process.

Potassium acid phthalate ($KHC_8H_4O_4$) is the most widely used primary standard for the standardization of base solutions. Substances that are used to a lesser extent are potassium acid tartrate ($KHC_4H_4O_6$), potassium biiodate [$KH(IO_3)_2$], and sulfamic acid ($HSO_3 \cdot NH_2$).

For the direct determination of the normality of an acid, the recommended primary standard is anhydrous sodium carbonate with a purity of 99.95%. Since it is impossible to purchase sodium carbonate commercially of this purity, it is general practice for analysts to prepare their own pure sodium carbonate from reagent grade sodium bicarbonate by prolonged heating at high temperature. Other substances which may be used as primary standards for the standardization of acid solutions are reagent grade calcium carbonate and mercuric oxide. These latter substances have the disadvantage of being insoluble in water; consequently, they must be dissolved in excess acid and back-titrated with standard base.

I have found that students secure more reliable results for the standardization of the HCl solution with the use of acid–base ratios (using NaOH as a *secondary standard*) than of the primary standards mentioned above. However, you should be aware of the possibility of direct titration of the HCl against a *primary standard* as a method for the standardization of this solution.

12·10 Standardization of Base Solutions

The essential steps necessary for the precise determination of the strength of the base solution are described as follows.

Preliminary Operations. Pure potassium acid phthalate (frequently labeled potassium biphthalate) is to be found on the side shelf. Weigh approximately 4 grams of this substance, on a piece of glazed paper, and transfer to a clean weighing bottle. Your name should be marked with a *graphite* pencil on the ground glass portion of the bottle, with the letters KHP to identify the contained substance. Put the weighing bottle and material (with the bottom of the bottle cradled in the up-turned top) in the drying oven, and dry at 110°C for at least one hour. Stopper the bottle and place in the desiccator until needed.

Wash three wide-mouth 500 ml Erlenmeyer flasks, rinse with *small* portions of distilled water, and label the flasks A, B, and C. The flasks need not be dried. Prepare a right-hand page of the notebook for recording the data of this experiment following the general form shown in Table 12·3. Proceed to the balance room with flasks, notebook, and desiccator containing the weighing bottle with dry potassium acid phthalate.

Weighing the Samples. It is customary to handle a weighing bottle with a paper shield, that is, a folded strip of paper encircling the bottle and held tightly to the body of the bottle by means of the ends of the strip.

First, weigh the weighing bottle containing the dry material, record the weight as in Table 12·3, and then transfer the estimated amount of material, by cautious pouring, into the flask in which it is to be treated. Close the weighing bottle, and weigh again; the loss, or difference, in weight represents the sample taken. Four weighings are required for three samples. In the transfer of material, it is necessary that all of the sample leaving the bottle should be caught in the container. It is poor practice to return material from the receiving vessel to the weighing bottle since the receiving vessel is usually not dry. You must

TABLE 12·3   Sample data of NaOH solution with potassium acid phthalate

Weighings:		Sample *A*		Sample *B*		Sample *C*	
Wt of KHP + bottle (before)		26.0297		25.3241		24.5764 g	
Wt of KHP + bottle (after removal of sample)		25.3241		24.5764		23.7941	
Weight of samples:		0.7056		0.7477		0.7823	

Buret readings:

	A		*B*		*C*	
HCl	NaOH	HCl	NaOH	HCl	NaOH
0.44	31.78	0.44	33.68	1.73	36.71 ml
0.44	0.26	0.44	0.36	0.44	0.56
0.00	31.52	0.00	33.32	1.29	36.15

(Hypothetical ratio of HCl/NaOH = 1.014 with phenolphthalein from Table 2·1)

General formula:

$$N = \frac{\text{wt of sample of KHP}}{\text{ml of NaOH} \times \text{meq wt of KHP}}$$

In case of back-titration:

$$N = \frac{\text{wt of sample of KHP}}{[\text{ml of NaOH} - (\text{ml of HCl/ratio})] \times \text{meq wt of KHP}}$$

Calculations: Deviations:

Sample *A* $N = \dfrac{0.7056 \text{ g}}{31.52 \text{ ml} \times 0.2042 \text{ g/meq}} = 0.1096$ meq/ml   0.0002

Sample *B* $N = \dfrac{0.7477}{33.32 \times 0.2042} = 0.1099$ 0.0001

Sample *C* $= \dfrac{0.7823}{[36.15 - (1.29/1.014)] \times 0.2042} = 0.1098$ 0.0000

Average $= 0.1098$ *N*   0.0001

Therefore the normality of the hypothetical base = 0.1098 *N*

---

learn by practice how to estimate the approximate sample weights.

Accurately weigh three samples to four significant figures (about 0.7–0.9 g each) of pure potassium acid phthalate into the three labeled titration flasks. Record these weights and subsequent data in the notebooks as indicated in Table 12·3.

Titrations.   To each of the flasks add 75 ml of distilled water (measured in graduated cylinder) and 2–3 drops of phenolphthalein solution. Rotate flasks gently to hasten complete mixing. Drain the water from the burets, rinse four times with small amounts of the respective solutions to be used, as instructed in Section 12·5, fill the burets properly, take readings, and record as indicated in Table 12·3. The acid solution is not used except for back-titration. However, the buret must be filled for each titration to avoid errors arising from lack of uniformity in the buret tube.

Place flask A under the NaOH buret, and add the base solution, rotating the flask to ensure mixing. The NaOH may be added rather rapidly as long as the pink color, produced on the first contact, fades quickly on mixing. When the color begins to disappear more slowly, decrease the rate of titration and finally complete it drop by drop, seeking a complete-stop end point. Stop at the usual phenolphthalein end point, that is, when a single drop produces a faint pink which lasts for as much as 15 seconds. Take the buret reading and record.

Back-titration.   Should too much NaOH be added (over-titration), the situation may be saved by adding a measured volume of HCl for which you have an acid–base ratio from Section 12·6. After enough HCl has been added to make the solution colorless, proceed to the usual end point by drop-by-drop addition of the base. Sample C in Table 12·3 illustrates a hypothetical over-titration with proper calculations. The acid–base ratio is also hypothetical, and is the one given as an example in Table 12·1.

Results involving back-titrations are not as reliable as those secured with complete-stop end points, because the former involves four buret readings for each result, whereas the latter involves only the two readings on the base buret. Furthermore, overly prolonged titrations allow carbon dioxide from the air to react with sodium hydroxide in the buret.

12·11 Standardization of Acid Solution by Comparison with Base Solution

The most convenient method of standardizing the HCl solutions is through the use of the NaOH as a *secondary standard* in combination with the acid base ratios. The hypothetical values indicated in Sections 12·6, 12·7, and 12·10 will be used to illustrate this method.

In Section 12·6 and Table 12·1 the acid–base ratio (HCl/NaOH) with phenolphthalein as an indicator was assumed to have the value in 1.014. No value for the acid–base ratio with the mixed indicator (Section 12·7) was conjectured, but for the purpose of this explanation we will assume that a value of 1.019 was obtained. As was explained in Section 12·7, these ratios are necessarily different. Consequently, the normality of the HCl will have two values: one value when used with phenolphthalein as an indicator, and the other value when used in titrations in which the mixed indicator solution is involved. Using the values assumed in the preceding sections, we can arrive at the following normalities for HCl:

with phenolphthalein (see Section 12·6):

$$\frac{\text{normality of base}}{\text{HCl/NaOH ratio}} = \frac{0.1098}{1.014} = 0.1083 \, N$$

with mixed indicators (see Section 12·7):

$$\frac{\text{normality of base}}{\text{HCl/NaOH ratio}} = \frac{0.1098}{1.019} = 0.1078 \, N$$

The student *must realize* that all of the values given in the preceding sections are merely illustrative for the purpose of explaining how

acids and bases may be standardized. The student's experimental values may vary considerably from those used in the examples.

Important. In the exercises involved in the analyses of unknown acids and bases, which are given in this chapter, the student will use the normality of the acid (with phenolphthalein) only in back-titrations involving the determination of the percentage composition of an unknown acid sample. The normality of the acid (with mixed indicator) is used in the analysis of impure soda ash.

ANALYSES OF ACID AND BASE SAMPLES

12·12 Preliminary Comments

For the analyses of acid samples, either you may be given pure solid acids or acid salts for the determination of equivalent weights, or you may be required to analyze impure acids for percentage compositions.

Among the solid acids and their acid salts that may be obtained in a degree of purity necessary for equivalent weight determination, a few may be listed as follows: succinic acid, benzoic acid, oxalic acid, sulfamic acid, and potassium bioxalate. Any of these substances can be issued as solid unknowns from which weighed samples may be titrated with standard NaOH to secure equivalent weight values. The general formula is as follows:

$$\text{eq wt} = \frac{(\text{wt in g of the acid}) \times 1000}{(\text{ml of NaOH used}) \times (N \text{ of NaOH})}$$

However, because these substances have widely differing equivalent weights, it is difficult for an instructor to write up laboratory procedures that can be used in highly precise equivalence determinations for a variety of the compounds. For example, a sample of oxalic acid would require approximately six times the volume of standard base as would be required by the same weight of potassium biiodate.

The instructor usually finds the most satisfactory solution to the problem of acid analysis is the use of samples of impure potassium acid phthalate. Such samples are commercially available, in varying percentage compositions, which have been determined by professional analysts. A procedure for the analysis of potassium acid phthalate samples containing known quantities of inert materials is given in Section 12·13.

Very few substances are satisfactory as unknowns for student analysis of base constituency. However, certain carbonates, particularly sodium carbonate and calcium carbonate, have been used quite successfully as unknown samples in elementary quantitative analysis. Sodium carbonate is more desirable since it is water soluble, whereas calcium carbonate is not. Commercial samples of sodium carbonate are available containing known quantities of inert materials. Usually the combination of sodium carbonate with an inert material is designated as soda ash, and the procedure for the analysis of such samples is contained in Section 12·14.

Although the particular choices of acid–base unknowns used in the exercises of Sections 12·13 and 12·14 are dictated by the academic considerations above, it should be emphasized that comparable acid–base analyses are encountered in research and quality control laboratories throughout the world. There, the analytical chemist is faced with the additional problems of selecting the size of sample to use, the best indicator for the analysis, the appropriate concentrations for the titrants, and so on—problems which, for lack of student laboratory time have already been answered for you in these particular exercises.

12·13 Analysis of an Acid Sample

The details in an acid analysis will vary with the type of unknown, but the general features are as follows.

Principles. The equivalent weight of any acid may be calculated by dividing the formula weight of the acid by the number of protons (customarily referred to as hydrogen ions) furnished in the reaction. For potassium acid phthalate, which yields one proton per hydrogen–phthalate ion, the equivalent weight is the formula weight.

A *gram equivalent weight* (or an *equivalent*) of an acid is that quantity of acid which furnishes a fw of protons (1.008 g of protons). An equivalent of acid will neutralize an equivalent of base (for example, 40.01 g of solid sodium hydroxide or one liter of a *normal* solution of sodium hydroxide). Similarly, a milliequivalent of acid will neutralize a milliequivalent of base, for example, one ml of a *normal* solution of base). A general relationship may be written:

$$N \times \text{ml} = \frac{\text{g of solute}}{\text{meq wt}} = \text{meq}$$

The present exercise is concerned with the analysis of a known acid (potassium acid phthalate with an equivalent weight of 204.2, or 0.2042 meq wt) of unknown purity. The general formula for calculating the % purity of the acid is:

$$\% \text{ pure acid} = \frac{\begin{array}{c}\text{wt of pure acid as}\\\text{determined by titration}\end{array}}{\text{total wt of impure acid}} \times 100$$

$$= \frac{(\text{ml base} \times N \text{ base}) \times \text{meq wt of acid}}{\text{wt of impure sample in g}} \times 100$$

Therefore,

$$\% \text{ KHP} = \frac{(\text{ml NaOH} \times N \text{ NaOH}) \times 0.2042 \times 100}{\text{g of impure KHP}}$$

Note that this formula is merely a modification of that given in Section 12·12 for determining an unknown equivalent weight for a pure acid.

Procedure. Secure an unknown sample from the instructor, and dry it in an oven at 110°C for a period of at least one hour. The dried sample is permitted to cool (in desiccator) for approximately 20 minutes. It is acceptable practice always to store dried samples (in weighing bottles) in the desiccator except when portions are being weighed.

Proceed to the analytical balance with the dried unknown (in desiccator), and weigh accurately (to 0.1 mg), three samples of approximately one gram each, into plainly marked 500 ml wide-mouth Erlenmeyer flasks. Record the weighings and subsequent titrations, as well as the calculations pertaining to them, in much the same manner as the sample data shown in Table 12·3.

Using a graduate, add 75 ml of distilled water to each sample, and dissolve the samples by gentle rotation of the flasks. Add 2 or 3 drops of phenolphthalein indicator solution, and titrate with your standard NaOH solution, seeking a complete-stop end point. Calculate the results according to the formula given in this section for obtaining the percentage composition of impure potassium acid phthalate.

If the end point is overstepped, add a measured volume of standard HCl from the acid buret until the solution is colorless, and then complete the titration with NaOH. Subtract the milliequivalents of HCl used in the back-titration from the total milliequivalents of NaOH necessary for the titration. In a calculation of this kind the preceding formula is modified to the following:

$$\frac{[(\text{ml of NaOH} \times N) - (\text{ml of HCl} \times N)] \times 0.2042 \times 100}{\text{wt of sample in g}} = \% \text{ KHP}$$

Precision. With careful work a student should be able to attain a precision of 2 or 3 parts per thousand for the analytical results of the foregoing analysis. A precision of more than 5 parts per thousand would indicate that additional samples should be analyzed. Note that errors in weighing, as well as errors in titration, may produce imprecise and inaccurate results.

Report. The report (for this unknown and all future analyses) should be submitted on a loose sheet of paper, $8\frac{1}{2} \times 11$ inches in size. It should be a *copy* of the weighings, titrations, and the formula for the calculations pertinent to the analysis. These details should be *entirely* complete, and in a neat, tabular form. All such reports will be retained by the instructor in case future reference to them is necessary.

12·14 Analysis of a Base Sample

For an example of an analysis of a base sample, we shall determine the percentage composition of soda ash.

Principles. Soda ash (impure sodium carbonate) is used as an unknown substance for student determinations because it is a typical solid of an alkaline material, and offers an exercise in the use of mixed indicators and in back-titration. Because of the physical nature of soda ash, it is sometimes difficult to obtain samples of a high degree of homogeneity. Even so, it seems preferable to weigh out individual samples of this material rather than large samples for use in "aliquot portion" techniques as described in many laboratory procedures.

The formula weight of sodium carbonate is 105.99. Since each carbonate ion reacts with two hydronium ions, according to the equation:

$$CO_3^{--} + 2H_3O^+ \rightleftharpoons 2H_2O + CO_2,$$

the equivalent weight of sodium carbonate is one-half of the formula weight, or 53.00. Consequently, the milliequivalent weight of sodium carbonate is 0.05300. The end point in a soda ash titration is reached by first adding an excess of acid titrant and then back-titrating with base titrant; therefore, it is necessary to subtract the milliequivalents of base titrant from the total milliequivalents of acid titrant used for the soda ash. The resulting formula for analysis of an impure sample of Na_2CO_3 is the following:

$$\% \; Na_2CO_3 = \frac{\text{determined wt of pure } Na_2CO_3}{\text{total wt of impure } Na_2CO_3} \times 100$$

$$= \frac{(\text{net meq acid used}) \times 0.05300 \times 100}{\text{g of impure sample}}$$

$$\% \; Na_2CO_3 = \frac{[(\text{ml HCl} \times N \text{ HCl}) - (\text{ml NaOH} \times N \text{ NaOH})] \times 0.05300 \times 100}{\text{g of impure sample}}$$

The normality of the HCl used here is that obtained with mixed indicator solution.

Procedure. Weigh, to the nearest tenth of a milligram, three samples of soda ash (properly dried in oven) of about 0.4 gram each, and place each sample in a clean 500 ml Erlenmeyer titration flask, marking each flask for identification as Λ, B, and C. Dissolve each sample in about 50 ml of distilled water. *Get initial buret readings here!*

Add 1 drop of methyl orange indicator to sample A and titrate with standard HCl until a definite color change is obtained, and then add 1 ml in excess. Record original buret readings, but do not make a final buret reading at this point since more acid is required to complete the titration.

Place the flask on a wire gauze, and heat over a free flame until the solution barely comes to a boil, then continue to boil gently for one minute to evolve the carbon dioxide liberated from the sample. Cool, by running tap water on the side of the flask, add 2 or 3 drops of bromcresol-green–methyl-red indicator, and titrate slowly with standard NaOH solution until a green color is obtained. Now back-titrate with acid to a pink color, then add base dropwise until one **drop** gives a sharp change to green. It is advisable to approach the end point by addition of base in each titration. Read both burets, and record final readings. Keep records as in preceding sections.

Titrate samples B and C in like manner, and record the results of your titrations. From the net milliequivalents of acid used, and the weights of the unknown samples, calculate the percentage of Na_2CO_3 in the original material, as indicated under *principles* in this section. To repeat, it is essential that the normality of the acid be that obtained with *mixed indicator* solution.

Precision and Report. The agreement of values between samples A, B, and C should be within 4 parts per thousand. If not, additional samples should be determined. Fill out a complete report sheet as was performed with unknown No. 1 (Section 12·13).

SUMMARY

Neutralization methods, frequently designated as acidimetry and alkalimetry, are used throughout the entire field of chemistry. It is impossible to find any segment of chemistry, be it organic, inorganic, analytical, physical, biochemical, industrial, or any other branch of chemistry, which does not depend upon neutralization methods for some purposes of control and research.

Most acid–base reactions take place in water solutions; however, many other solvents have been used as reacting mediums for such reactions. The equilibrium of the hydroxide and hydronium ions in water is a controlling factor of many chemical processes. Topics relating to acidimetry and alkalimetry are too numerous to be listed in this summary; however, any chemistry-oriented student is urged to supplement his experiences in these exercises by additional readings in more advanced texts.

This chapter includes only the minimum essentials for the introduction of neutralization methods for a one-term course in quantitative chemistry. The important analytical tools, the volumetric buret and the pH meter, have been introduced with appropriate experimentation to indicate their importance. The buret is the most widely used implement of titrimetric methods, whereas the pH meter has a variety of uses in graphical analyses. Both tools will be involved in later experiments.

CHAPTER 13

Redox Methods

PERMANGANATE PROCESSES

13·1 Preliminary Comments

Many ions may be quantitatively oxidized or reduced in a number of analytical processes. The variety is so great that it is impossible to examine many in an elementary course. It may suffice to state that redox methods are more widely used than all other volumetric methods combined.

Important oxidizing agents, which are frequently used in standard solutions, include potassium permanganate, potassium dichromate, ceric ammonium sulfate, iodine, and potassium iodate. Reducing agents used in the form of standard solutions include sodium oxalate, sodium arsenite, ferrous ammonium sulfate, and sodium thiosulfate.

Potassium permanganate is probably the most widely used of all volumetric agents. It is a versatile reagent in that it may be used to determine many substances by direct or indirect titrations, and may be controlled to produce products in several oxidation states. It also has the peculiar advantage over most other titrants in that the highly colored MnO_4^- ion serves as its own indicator.

Several disadvantages attend the use of permanganate solutions, which are listed as follows:

(1) In preparing $KMnO_4$ solutions it is necessary to remove MnO_2, by filtration, which may be present as an impurity in the reagent, or which may be formed by a reaction between MnO_4^- and impurities in distilled water. The presence of MnO_2 catalyzes the decomposition of $KMnO_4$ solutions.

(2) An aqueous solution of permanganate is not completely stable. There is a tendency for the ion to oxidize water, and traces of organic material in distilled water, to produce MnO_2 and oxygen. Although this decomposition reaction is fairly slow it may necessitate the occasional restandardization of the solution.

(3) There is a tendency for the permanganate ion to oxidize the chloride ion when hydrochloric acid occurs in a titration mixture as a preliminary solvent.

(4) There are numerous possible reaction products in permanganate oxidation unless the titrations are carried out under rigorous, specified conditions.

In spite of the disadvantages recognized as occurring with the use of MnO_4^- as an oxidizing agent, many determinations have been extensively studied and have been found to be highly accurate when properly performed.

13·2 Preparation of Approximately 0.1 *N* KMnO₄

In a strongly acid solution the permanganate ion is reduced to the manganous ion according to the following half-reaction:

$$5e + MnO_4^- + 8H^+ \rightleftharpoons Mn^{++} + 4H_2O$$

FIGURE 13·1   Filtration of permanganate solution through Gooch crucible

Such a solution is a powerful oxidizing agent, with a standard electrode potential of 1.50 volts. Five electrons are required in the redox process, indicating that the equivalent weight of potassium permanganate is one-fifth of its formula weight.

$$KMnO_4/5 = 158.04/5 = 31.61$$

To prepare a 0.1 N $KMnO_4$ requires one-tenth of the equivalent weight in grams of the reagent dissolved in sufficient water to produce one liter of solution.

Using a laboratory trip-beam balance, weigh on a watch glass a 3.25 g portion of $KMnO_4$ crystals. Divide roughly, half-and-half, between two *clean* 600 ml beakers, and add approximately 500 ml of distilled water to each. Cover beakers with 6-inch watch glasses and heat to boiling. Place the hot solutions, still covered, inside the desk for a period of at least 24 hours.

As part of the same work, clean with dichromic acid cleaning solution a 500 ml side-arm filtering flask, and a 1 liter glass-stoppered bottle. After cleaning and thoroughly rinsing, leave these containers in a suitable position to drain for 24 hours.

Before storing away the filter flask prepare a Gooch crucible* with an asbestos mat of approximately 1.5 mm thickness. To prepare the mat, shake the shelf-reagent bottle containing asbestos soup (a mixture of water and a small amount of acid-washed asbestos fibers), then pour the soup into the crucible, with suction partly on, as indicated in Figure 13·1. While the soup is passing through the crucible, tamp the mat into formation with a stirring rod whose ends have been well-rounded by fire polishing. A properly prepared mat can be distinguished by looking through the crucible toward a strong light. If no light can be seen, the mat is too thick. If the holes of the crucible are visible the mat is

---

* A glass-fritted crucible is easier to use, but more difficult to clean.

too thin. If a small amount of light can be observed through the mat it is of suitable thickness.

After 24 hours or more have elapsed, the permanganate solution may be filtered. In filtering the solution, arrange the Gooch crucible as shown in Figure 13·1. Be certain to insert a safety bottle between the filter flask and the aspirator, to prevent tap water from being sucked accidently into the filtered permanganate. It will be necessary to empty the filtering flask several times during the filtration.

Store the filtered permanganate solution in a clean one-liter glass-stoppered bottle, and keep in the dark when not in use. When properly prepared, and protected from heat and light, the solution is sufficiently stable for quantitative work for a period of several weeks. Any decomposition can be recognized by the deposition of a brown coating of MnO_2 on the sides and bottom of the storage bottle.

13·3 Standardization of KMnO₄ Solution

Potassium permanganate is not available in sufficient purity to be weighed directly as a standard. It is customary to prepare an approximately $0.1\ N$ solution, and then titrate the solution against a primary standard (see Section 12·9). The most frequently used standards are sodium oxalate, arsenious acid, and iron wire. Of these, sodium oxalate, of a high degree of purity, is the most satisfactory for student use. The overall reactions of permanganate with sodium oxalate are given in the following equations:

$$2Na^+ + C_2O_4^{--} + 2H^+ \rightleftharpoons$$
$$H_2C_2O_4 + 2Na^+$$
$$2MnO_4^- + 5H_2C_2O_4 + 6H^+ \rightleftharpoons$$
$$2Mn^{++} + 10CO_2 + 8H_2O$$

Weigh out to the nearest 0.1 mg three samples, between 0.2500 and 0.3000 g, of previously dried $Na_2C_2O_4$, and transfer the samples to wide-mouth 500 ml Erlenmeyer flasks (properly identified). Dissolve each sample in a solution of 100 ml of $1.5\ N\ H_2SO_4$ (prepared by dis-

solving 13 ml of concentrated H_2SO_4 in 300 ml of water).

Heat the solution to a temperature between 80 and 90°C, measured with a thermometer. Once introduced into the flask, the thermometer must not be removed until the titration is completed. Titrate with permanganate with constant swirling, but be certain that the temperature is maintained above 70°C at all times. At first the permanganate should be added drop by drop, with swirling after each addition, until the color disappears. After several drops have been added the permanganate may be added fairly rapidly with constant swirling. Toward the end of the titration particular care must be taken to allow the color, produced by each drop, to disappear before the addition of the next, in order to avoid overstepping the end point. Titrate to the first permanent *faint* pink color, which should last about 15 seconds.

From the data obtained, calculate the normality of the permanganate solution. Duplicate values should check within two parts in one thousand. If the average deviation is more than four parts per thousand the exercise should be repeated.

The half-reaction of oxalic acid (derived from sodium oxalate) involves two electrons for each formula weight of the oxalic acid, as is indicated in the following equation:

$$H_2C_2O_4 \rightleftharpoons 2H^+ + 2CO_2 + 2e$$

Consequently, the equivalent weight of sodium oxalate is one-half of its formula weight, and the normality of the permanganate may be obtained from the following formula:

$$\frac{\text{Wt of dried sodium oxalate in g}}{\text{ml of } KMnO_4 \times 0.06700}$$
$$= N \text{ of } KMnO_4 \text{ solution}$$

13·4 Analysis of an Iron Ore with Permanganate

The principal iron ores occur in some form of an oxide. Treatment of the ore in the determination of iron consists of three separate steps: (1) solution of the sample; (2) reduction of the dissolved iron to the ferrous state; (3) titration of the ferrous ions to the ferric state with standard permanganate. A number of precautionary measures are necessary within each step. These measures are included and explained within the detailed procedure for the analysis.

After a portion of the ore has been dried for an hour or more, at 110°C, weigh three samples accurately on the analytical balance. The weights of the samples should fall between 0.3000 and 0.4000 g. Transfer the weighed samples to 500 ml wide-mouth Erlenmeyer flasks, and dampen each sample with a few drops of distilled water. The purpose of the moistening is to prevent any loss of the finely divided powder.

Add 15 ml of concentrated hydrochloric acid to each sample; cover with watch glasses, and heat gently with an open flame (*perform this operation in a hood*) until the samples are disintegrated, as indicated by the disappearance of all dark particles of the original sample. Keep the volume fairly constant by additions of distilled water, if necessary. The solution process usually requires about 30 minutes. After this interval 99.9% of the iron is in the solution regardless of the outward appearance of the solution. The resulting solution is yellow in color and contains a mixture of ferrous and ferric ions. The latter is probably a chloride complex with a formula of $FeCl_4^-$.

The following procedure is to be followed for each sample individually, from stannous reduction through the permanganate titration, because ferrous ions are not stable for an extended period:

Heat the individual sample nearly to boiling, and add stannous chloride solution dropwise until complete reduction of ferric ions is accomplished as is indicated by the disappearance of the *yellow* color. However, there may remain a faint greenish-blue color which is characteristic of the ferrous iron. Avoid an excess of more than *two* drops of stannous chloride. The equation for the stannous reduction is as follows:

$$2FeCl_4^- + Sn^{++} \rightleftharpoons$$

$$2Fe^{++} + SnCl_6^{--} + 2Cl^-$$

Cool the flask completely by running cold water over the outside of the container, and without delay add rapidly 25 ml of mercuric chloride solution. (If the reagent is not added rapidly, a local excess of stannous ions may cause the reduction of $HgCl_2$ to Hg instead of to Hg_2Cl_2.) A small white silky precipitate of mercurous chloride should result. If the solution is gray or dark in color it should be discarded. If there is no precipitate, insufficient stannous chloride was used, and such a solution too should be discarded. The purpose of adding the mercuric chloride is to destroy excess stannous ions:

$$Sn^{++} + 2HgCl_2 + 4Cl^-$$

$$SnCl_6^{--} + Hg_2Cl_2$$
$$\downarrow$$

Dilute the solution to approximately 300 ml with distilled water, and then add 25 ml of preventive (Zimmermann–Reinhardt) solution. This solution contains a mixture of manganous sulfate, phosphoric acid, and sulfuric acid. The manganous sulfate lowers the oxidizing potential of the permanganate–manganous system and thereby inhibits the possible oxidation of the chloride ion. Phosphoric acid forms a colorless complex phosphate with ferric ions, possibly $Fe(PO_4)_2^{3-}$, and thus prevents the formation of yellow $FeCl_4^-$ ions during titration, which would obscure the permanganate end point.

Titrate immediately with standard permanganate solution, with constant whirling of the flask, to the appearance of a faint pink color which remains for 15 seconds. Rapid titrations give more accurate results by decreasing the likelihood of side reactions.

Compute and report the average percentage of Fe in the iron ore. The results should agree within three parts per thousand. The formula for the calculation is as follows:

$$\frac{(\text{ml of } KMnO_4 \times N) \times 0.05585 \times 100}{\text{wt of sample in g}}$$

$$= \% \text{ Fe in ore}$$

13·5 Techniques of Filtration with Filter Paper

The use of filter paper is usually restricted to gravimetric procedures in which a substance is precipitated, washed, and then ignited to constant weight. However, the determination of calcium by indirect analysis with standard permanganate, as given in Section 13·6, is a useful volumetric determination, which requires quantitative precipitation and subsequent quantitative filtration of calcium ions as calcium oxalate, prior to volumetric titration.

Most quantitative filtrations require 11 cm, acid-washed papers* that may be procured in coarse, medium, or fine porosities. This size of filter paper is customarily fitted into a fluted glass funnel, which has an angle of 60 degrees and a diameter of 65 mm.

The process of folding the filter paper and fitting it into a funnel involves essential but simple techniques. As indicated in Figure 13·2,

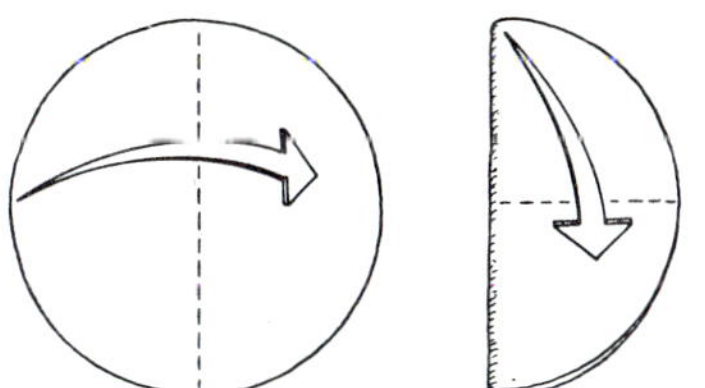

FIGURE 13·2 Folding a filter paper

the filter paper is folded along its diameter, and again along the radius at nearly right angles to the first fold. When the paper is opened, it forms a cone, which has an angle of 60 degrees. Moisten the cone and insert it securely into a fluted funnel. Using the fingers, press the top part of the paper into close contact with the funnel. (Older procedures have suggested that the outer top corner of the cone be torn off to speed up filtration. This is a questionable procedure, which is quite unnecessary with *fluted* funnels.)

Place the properly prepared funnel and filter paper in a slot of a wooden funnel support, with the receiving beaker underneath. It is good technique for the stem of the funnel to touch the receiving vessel to avoid spattering.

In a quantitative precipitation the precipitate usually is aged by standing to increase crystal size; the liquid above the precipitate is called the supernatant liquid. After the supernatant

---

*A glass-fritted crucible may be used in place of filter paper, but such a crucible is difficult to keep clean.

liquid has been properly aged (or digested), pour it cautiously down a stirring rod (Figure 13·3), taking care that the level of the liquid does not come within $\frac{1}{4}$ inch of the top of the filter cone. Keep the bulk of the precipitate in the beaker as much as possible. A precipitate on a filter paper cannot be washed as thoroughly, or as rapidly, as in the beaker.

FIGURE 13·3 Technique of quantitative filtration

To minimize solubility losses, the precipitate should be washed with the smallest practical quantities (usually 10 to 15 ml) of liquid. During each washing, stir the precipitate with the liquid in the original precipitation beaker, and allow to settle; then pour the supernatant liquid through the filter, still attempting to keep the bulk of the precipitate in the beaker. Usually five washings are sufficient.

13·6 The Determination of Calcium in CaCO₃ with Permanganate Solution

A weighed sample of impure calcium carbonate is dissolved in acid, and the calcium is precipitated as insoluble calcium oxalate by the addition of ammonium oxalate and ammonia. The precipitated calcium oxalate is filtered quantitatively and washed free of excess oxalate. It is then dissolved in dilute sulfuric acid, and the resulting solution titrated with standard permanganate. The amount of permanganate required to oxidize the oxalate is a measure of the total calcium. The redox reaction is essentially the same as that given in Section 13·3. *Note:* The samples furnished in this experiment contain no material insoluble in acid and no ion, other than calcium, that can be precipitated as the oxalate. When the method is used for the analysis of native limestone, interfering substances must be removed by lengthy preliminary procedures.

Weigh accurately on the analytical balance three samples between 0.3000 and 0.4000 g each. Transfer the weighed samples to 400 or 600 ml beakers (properly identified, for example as A, B, and C), and dampen each sample with distilled water from wash bottle. Cover each beaker with a watch glass, and, taking care to avoid loss by effervescence, introduce *approximately* 10 ml of 6 N HCl under each watch glass with the aid of a pipet. When the evolution of carbon dioxide has ceased, add 10 ml of distilled water and warm gently if necessary to effect complete solution. Wash down, into the beakers, anything on the underside of the cover glasses, and then dilute to about 200 ml with distilled water.

Using a graduate, add to each beaker 20 ml of 0.25 F ammonium oxalate solution. Heat to gentle boiling and, if any precipitate or turbidity remains, dissolve by the cautious addition of 6 N HCl. To the hot solution add 6 N ammonia gradually until a *strong* odor of ammonia persists after stirring. Leaving the stirring rods in the beakers and covering each beaker with a watch glass, set aside until the next period or for at least 18 hours. *Note:* In every quantitative precipitation, it is customary technique to provide each beaker with a stirring rod having fire-polished ends. Once placed in the solution, a stirring rod *must remain* there during the remainder of the analysis except *when held in the hand.* If laid down, adhering material will be lost and impurities may be introduced.

The technique of quantitative filtration is discussed in Section 13·5, which should be read carefully before continuing the calcium determination.

Only quantitative filter paper of medium porosity, such as Whatman No. 40, should be used for filtering the calcium oxalate precipitate. Obtain this paper from your instructor. (Quantitative filter paper is quite expensive and, to avoid waste, is not issued in the desk kit.)

Observing all the precautions explained in Section 13·5, filter the solution from the precipitation sample A through 11 cm quantitative filter paper in a fluted glass funnel. (Each funnel is identified for each precipitate, for example, as A, B, and C, and each is supported on a funnel rack over containers to receive the supernatant liquids and washings. *Never use suction with filter paper.*) Samples B and C may be filtered simultaneously with A through two other filters. Leave the precipitates in the original beakers as much as possible. When all of the liquids have passed through the filters, discard the water-clear filtrates. Wash the precipitates repeatedly with small portions (10–15 ml) of cold distilled water containing a few drops of ammonia water. After three or four washings, test the filtrate for the oxalate ion. Continue the washing until a 10 ml portion of the last filtrate gives no test for the oxalate. (Too many washings will dissolve a precipitate and cause the analysis to run low.) *Note:* Test for the oxalate ion in the washings as follows: Collect approximately 10 ml of the filtrate in a test tube, add 1 ml of 3 N H_2SO_4, and heat nearly to boiling; then add *one drop* of potassium permanganate solution. If the pink color remains, the absence of the oxalate is confirmed. As long as the pink color is bleached, the washing should be continued.

When the test for the oxalate ion is negative, remove the filter paper from the funnel and place it, with its contents, in the original beaker containing the bulk of the precipitate. By means of a graduated cylinder, add 100 ml of 3 N sulfuric acid* and warm gently with stirring to dissolve the calcium oxalate. Dilute with 100 ml of distilled water, and heat nearly to boiling. Titrate the hot solution with standard permanganate. The presence of the quantitative filter paper will not interfere with the end point reading.

It is customary to calculate calcium in terms of CaO, which has a millequivalent weight of 0.02804. The formula for the calculation is given as follows:

$$\frac{(\text{ml of KMnO}_4 \times N) \times 0.02804 \times 100}{\text{wt of sample in g}}$$

$$= \% \text{ CaO}$$

SOME IODINE PROCESSES

13·7 Preliminary Comments

Iodine is a weaker oxidizing agent than potassium permanganate. It can react quantitatively only with substances that are active reducing agents. Among the limited number of substances that can be oxidized quantitatively with iodine are H_2S, $H_2AsO_3^-$, and H_2SO_3. On the other hand, the iodide ion may be readily oxidized to molecular iodine by most strong oxidizing agents. The use of iodine methods is largely due to the fact that starch in an iodine–iodide system is the most sensitive visual indicator in quantitative chemistry.

Since iodine may be either reduced or produced by oxidation, there are two different types of iodine methods—direct and indirect. The use of iodine as a direct titrating agent is known as the *iodimetric* method, and the production of iodine by iodide ion to iodine is designated as the *iodometric* method. Typical examples are the following:

iodimetric method:

$$H_2AsO_3^- + I_2 + H_2O \rightarrow$$

$$H_2AsO_4^- + 2I^- + 2H^+$$

iodometric method:

$$2Cu^{++} + 4I^- \rightleftharpoons 2CuI + I_2$$

* *Note:* If the sulfuric acid is a great deal stronger than 3 *N*, it will cause the potassium permanganate to oxidize the filter paper; consequently, the analysis will not be reliable.

However, neither of the foregoing reactions are necessarily quantitative. The first reaction must be driven to completion by the addition of bicarbonate ions to remove the hydrogen ions; and in the second reaction, the liberated iodine must be titrated by standard thiosulfate solution.

The titration of iodine against sodium thiosulfate, with starch as an indicator, is extremely accurate. In fact, the thiosulfate ion cannot be titrated quantitatively by any reagent other than iodine. This fundamental reaction takes place only in slightly acid or neutral solutions. The equation is as follows:

$$2S_2O_3^{--} + I_2 \rightleftharpoons S_4O_6^{--} + 2I^-$$

Because the titration is reversible, it is possible to back-titrate an overstepped end point with the opposite solution. The reversibility also permits the secondary standardization of one solution when the other solution has been standardized against a primary standard. It is usually customary to standardize iodine solution against arsenious oxide, and then to standardize the thiosulfate solution against the iodine solution by the ratio value for the two solutions.

13·8 The Preparation of Approximately 0.1 *N* Iodine Solution

Iodine is relatively insoluble in water, but dissolves readily in potassium iodide solution due to the formation of the complex I_3^- ion. The highly reversible reaction is indicated by the equation,

$$I_2 + I^- \rightleftharpoons I_3^-$$

However, the rate with which iodine dissolves in KI is slow, and particularly so if the iodide concentration is low. It is common practice to dissolve the iodine in a concentrated iodide solution, and then dilute to the desired volume. *Extreme care* must be taken to be certain that all of the iodine is dissolved before dilution; if it is not, the normality of the resulting solution will increase as the undissolved iodine eventually dissolves.

Procedure. Using a triple-beam balance, weigh out 25 g of potassium iodide on glazed paper, and dissolve in 20 ml of distilled water contained in a 150 ml beaker. Also weigh out 9.8 g of reagent-grade iodine on a watch glass, using the rough balance. Add the iodine to the potassium iodide solution, and stir until the iodine is completely dissolved. (Examine the surface of the solution for undissolved particles.) After the solution is *complete*, transfer to a glass-stoppered one-liter bottle, and dilute to *approximately* 750 ml. (A 250 ml volumetric flask may be used for measuring 750 ml of distilled water in the dilution.) Mix thoroughly, store in the dark, and let the solution stand for at least 24 hours before standardizing.

13·9 Preparation of Approximately 0.1 *N* Thiosulfate Solution

Heat 1 liter of distilled water (500 ml in each of two 600 ml beakers covered with watch glasses) to boiling, and continue to boil for five minutes to remove CO_2. While it is *warm*, transfer it to a *clean* one-liter glass-stoppered bottle. By means of a paper cone, add 25 grams of $Na_2S_2O_3 \cdot 5H_2O$ (weighed on triple-beam balance) and 0.1 g Na_2CO_3 (a level micro-spatulaful). Shake thoroughly, and then store in the dark. The sodium carbonate is added to prevent disintegration of the thiosulfate ion.

13·10 Preparation of Starch Indicator

Stir approximately 2 g of "water-soluble" starch in 10 ml of water (in a small beaker) to produce a paste. Slowly add the paste to 200 ml of boiling water, and continue to boil for two minutes. Cool, and store in a *clean* stoppered bottle.

The starch indicator should be freshly prepared before it is used, since it is susceptible to bacterial action. However, even without the addition of preservatives, it is usually satisfactory for use for a period of about a week. Discard the suspension when it becomes definitely cloudy or when it gives a reddish color with iodine solution.

13·11 Standardization of Iodine Solution

Pure, primary-standard-grade arsenious oxide is to be found on the side shelf. Weigh out on the triple-beam balance, on glazed paper, a portion of about 1.5 g, and place in a clean dry weighing bottle. Dry in the oven at 110°C for at least one hour. Store in desiccator until needed.

Arsenious oxide is insoluble in water, and dissolves only slowly in acids. On the other hand, it is readily soluble in a high concentration of hydroxide ion as is indicated by the equation,

$$As_2O_3 + H_2O + 2OH^- \rightleftharpoons 2H_2AsO_3^-$$

The titration of the arsenite ion is not quantitative except in a buffered solution containing bicarbonate ions. In preparing the samples for titration it is essential that the directions in the next paragraph be followed in exact detail.

Weigh accurately (to 0.1 mg) three samples of arsenious oxide of about 0.2 g each, and transfer to 500 ml titration flasks. Add to each flask 10 to 15 ml of 1 *N* NaOH (prepare by adding 5 ml of 50% NaOH to 95 ml of water), and warm to hasten the solution of the arsenious oxide. Cool by immersing in cold water, add a piece of blue litmus, and then neutralize with 1 *N* HCl (prepare by diluting 8 ml of concentrated HCl to 100 ml with water) until the solution is distinctly acid. Add approximately 3 g of sodium bicarbonate to each flask. Dilute to about 100 ml, add 5 ml of starch indicator, and titrate with iodine solution to the *first* permanent appearance of a deep blue color.

Caution. The solution should not become acid during the titration. If the litmus paper turns red during the titration, add more $NaHCO_3$. Also, iodine tends to bleach litmus paper to a white color, and it may be necessary to add another piece of litmus paper. However, for best results, titrations should proceed at a pace such that bleaching is not permitted to occur.

The titration of the arsenite ion with iodine is not complete, except in the presence of excess bicarbonate ions. The bicarbonate ions react with hydrogen ions to cause the liberation of carbon dioxide. The two reactions are indicated in the following equations:

$$I_2 + H_2AsO_3^- + H_2O \rightleftharpoons$$

$$2I^- + H_2AsO_4^- + 2H^+$$

$$HCO_3^- + \rightleftharpoons H_2O + CO_2$$

Calculate the normality of the iodine solution. The average deviation should be less than 2 parts per thousand. The redox reaction involves the exchange of two electrons per arsenic atom; consequently, the milliequivalent weight of arsenious oxide, which contains two arsenic atoms, is 0.04946. This is obtained by dividing the formula-weight by 4000. The normality of the iodine is calculated from the formula,

$$\frac{\text{wt of dried arsenious oxide in g}}{\text{ml of iodine solution} \times 0.04946}$$

$$= N \text{ of iodine solution}$$

13·12 Standardization of Thiosulfate Solution

Several procedures have been developed for the standardization of thiosulfate against a number of primary standards, for example, KIO_3, $K_2Cr_2O_7$, Cu, and so on. However, it is not necessary to standardize both iodine and thiosulfate against primary standards. The normality of the thiosulfate can be determined satisfactorily by running a series of ratios against standard iodine solution. This is the same type of procedure as that performed in Section 12·6 in the acid–base exercise.

Procedure. Place the thiosulfate and iodine solution in burets, following precautions given in earlier sections. Run out between 30 and 35 ml of thiosulfate into a titration flask, add 5 ml of starch solution, and dilute with 100 ml of distilled water. Titrate with standard iodine to the *first* permanent appearance of a deep blue color. Repeat until the average ratio of the two solutions is determined with a deviation of less than 2 parts per thousand.

Calculations.

$$N \text{ of iodine} = N \text{ of thiosulfate}$$

$$\times \frac{\text{vol of thiosulfate}}{\text{vol of iodine}}$$

13·13 The Iodimetric Determination of Antimony in Soluble Antimony

Antimony undergoes reactions with iodine that are similar to those described in Section 13·12. The principal difference is the fact that antimony tends to precipitate from dilute acid solutions in the form of insoluble SbOCl, as indicated by the equation,

$$Sb^{3+} + H_2O + Cl^- \rightleftharpoons SbOCl\downarrow + 2H^+$$

To avoid this possibility, tartaric acid is added to produce a stable, yet soluble, complex with antimony. The equation for the complex formation is usually written as,

$$SbCl_4^- + H_2C_4H_4O_6 + H_2O \rightleftharpoons$$

$$H(SbO)C_4H_4O_6 + 4Cl^- + 3H^+$$

The potassium salt of this antimonyl tartrate complex is a soluble substance known as tartar emetic. It is available commercially and may be represented by the formula $KSbOC_4H_4O_6$.

Antimony in the form of an ore is somewhat difficult to get into solution; consequently, the analysis of stibnite—an ore of antimony—involves a painstaking procedure. On the other hand, commercially prepared samples of impure tartar emetic (the "soluble-antimony" samples referred to in this section) can be analyzed without difficulty. They are fairly soluble in water, and since they already exist in the form of the tartrate complex, only a small amount of additional tartrate is necessary to prevent the possibility of precipitation through hydrolysis.

Procedure. Dry the sample containing soluble antimony at 110°C for at least one hour. After cooling, weigh accurately three portions on the analytical balance (between 0.7 and 1.0 g) to 0.1 mg. Transfer each portion to a 500 ml titration flask, and add 100 ml of distilled water.

To the solution (or suspension) add approximately 2 g of potassium sodium tartrate, 3 g of sodium bicarbonate, and 5 ml of starch indicator; whirl until the mixture is uniform. Titrate with standard iodine to the *first* permanent appearance of a deep blue color. The titration reactions are usually designated as,

$$(SbO)C_4H_4O_6^- + I_2 + H_2O \rightleftharpoons$$
$$(SbO_2)C_4H_4O_6^- + 2I^- + 2H^+$$

$$H^+ + HCO_3^- \rightleftharpoons CO_2 + H_2O$$

The redox reaction involves a change of two electrons per antimony atom; therefore, the milliequivalent weight of antimony is the atomic weight divided by 2000. Calculate the percentage of antimony in the sample, using the following formula:

$$\frac{\text{ml of iodine} \times N \times 0.06088 \times 100}{\text{wt of sample in g}} = \% \text{ Sb in sample}$$

Results should agree within less than 3 parts per thousand.

13·14 The Iodometric Determination of Copper in Impure Copper Oxide

The number of substances that can be determined by iodometry is greater and more varied in nature than the number determinable by iodimetry. As was stated in Section 13·7, an iodometric analysis is an indirect iodine method in which an oxidizing agent liberates iodine from an iodide. Invariably the liberated iodine is titrated with thiosulfate solution, although solutions of arsenious acid or other reducing agents are possibilities.

The determination of copper is a typical, and widely used, example of the indirect iodine method. As in other analyses, it is more convenient to use a commercially prepared sample* rather than a copper ore as the substance to be analyzed. It is generally true that the analysis of any ore involves a more complex procedure than for an impure substance.

The cupric ion is an oxidizing agent that reacts with a soluble iodide according to the following equation:

$$2Cu^{++} + 4I^- \rightleftharpoons 2CuI\!\downarrow + I_2$$

Actually, the copper ion is a poor oxidizing agent, and the reaction is quantitative only because of the low solubility of cuprous iodide. Thus, the reaction must be regarded as a precipitation process as well as an electron interchange. The liberated iodine is titrated, in the presence of the precipitated CuI, by standard thiosulfate solution.

$$I_2 + 2S_2O_3^{--} \rightleftharpoons 2I^- + S_4O_6^{--}$$

The two foregoing reactions are carried out in a highly buffered acetic acid solution since neither the iodine nor the thiosulfate solution is stable in strongly acid or basic solutions.

Experimental results indicate that cuprous iodide tends to adsorb iodine sufficiently to cause low analytical results. This difficulty is largely eliminated by the addition of the thiocyanate ion, which tends to displace adsorbed iodine on the CuI particles.

Procedure. Dry the impure sample at 110°C for at least one hour. Weigh out (to the nearest milligram) three samples, each between 1.5 and 2.0 g, into 500 ml titration flasks. Add 10 ml of 6 N nitric acid (concentrated nitric acid is 15 N) and dissolve by warming gently. Dilute to 20 ml and add 5 ml of urea† solution (4 g in 100 ml). Boil for one minute and then cool in tap water. Dilute to 100 ml and add NH_3 until a deep blue color prevails. Boil to remove excess ammonia, then neutralize with acetic acid, and add 3 ml in excess. Boil another minute, and then cool to room temperature. Add 5 g of potassium iodide, dissolved in 10 ml of water, to each sample. Titrate with standard thiosulfate until the yellow-brown color of iodine is almost gone (*but not quite*). Add 5 ml of starch solution and

---

* Analyzed (commercially prepared) samples are designed for training purposes and avoid impurities which might cause undue complications.

† The urea reacts with any oxides of nitrogen which may remain from the nitric acid treatment.

continue titration, drop by drop, until the deep blue color changes to gray. Add 2 g of KSCN (or NH_4SCN) and swirl until the crystals are dissolved. Titrate just to the disappearance of the blue color, leaving a creamy white suspension of cuprous iodide.

Calculations. Since the redox reaction in the titration involves the interchange of only one electron, the milliequivalent weight of copper is the atomic weight divided by 1000, or 0.06355. The percentage of copper in the sample may be obtained from the following formula:

$$\frac{\text{ml of thiosulfate} \times N \times 0.06355 \times 100}{\text{wt of sample}} = \% \text{ Cu}$$

The agreement among values should be within 3 parts in a thousand.

SUMMARY

Because the number of redox methods is very large, a fair sampling of such methods is impossible in this manual. The four determinations included in this chapter are representative, but they do not provide a comprehensive perspective of the many oxidation–reduction possibilities in quantitative analysis. Such a survey is beyond the limitations of time within a one-term course. Outside readings pertaining to this area are recommended and urged for those students with inquiring minds. Your instructor will be glad to suggest references in this field.

The calculations involved in redox methods of quantitative analysis depend on the relationship between electron charge and the equivalent weight. The student should note these relationships, which are emphasized in the experimental procedures.

Filtration techniques involving the use of a Gooch crucible and filter paper have been introduced. These techniques become increasingly important in later gravimetric determinations.

CHAPTER 14

Colorimetric Methods

INTRODUCTION

14·1 Preliminary Explanations

When light passes through a transparent medium, the amount of visible radiation decreases because of absorption during the passage. Colorless materials such as window glass absorb all wavelengths of visible light equally; therefore, the color of the light emerging is the same as the color entering. Other substances absorb certain wavelengths preferentially, and in such a case, the color observed is complementary to the color absorbed (see Table 7·1). If white light is passed through a $FeSCN^{++}$ solution, green is absorbed, whereas the transmitted light is red. Similarly, a chromate solution appears yellow, a permanganate solution is purple, and a $Cu(NH_3)_4^{++}$ is seen as deep blue, because the complementary colors have been absorbed by these ions.

The theoretical aspects of the quantitative absorption of light have been discussed in Sections 7·3, 7·4, and 7·5. The mathematical relationship associated with the absorption of light, known as the Bouguer–Beer Law, is denoted as

$$\log \frac{I_0}{I} = abc$$

The terms making up this formula were defined in Section 7·3; however, the variables contained in abc may be more effectively indicated by the following illustrations:

Variable a. Green light rays are absorbed more strongly by red wine than by white wine of similar concentration; in like manner, yellow light is absorbed more strongly by a MnO_4^- solution than by a CrO_4^{--} solution of equal concentration.

Variable b. Light passes easily through a thin layer of colored glass such as a sun shade, but the thick stained-glass windows of a cathedral keep the interior quite dark.

Variable c. Light passes readily through a dilute solution of $FeSCN^{++}$ (which is weakly colored), but very slightly through a concentrated solution.

Quantitative measurements of light absorbed are frequently expressed in terms of a logarithmic quantity designated as absorbance (A); the mathematical relationship of absorbance to transmittance (T) is as follows:

$$A = -\log_{10} T = abc$$

Usually the values a and b are constants; hence, the concentration, c, can be determined by measuring the absorbance A. For a series of solutions having equal values for the products $a \times b$, the quantities A and c are directly proportional. Therefore, according to Beer's Law, the concentration of the absorbing substance in solution is a linear function of absorbance. From $\% \, T$ readings, the corresponding A values can be calculated with the use of log tables, or even better, $\% \, T$ data may be plotted directly on semilog paper (see Figure 7·3).

14·2 General Method

Quantitative spectrophotometry is concerned with the problem of analyzing a solution, which contains a known light-absorbing substance in *unknown concentration*, designated for convenience as C_x. The analyst prepares a series of solutions containing the same light-absorbing material, but in a variety of *known concentrations*, which may be indicated as C_1, C_2, C_3, and so on. The absorbance of each of the solutions is measured, and the variables a and b are kept constant by selecting a single wavelength of light for analysis and by using the same cuvette for each measurement. When absorbance is plotted (vertically) against known concentration (horizontally), a linear graph is obtained, from which the concentration of the unknown can be determined by interpolation. The graphical method eliminates the necessity of knowing the exact values of a and b.

If the unknown solution happens to contain *two* light-absorbance components to be analyzed simultaneously, then the wavelengths must be selected, two sets of known solutions must be measured, and two absorbance equations must be solved simultaneously for the unknowns. The accuracy of the method generally diminishes as the number of simultaneous unknowns increases, but it is significant that two unknowns may be determined simultaneously by spectrophotometric measurements.

14·3 Instrumentation

A spectrophotometer is an instrument designed for measuring the absorption of radiant energy. A few instruments may be used for choosing the proper wavelength for absorption, and also for measuring the transmittance (at this wavelength) through a particular solution (see Section 7·13). In the visible spectrum, white light emanating from a tungsten lamp is split into its component wavelengths by a grating (or by a

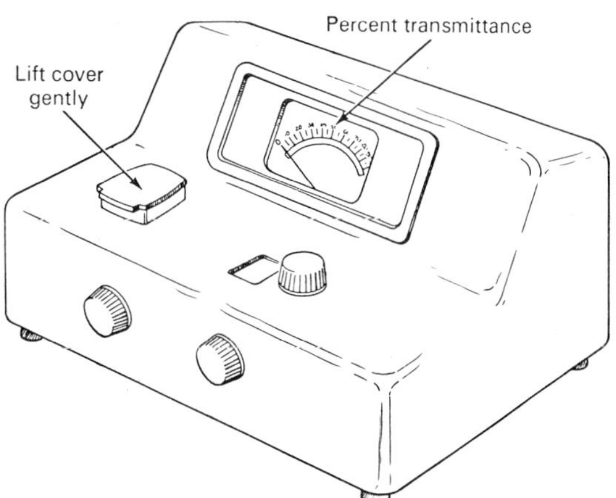

FIGURE 14·1 The Bausch and Lomb "Spectronic 20" Spectrophotometer

prism). The latter is oriented through various positions, so that only the desired wavelengths pass through an exit slit into the sample area. The absorbing solution is placed into a glass cuvette (much like an ordinary test tube, but especially constructed to be uniform in diameter), and lowered into the light path. The transmitted light is measured by means of a phototube, and recorded on a large meter. Although the absorbance value is more useful chemically, most spectrophotometers measure $\% \, T$ directly. The latter value changes linearly with the intensity of the light passing through the solution, and can be read more accurately on the meter scale.

As stated previously, a few spectrophotometers can be used to measure all the various wavelengths of the visible spectrum transmitted through a solution. From this spectrum (see Figure 7·2), it is possible to determine the wavelength for maximum absorption. Usually, the wavelength of maximum absorption is the best wavelength for the analysis of a particular solution.

Figure 14·1 provides a sketch of the Bausch and Lomb "Spectronic 20" Spectrophotometer, which will be used in the experiments to follow.

ESSENTIAL TECHNIQUES IN SPECTROPHOTOMETRY

Several techniques, some of which are also common in general analytical work, are especially important in spectrophotometric analysis and are introduced in the present experiment.

14·4 Use of Volumetric Pipets

A pipet is a device, fabricated from glass, which has been factory-calibrated *to deliver* (T.D.) a definite volume of liquid from one container to another.

The pipet should have been thoroughly cleaned according to the directions given in Section 11·6. It should not contain any moisture from the cleaning process; if it does, rinse it at least twice with the solution to be used.

Insert the pipet into the liquid to be used, permitting the tip to reach nearly to the bottom of the container; then, by suction, draw up the liquid until its level is well above the calibration mark on the upper stem. Suction may be attained by the mouth or by a squeeze bulb. Unless the liquid is corrosive or poisonous, it is better to omit the squeeze bulb, which is awkward to handle and difficult to adjust. When the liquid is above the calibration mark, quickly cap the pipet with the index finger (never the thumb). Tilt the pipet slightly and wipe the exterior free of adhering liquid. To adjust the liquid level, hold the pipet in a vertical position over a *waste* beaker, and cautiously release the pressure upon the index finger, thereby allowing the meniscus (see Section 12·5) to fall until it is exactly even with the calibration mark. Remove any hanging droplet of liquid by touching the tip to the side of the *waste* beaker. Finally, allow the liquid to drain freely from the pipet into the *desired* container (with the pipet almost vertical and its tip in contact with the glass wall), *without forcing or blowing out* the liquid remaining within the tip. This partial drop of liquid remaining in the pipet is included in the factory-calibration of any T.D. pipet. If the pipet is chemically clean, no droplets will remain on the inner surfaces after the liquid has been drained.

Use mouth suction or a squeeze bulb only to draw liquid into the pipet and not to adjust the level to the calibration mark. Some practice may be necessary before you become adept at removing the bulb (if used) with the free hand and quickly capping with the index finger of the hand holding the pipet; yet, this is essential for good results.

Two additional precautions are necessary in the proper use of the pipet:

(1) In grasping the pipet, use only the thumb and first two fingers on the stem above the graduation mark. Never handle the bulb of the pipet, since it is calibrated at a definite temperature.

(2) As in the use of the buret, the liquid in the pipet must be completely free of air bubbles.

14·5 Use of Volumetric Flasks

A volumetric flask is calibrated to contain (T.C.) a specified volume of liquid at a specified temperature (generally 20°C). The *bottom* of the liquid meniscus should coincide exactly with the calibration mark. The flask can be used for the preparation of standard solutions; quantitative dilutions, and the preparation of unknowns to a known volume.

Preparation of Standard Solutions. Analytical grade reagents may be weighed out carefully, dissolved, and diluted properly to yield a solution of known formal concentration without need of further analysis. Thus, 1.824 mfw of pure reagent diluted in a 10.00 ml volumetric flask produces a solution that is 1.824 mfw/10.00 ml, or 0.1824 *F*.

Quantitative Dilutions. Quantitative dilution is accomplished by pipetting a known volume of a solution into a volumetric flask, and diluting exactly to the new volume. Thus, a 15.00 ml pipetted sample can be diluted to 50.00 ml to yield a solution of formality exactly 15/50 times as great as the original solution.

Preparation of Unknowns to Known Volumes. Volumetric flasks are used also to dissolve samples of impure substances to a specified volume (do not heat a volumetric flask). From subsequent analysis of the formality of the solutions and the known volume, it is possible to determine the number of fw of pure substance present in the impure sample.

14·6 Use of Cuvettes

Two matched cuvettes are used in the measurement of the $\% T$ of a solution. In this experiment, one cuvette will contain distilled water (pure solvent) and the other will be used for the various solutions of $Cu(NH_3)_4^{++}$, which you will prepare. Matched cuvettes have similar size and curvature, and show nearly identical effect upon a beam of light. The following *rules for handling cuvettes* should be observed:

(1) Never touch the lower half of the cuvette, through which the light beam will pass.

(2) Lightly wipe off any liquid or smudges from the outer surfaces, using *only the special tissue provided.*

(3) Always rinse the cuvette with *several small portions* of solution before filling it for a measurement. (Only about 3 ml is needed for the measurement.)

(4) Notice the *index line* near the top of the cuvette. The cuvette should be placed in the spectrophotometer with the index line toward the front. After the cuvette is firmly seated, rotate the cuvette gently until the mark is aligned with that on the instrument. It is very important that the cuvette be aligned precisely the same way for each separate measurement.

(5) Dirty cuvettes should be rinsed several times with pure water, and the outside surfaces lightly wiped with tissues, after final use. Under no circumstances may cleaning solution or test tube brushes be used on cuvettes.

14·7 Use of a Calibration Graph

A calibration curve (see Figure 7·3) is prepared from measurements of a series of solutions of known concentration. The measurement of a solution of unknown concentration is subsequently compared with the known measurements on the graph. In this type of graphical analysis, the accuracy of the standard curve is of critical importance for obtaining accurate results.

In this laboratory exercise the calibration graph should be a straight line (since Beer's Law is obeyed under the conditions of the experiment). Also, it should pass through the origin of the graph (that is, the point corresponding to zero absorbance versus zero concentration, or 100% T versus zero concentration). Any deviations of the measurements from a perfect straight line are indications of lack of precision. In such cases, measurements should be repeated. Deviations may be due to faulty experimental technique

(most likely), severe fluctuations in the AC line voltage (occasionally), and (to a small extent) inherent limitations imposed by the spectrophotometer itself. Occasionally, a point is simply misplotted.

Common technical errors, which may distort colorimetric data, are listed as follows:

(1) Failure to keep the outer surface of the cuvette clean.

(2) Failure to use the same cuvette for all measurements.

(3) Failure to rinse the cuvette thoroughly enough with the new solution before taking measurements.

(4) Failure to orient the cuvette exactly the same way for each measurement.

(5) Failure to account for slight "play" of the plastic cuvette holder within its cavity in the instrument.

(6) Failure to set 0% T and 100% T carefully.

(7) Inaccurate pipetting.

(8) Inaccurate dilution within the volumetric flask.

(9) Insufficient mixing of the solutions.

(10) Use of unclean volumetric glassware.

The student is urged to take necessary precautions in avoiding the various pitfalls in colorimetric analysis. Indications of such errors may sometimes be observed from the graphical representation of the data.

ANALYSIS OF COPPER IN COPPER OXIDE

14·8 Preliminary Preparations

Special equipment provided for the experiment includes the following: a Bausch and Lomb "Spectronic 20" Spectrophotometer with constant voltage transformer, a cuvette rack with two matched cuvettes and a third cuvette that has a piece of chalk with a diagonally-ground surface, a box of tissue, a marking pencil, and a box of plastic wrap.

From the equipment in your desk, you will need the following items: a 25 ml graduated cylinder, a 250 ml volumetric flask, a 25 ml volumetric pipet, the wash bottle containing distilled water, two 250 ml beakers for preparing solutions, and five one-pint bottles (the Thro-A-Way type are excellent for this experiment).

Two sheets of semilog graph paper (preferably one-cycle) are needed for plotting experimental data.

The sample of copper oxide (obtained from the instructor) must be dried in the oven at 110°C for at least one hour. The procedures of the experiment must be read carefully beforehand in order to make efficient use of laboratory time.

Connect electrically, and turn on the spectrophotometer by rotating the *left* front knob clockwise. Use this knob to adjust the meter needle to % T. While the instrument is warming up (about 20 minutes), complete the preliminaries and prepare the solutions needed in the experiment.

To observe the diffracted light, which passes through the sample cell, carefully insert the test tube containing the chalk into the sample holder. Make certain the chalk's diagonal surface slopes downward toward the *right*. Rotate the 100% T adjust dial (*right* front knob) clockwise; a band of color should appear upon the diagonal surface. Now rotate the wavelength dial (top right), and observe the variation in the color of the band, from violet to blue near 400 mμ to red near 700 mμ. If the rotation continues beyond 700 mμ to 1000 mμ, a repetition of the violet, blue, and green colors will be seen, owing to second order diffraction from the grating. When the spectrophotometer is used at wavelengths greater than 700 mμ, a red filter must be inserted into the instrument to counteract this effect.

Remove the tube, with the chalk, from the sample holder.

14·9 Preparation of Standard Copper Solutions

A stock of solution of standardized cupric nitrate will be available in the laboratory. The standard solution is prepared by dissolving 2.000 g of 99.90% copper metal in nitric acid, and diluting quantitatively to exactly one liter of solution. Such a solution contains 2 mg of Cu per ml.

Obtain approximately 200 ml of the stock solution in a *clean, dry* 250 ml beaker.

Using the volumetric pipet, transfer exactly 25.00 ml of the standard solution from the beaker to the *clean* 250 ml volumetric flask (it need not be dry). Fill half of the volumetric flask with distilled water, to dilute the copper solution. Operating in the fume hood, add small portions of concentrated ammonia (from a 25 ml graduated cylinder) with swirling, until a *clear* deep-blue solution is obtained (after formation and redissolution of a milky-blue precipitate). Add 30 ml of concentrated ammonia in excess, and cool to room temperature. Dilute to the graduated mark with distilled water. Transfer the solution into a clean, dry, one-pint Thro-A-Way bottle, which is stoppered with a plastic screw top. Stopper tightly, and shake the solution thoroughly for several minutes. Label it solution I.

Repeat the solution process, to produce a second known copper standard, containing *50* ml of stock solution (use pipet twice), and label solution II.

Continue the solution procedure once again, this time pipetting a total of 100 ml of stock solution into the clean volumetric flask, to obtain a third known copper solution, which is labeled solution III.

Prepare a "blank" solution (labeled solution 0) by diluting 30 ml of concentrated ammonia to approximately 250 ml in a clean one-pint Thro-A-Way bottle. Mix thoroughly, as you did solutions I, II, and III.

In your notebook make a table, like the example below, for the solutions you have prepared:

Concentration of Cu in the standard stock solution: __ mg/ml

Solution	Ml of stock solution used	Final volume	Final concentration (calculated)
0	0.00	250.0	—
I	25.00	250.0	—
II	50.00	250.0	—
III	100.00	250.0	—

14·10 Preparation of Unknown Solution from Copper Oxide

Weigh accurately one sample of *dry* copper oxide totaling about 2.5 g. Place the sample in an empty 250 ml beaker (clean). Add 30 ml of concentrated HNO_3 and 30 ml of distilled water, and warm gently in the hood to dissolve sample completely. Allow to cool. Again in the hood, add small portions of concentrated NH_3 from the 25 ml graduated cylinder, swirling until a clear deep blue solution is obtained (no cloudiness). Add 30 additional ml NH_3, then transfer the solution quantitatively (see Section 15·2) to an empty (clean) 250-ml *volumetric* flask (fitted with plastic stopper), using distilled water for the rinsing. Finally dilute exactly to the 250-ml mark with distilled water. Mix the solution well.

Record the essential details regarding the "unknown" solution in your notebook.

14·11 Standardization of the Spectrophotometer

Be sure to observe the following for standardization of the spectrophotometer.

(1) If necessary, rezero, using the left front knob, with no cuvette in the instrument.

(2) Rotate the right front knob far counterclockwise.

(3) Rinse one of the cuvettes several times with the "blank" solution 0, fill half full with the solution, dry the outside surfaces with tissue, cap with plastic wrap, and align the cuvette in the instrument as described previously. *Close the top* of the sample holder.

(4) Rotate the right front knob clockwise, until the meter needle reaches exactly 100% T. (Do not allow the reading to exceed 100% T.) Now the spectrophotometer is standardized for the wavelength at which the instrument is set, using any sample containing dilute ammonia solution as the solvent.

Retain the cuvette, half full, for later standardizations (henceforth called the "reference cuvette").

Important: The spectrophotometer must be restandardized (following the four steps outlined above) for *each new* wavelength setting. Furthermore, you must perform the exercise in Section 14·12 before you can make any quantitative measurements upon the four known and one unknown copper solutions (which you prepared from the instructions in Sections 14·9 and 14·10).

14·12 Measurement of the Visible Spectrum of $Cu(NH_3)_4^{++}$

A fairly complete visible spectrum is measured on one of the known solutions (prepared in Section 14·9) for the purpose of obtaining the most suitable wavelength for maximum absorption. The value so obtained is used in the quantitative determination of the cupric ammine complex. Either solutions I, II, or III may be used, but solution III is preferred since it is the most concentrated in respect to the copper complex-ion.

Procedure. Set the wavelength at 450 mμ (upper right dial) and standardize (Section 14·11) with the reference cuvette (solution 0) in the sample holder. After the procedure is completed, remove the reference cuvette.

Fill the second cuvette (designated as the "sample" cuvette) halfway with solution III; be certain that the cuvette has been prerinsed with the same solution. Wipe the cuvette dry with tissue, cap with plastic wrap; then insert cuvette into sample holder, and align properly. Record the % T reading at this wavelength. Readjust "0" and "100" as described in Section 14·11, and remeasure the % T of the sample III solution several times, at the 450 mμ wavelength, until the average % T appears to be satisfactory.

Follow the same procedure at new wavelength settings of 475, 500, 525, 550, 575, 600, 625, 650, and 675 mμ. Record the readings and other data in your notebook in a tabular form similar to the following:

Wavelength (mμ)	% T reading	Solution measured
450	—	solution III
475	—	solution III
—	—	—
—	—	—

Graphic Analysis. After obtaining the data indicated above, plot the values on semilog paper, with % T on the vertical (log) scale, and wavelength on the horizontal (nonlog) axis. The curve, so obtained, indicates a relatively flat region (between 600 and 625 mμ) of the $Cu(NH_3)_4^{++}$ spectrum, where the solution absorbs light most strongly. A wavelength of 600 mμ has been selected as a suitable value to measure the absorption on a quantitative basis.

14·13 Analysis of Absorption Data Using Relationships of Beer's Law

The absorption of each prepared solution (knowns and unknown) is measured (several times) at the selected (600 mμ) wavelength. The method of analysis has been discussed in Section 14·2.

Procedure. Empty and rinse thoroughly the "sample cuvette", and fill it with solution 0.

Set the spectrophotometer at a wavelength of 600 mμ.

Prepare a table for recording analytical data, similar to the following (the initial values in the table may be identical with those indicated):

Solution	Concentration (mg Cu/ml)	% T measured
0	0.00000	100
—	—	—
—	—	—

% T average	A calculated	Wavelength mμ
	0	600
—	—	—
—	—	—

(continue in length as necessary)

Restandardize the instrument, using the "reference cuvette"; then, measure % T of the "sample cuvette" containing solution 0. Repeat several times, and record in the table *each % T* measurement; the first values may be those given in the form for the table.

Rinse the sample cuvette repeatedly with solution I; then measure the transmittance of the solution at 600 mμ. It is desirable to restandardize the spectrophotometer for each measurement.

Similarly, measure the % T of solutions II and III, and the unknown solution. Appropriate rinsing of the cuvette is necessary before each measurement.

After constructing the Beer's Law curve, locate the measurement made with the unknown solution in mg/ml. Enclose the value in a square box, to distinguish it from other measurements used to prepare the graph.

Calculate the % Cu in the copper ore sample. The mathematical relationship is as follows:

$$\% \text{ Cu} = \frac{(\underline{\quad}\text{mg/ml Cu in vol flask}) \times (250.0 \text{ ml}) \times 100}{(\underline{\quad}\text{g. Cu unknown sample placed in flask}) \times 1000}$$

Termination of Experiment. After the procedure has been completed, empty the cuvettes, *thoroughly rinse* them with water, and return them to the cuvette rack. Return individual equipment to your desk. Neatly arrange the items of special equipment for the convenience of the next student to perform the colorimetric experiment; also leave desk space clear. Unplug the spectrophotometer.

Submit the following as a special laboratory report *within one week* after the experiment has been completed:

(1) the semilog plot of the spectrum of the ammoniacal Cu^{++} solution, prepared in Section 14·12.

(2) the graph of % T versus concentration for the known and unknown copper solutions, as prepared in this section.

(3) the calculation of the % Cu in the solid unknown, as computed in this section.

The two graphs must be labeled with all details, so that each graph is self-sufficient for examination by the instructor.

14·14 Additional Commentary

Various diverse methods are available for analyzing solutions containing copper. Section 13·14 contains a procedure for the iodometric determination of copper. A very accurate determination may be made by electroanalysis. The foregoing spectrophotometric analysis in aqueous ammonia gives less precise results than most of the many possible determinations of the cupric ion. However, it has been included in this manual for the following reasons: it is a comparatively simple procedure involving readily available chemical substances; it provides first-hand experience with some of the important techniques of colorimetric analysis; finally, it gives the student an opportunity to use a reliable spectrophotometer.

Colorimetric methods of copper analysis are used most frequently in the determination of *trace* amounts of copper. A wide variety of organic complexing agents produce distinctive copper complexes with high absorbancy indexes and strong light-absorbing properties. However, to attain greater precision it is necessary to utilize more sophisticated spectrophotometers. Unfortunately, such instruments are quite expensive, and may not be suitable for an elementary course in quantitative chemistry.

COLORIMETRIC DETERMINATION OF FERROUS IRON

14·15 Preliminary Comments

One of the most sensitive methods for the determination of iron is based upon the formation of the orange-red ferrous-orthophenanthroline complex ion. Three molecules of 1,10-phenanthroline react with one ferrous ion in the following manner:

$$Fe^{++} + 3 \text{ (phen)} = [Fe(\text{phen})_3]^{++}$$

The chelate complex, which follows Beer's Law closely in its absorbance, is very stable and remains unchanged over a long period of time.

Orthophenanthroline is a weak base, behaving similarly to ammonia, in reacting slightly in water solution to give the phenanthrolinium ion and hydroxide ion,

$$\text{phen} + HOH = \text{phen } H^+ + OH^-$$

In an acid solution, the principal species present is the phenanthrolinium ion (analogous to the NH$_4^+$ ion); consequently, the complex formation in acidic solution may be indicated by the following equation:

$$Fe^{++} + 3 \text{ phen } H^+ \rightleftharpoons Fe(\text{phen})_3^{++} + 3H^+$$

The equilibrium constant for the reaction is 2.5×10^6, which strongly favors complex formation. However, the position of the equilibrium is somewhat dependent upon pH. If the pH is below 2, the reaction is incomplete, resulting in a weak color. Usually a pH of about 3.5 is recommended for the analysis, although the color is stable within a pH range of 2 to 9, and a more *careful* control of this variable is not necessary.

Ferric iron is reduced by adding an excess of hydroquinone. The ferrous iron remains stable as the ferrous-orthophenanthroline ion for many months.

14·16 Reagents

The reagents for analysis are usually prepared by the instructor, and are available in the laboratory. For informational purposes, the preparations are indicated as follows:

Hydroquinone solution. A 1% solution is prepared by shaking the compound in distilled water, and buffering it with sodium citrate to an approximate pH of 4.5.

Sodium citrate solution. 250 g of the dihydrate is added to sufficient distilled water to make one liter of solution.

1,10-Phenanthroline solution. 0.5% solution of the monohydrate in distilled water is used. Warm to effect solution and store in dark; discard the solution when it becomes colored. Two milliliters of this solution is required for each milligram of iron.

Standard Iron Solution. 0.3511 g of reagent grade ferrous ammonium sulfate is dissolved in 100 ml of distilled water; add 1 ml of concentrated sulfuric acid, and dilute exactly to one liter in a volumetric flask. One milliter of the standard iron solution contains 0.05 mg of iron.

pH Paper. This should cover the approximate range of 3.0 to 5.0, such as pHydrin paper.

14·17 Procedure

Review Sections 14·1 through 14·8; the *Use of Cuvettes* is of special importance. The special equipment provided for this experiment is the same as that specified in Section 14·8.

Preparation of Diluted Standard Iron Solutions. Obtain approximately 150 ml of standard iron solution (containing about 0.05 mg Fe/ml) in a *clean, dry* 250 ml beaker. The exact concentration of Fe is indicated on the bottle label.

Using the 25 ml pipette, transfer 25.00 ml of the stock solution (see Section 15·2) from the beaker to your *clean* 250 ml volumetric flask (it need not be dry). Fill the volumetric flask halfway, so as to dilute the iron solution. Add sodium citrate solution dropwise to the flask. Shake the flask after each addition (only a few drops will be necessary). The flask should be stoppered with a leak-proof plastic stopper. Touch the damp stopper against a piece of pH paper. Continue to add sodium citrate dropwise until pH paper indicates a pH between 3.0 and 4.5. (Avoid excessive testing with the pH paper, to avoid undue loss of solution.) Use a graduate to add 8 ml of hydroquinone solution and 8 ml of orthophenanthroline solution. Dilute to mark with distilled water. Transfer for storage to a *clean, dry*, one-pint Thro-A-Way bottle, fitted with a plastic screw top and label *Solution I*; mix solution thoroughly by shaking.

Repeat the instructions given in the previous paragraph, this time pipetting *two* 25.00 ml portions of the standard iron solution into the clean 250 ml volumetric flask, to produce a second known iron solution which is exactly twice as concentrated as the first one. Use a 250 ml Thro-A-Way bottle for storage, and label it as *Solution II*.

Repeat the instructions once again, this time pipetting *three* 25.00 ml portions of the standard iron solution into the clean 250-ml volumetric flask, to obtain a third known iron solution which is exactly three times as concentrated as the first solution. Store in a Thro-A-Way bottle and label as *Solution III*.

Prepare a "blank" solution, designated as *Solution 0*, by diluting 8 ml of hydroquinone and 8 ml of orthophenanthroline to approximately 250 ml in a clean Thro-A-Way bottle. No adjustment of pH is necessary.

In the notebook prepare a table for the solutions that have been diluted, calculated on the basis of the known concentration (____ mg Fe/ml) of the original standard iron solution.

Diluted Standard Iron Solutions

Solution	Ml of standard solution used	Final volume	Final concentration, mg Fe/ml (calculated)
0	0.00	250	0.0000
I	25.00	250.0	________
II	50.00	250.0	________
III	75.00	250.0	________

Preparation of Unknown Solution of Impure Ferrous Ammonium Sulfate. Obtain an unknown sample from instructor. This material is *not* to be dried, because at elevated temperatures the sample would lose water of hydration (part of its original composition) in a nonquantitative fashion, and the iron would be slowly oxidized (resulting in further change in composition).

Weigh out (to 1 mg) approximately 1 g into a 400 ml beaker. Add 100 ml of distilled water and 1 ml of concentrated H_2SO_4. Stir thoroughly until solution is effected; then transfer quantitatively (see Section 15·1) to a one-liter volumetric flask, and dilute to mark. Mix thoroughly for several minutes by repeatedly inverting the flask (fitted with a plastic leakproof stopper).

Using the 25 ml pipette, transfer 25.00 ml of the unknown solution to a *clean* 250-ml volumetric flask (it need not be dry). Fill the volumetric flask halfway with distilled water; add sodium citrate dropwise, and test for pH as described in *Preparation of Diluted Standard Solution* in this section. When the pH is between 3.0 and 4.5, add 8 ml of hydroquinone solution and 8 ml of orthophenanthroline solution, and dilute to mark with distilled water. Mix the solution thoroughly; this is your prepared *Unknown Solution.*

14·18 Operating Instructions for the Spectronic Colorimeter

Detailed instructions for the operations of the "Spectronic 20" colorimeter are given in Sections 14·11 and 14·12. These instructions still apply, but will not be reiterated in detail. Condensed directions are as follows:

Turn on the instrument with left front knob and allow to warm up for about 20 minutes. Since the absorbancy spectrum exercise is given in Section 14·12, for the $Cu(NH_3)_4^{++}$ complex, it is not necessary to repeat this exercise for the ferrous-orthophenanthroline complex. It has been determined, experimentally, that the wavelength of maximum absorbance for the ferrous complex is 508 mμ. Set the wavelength dial (top, right-hand knob) at 508. With sample com partment empty, adjust the left front knob so that the meter pointer reads 0% T. Insert *Solution 0*, and adjust the right front knob so that the meter pointer reads 100% T. Replace the *Solution 0* with *Solutions I, II,* and *III*, and with the *unknown solution*. Use the same cuvette for all measurements, rinsing it well with each solution before testing. Determine duplicate or triplicate % T measurements on separate portions of each solution. From the data of the four standards (Solutions 0, I, II, and III) prepare a calibration curve, which should be a straight line on semilog paper, in terms of (log)% T versus mg Fe/ml in the samples. From the % T data of the unknown, ascertain the mg of Fe/ml in the

unknown; then calculate the % Fe present. *Record all data.*

After the experiment has been completed, empty and *thoroughly rinse* the cuvettes with water, and return them to the cuvette rack. Return individual equipment to your desk. Neatly arrange the items of special equipment for the convenience of the next student to perform the experiment; also leave the desk space clear. Unplug the spectrophotometer.

14·19 Calculations

For clarification, part of the previous instructions are repeated: Plot % T versus mg of Fe/ml on semilog graph paper. If a straight line is obtained, it verifies Beer's Law. If all the measurements do not fall on a straight line, recheck your values and calculations, and consult your instructor. Always enclose each data point within a small circle; indicate the concentration of the unknown solution in a square box to distinguish it from the other measurements used in preparing the graph.

The mathematical relationship in calculating the percentage Fe in the sample of ferrous ammonium sulfate is as follows:

$$\frac{\text{mg Fe/ml} \times \text{dilutions} \times (1000 \times 10) \times 100}{\text{wt of unknown sample} \times 1000} = \% \text{ Fe}$$

CHAPTER 15

Precipitation Methods

VOLUMETRIC METHODS

15·1 Introductory Statements

Volumetric methods of analyses that make use of precipitate formations are usually designated as *precipitation titrations*. In order to be usable volumetrically, a precipitate must form rapidly and be fairly insoluble. In other words, the reaction should be complete at the equivalence point if possible. Moreover, an indicator must be available that will produce an observable change in the solution (or precipitate) near the equivalence point. Because of these limitations, relatively few precipitation reactions are satisfactory for titrimetric processes.

The most widely used precipitate-formation titrations are those involving the silver ion versus the various halides and pseudohalides. These procedures are frequently called *argentometric precipitation methods*. As typical examples of these methods, the Mohr method for the chloride, and the adsorption indicator method for the same ion, have been selected for laboratory exercises. Alternately, an emf graphical method is used, involving a pH meter as detector with a silver wire as the indicator electrode (see Sections 15·5 and 15·6).

15·2 Preparation of Approximately 0.1 *F* Silver Nitrate

Normality, molarity, and formality are the units most frequently used in designating con-

centrations of solutions. The term *formality* is probably the most unambiguous expression for the concentration of an ionic compound (see Section 2·3).

Procedure. Counterpoise a clean, dry 50 ml beaker on a triple-beam balance and weigh into it 17 g of pure, dry $AgNO_3$ (as furnished—do not dry in oven). Proceed to the analytical balance and weigh beaker plus silver nitrate to 0.001 g; record the weighing on a right-hand page of the notebook. Carefully transfer as much as possible of the crystalline $AgNO_3$, avoiding any loss, into a clean, but not necessarily dry, 250 ml beaker. Reweigh the original 50 ml beaker to 0.001 g, record, and take the difference from the first weighing, as the actual weight of the $AgNO_3$.

Working quantitatively, dissolve the $AgNO_3$ (now contained in the 250 ml beaker) in approximately 75 ml of water; then transfer the solution without loss to a clean one-liter volumetric flask. The procedure for *quantitative transfer* of the solution is illustrated in Figure 15·1. Place a clean analytical funnel in the flask. Hold a stirring rod vertically above the funnel and pour the solution from the beaker down the rod, keeping the spout of the beaker against the rod. After all of the solution is removed, rinse beaker and rod, as illustrated, with distilled water. Make certain that the stream of water reaches all of the inside surface of the beaker. Remove the beaker and rod; then withdraw the funnel slowly, meanwhile rinsing both the inside and outside of the stem. Now, direct a stream of

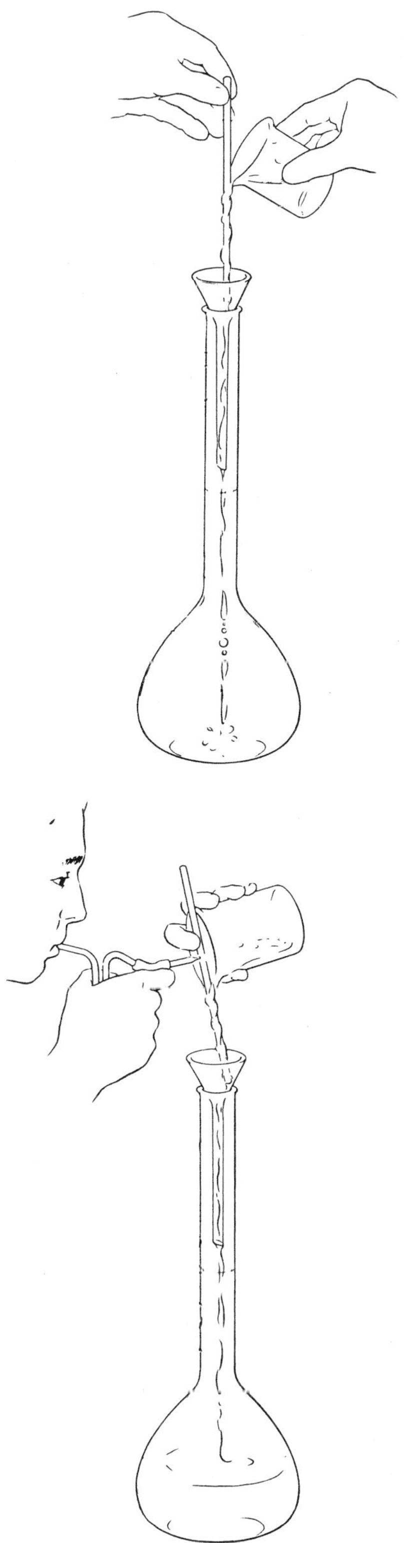

FIGURE 15·1   Quantitative transfer of solution to volumetric flask

water down into the volumetric flask to remove any solution which may have adhered near its mouth.

Next, pour distilled water, from a clean beaker, into the volumetric flask until the solution level is *just below* the graduation mark on the stem of the flask. Allow the solution to reach room temperature, and then add distilled water, drop by drop, from a pipet until the liquid level meniscus coincides with the graduation mark. Mix thoroughly by inverting the flask back and forth for several minutes. (The flask should be stoppered with a leak-proof plastic stopper.) Do not use the neck of the flask as a handle.

Calculation. From the known weight of $AgNO_3$ contained in the one liter (exactly measured) of solution thus prepared, calculate and record in notebook the formality of the solution

$$formality = \frac{wt\ of\ the\ AgNO_3\ in\ g}{ml\ of\ solution \times mfw\ of\ AgNO_3}$$

For example, if you weighed out 17.130 g of $AgNO_3$ and dissolved it in exactly one liter of solution, then

$$formality = \frac{17.130\ g}{1000\ ml \times 0.1699} = 0.1008\ F$$

This is a hypothetical example; your formality will be different.

15·3 Standardization of Silver Nitrate Solution by the Mohr and Adsorption Methods

The formality for the $AgNO_3$ solution obtained in Section 15·2 should be fairly accurate since the purity of reagent grade $AgNO_3$ is almost that of a primary standard chemical. However, the procedures for the Mohr and adsorption methods for the chloride contain slight inherent errors, which may be canceled by standardizing the $AgNO_3$ solution against pure sodium chloride.

Procedure. Dry a 2 g portion (triple-beam balance) of primary-standard sodium chloride for one hour at 110°C. Weigh out on the analytical balance to 1 mg, a sample of the dry NaCl of 1.4 to 1.5 g. Transfer quantitatively to a clean 150 ml beaker. Dissolve in about 60 ml of distilled water, transfer the solution quantitatively to a 250 ml volumetric flask, wash the beaker and stirring rod, and add washings to the flask taking care *not* to fill above the mark. (Follow the techniques described in Section 15·2 concerning the quantitative transfer of a solution.) When the solution has reached room temperature fill the volumetric flask exactly to the graduation mark. Mix thoroughly by inverting the volumetric flask (stoppered with a leak-proof plastic stopper) back and forth for several minutes. The formality of the sodium chloride is obtained from the following formula:

$$F = \frac{\text{wt of NaCl in g}}{250\,\text{ml} \times 0.05844}$$

Record the NaCl formality in the notebook.

The chloride solution just prepared is to be used in Parts I and II for the standardization of the $AgNO_3$ solution. Note carefully that both of the following titrations require good light to observe the end points; however, they should not be performed in direct sunlight. (Even in artificial light the precipitate will darken if the titrations are performed too slowly.) Conserve the NaCl solution as much as possible in order that it may suffice for 6 to 8 titrations, each involving 25 to 30 ml.

Part I. Standardization by Adsorption Indicator Method. Using the smallest quantities of reagents necessary for rinsing, fill one buret with $AgNO_3$ solution and the other with NaCl solution. Read and record both burets. Run into a clean 500 ml titration flask about 25 ml of the chloride solution and add approximately 25 ml of distilled water. Stir in 6 or 8 drops of dichlorofluorescein indicator and 0.1 g of dextrin (a level microspatulaful, no more). The dextrin stabilizes the colloidal state of the precipitate.

Titrate fairly rapidly with $AgNO_3$ solution, inasmuch as a slow titration may permit the precipitate to darken and thereby obscure the end point. At the end point there is a sudden change from white to pink in the color of the AgCl

particles.* It is necessary to whirl the solution constantly; otherwise, there is a tendency for the AgCl particles, which indicate the pink end point, to settle to the bottom of the flask. Since the end point is somewhat reversible, it is sometimes possible to run back and forth across the point at which the pink and white particles are differentiated. This is accomplished by adding alternately a slight excess of $AgNO_3$ or NaCl solution. Having determined duplicate titrations, calculate *at once* in order to ascertain whether it is necessary to make additional titrations to secure a precision of 2 parts in a thousand. Assume that the NaCl formality is more accurate than the $AgNO_3$ formality. The formality of the $AgNO_3$ may be obtained from the following formula:

$$(AgNO_3 \text{ solution}) \qquad (NaCl \text{ solution})$$
$$\text{ml} \times F \quad = \quad \text{ml} \times F$$

Record the $AgNO_3$ formality in the notebook, designating it as obtained by the adsorption method.

Part II. Standardization by the Mohr Method. After the burets have been filled (as directed in Part I) with $AgNO_3$ and NaCl solutions, respectively, and the readings recorded, measure approximately 25 ml of the NaCl solution into a medium-sized porcelain casserole (300 ml or larger). A flask or beaker does not permit a good observation of the end point. Add 25 ml of distilled water (as measured in a graduate) and 2 ml (as measured in a graduate, or 40 drops) of 0.1 F K_2CrO_4 indicator solution. Without further dilution titrate fairly rapidly with the $AgNO_3$ solution to the first permanent change in tint from lemon yellow to orange.† During the titration stir the solution constantly with a short

---

* The color change of the AgCl particles occurs when *negatively* charged, colored dichlorofluorescein ions become adsorbed on a positively charged colloidal surface. This change occurs at the equivalence point because the electrostatic charge on the surface of the colloidal AgCl particles goes then from negative (in the presence of excess untitrated chloride) to *positive* (in the presence of excess silver ions). See Figure 8·2.

† The color change from lemon yellow to orange occurs when the dissolved CrO_4^{--} ions (lemon yellow) first begin to be precipitated by excess Ag^+ titrant (as red Ag_2CrO_4 precipitate), indicating that the chloride ions present in the sample have already been precipitated (as less soluble, white AgCl).

stirring rod. Before taking a buret reading run back and forth across the end point by alternately adding a few drops of excess $AgNO_3$ and NaCl solutions. Make at least three independent titrations. From the average of concordant results, calculate and record the *formality* of the $AgNO_3$ as obtained by the Mohr method. The calculations are similar to those indicated for the adsorption method as given in Part I.

For your convenience and that of your instructor you should record, in tabular form, the *formality* of the silver nitrate solution as determined by each of the three methods of standardization. A suggested form, for your notebook, is given as follows:

Consolidated Record of Standardization
of Silver Nitrate Solution

Formality

A. From method of preparation (direct weighing of silver nitrate crystals)   ———————

B. From NaCl standardization with adsorption indicator (average)   ———————

C. From NaCl standardization by the Mohr method (average)   ———————

15·4 Analysis of Unknown Chloride Sample by Adsorption and Mohr Methods

Dry the sample of unknown chloride for at least one hour at 100°C. Weigh out on the analytical balance, to 1 mg, about 5 g of the material and record the weight in the notebook. Transfer the weighed portion to a clean, dry 150 ml beaker.

Working *quantitatively*, as described in Section 15·2, dissolve the weighed portion in about 60 ml of distilled water, and transfer the resulting solution to a 250 ml volumetric flask. (The flask should have been cleaned thoroughly with distilled water after its use for the experiment described in Section 15·3.) In filling the flask to the graduated mark, follow the procedure given in Section 15·2. Mix thoroughly for several minutes.

Analysis by the Adsorption Method. Fill the burets, respectively, with standard $AgNO_3$ solution and the solution of unknown chloride, taking care to waste as little as possible of the latter solution. Using the procedure of the Adsorption Indicator Method as described in Section 15·3 (Part I), complete three or four titrations. In each titration use about 20 ml of the chloride solution. From the formality of the $AgNO_3$ solution, as obtained by the adsorption method (Section 15·3, Part I), calculate and record the percentage of chloride in the sample. The formula for the calculation is as follows:

$$\frac{\text{ml of } AgNO_2 \times F \text{ of } AgNO_3 \times 0.03545 \times 100}{\text{wt of sample in g} \times (\text{ml of unknown chloride}/250 \text{ ml})} = \% \text{ Cl}$$

In the formula, the product "ml of $AgNO_3 \times F$ of $AgNO_3$" is the number of mfw of $AgNO_3$ used as titrant, which is equal numerically to the number of mfw of chloride ion precipitated. To convert from *mfw* of chloride to units of mass (*grams*), the conversion factor 0.03545 is used. Thus the formula compares the mass of chloride ion precipitated (numerator) with the mass of sample originally present in the particular volume of unknown solution used for the titration (denominator).

Analysis by the Mohr Method. Complete three or more titrations with about 20 ml of the unknown solution, using the technique of the Mohr Method as practiced in Section 15·3 (Part II). Calculate the percentage of chloride in the sample by the same formula as in the foregoing paragraph, but, in this calculation, use the formality of the $AgNO_3$ which was obtained by the Mohr Method (Section 15·3, Part II).

Consolidated Report on the Analysis of Unknown Chloride. The procedures in arriving at the final percentage of the unknown chloride are somewhat lengthy; consequently, a simple report on the analyses will suffice. Make the report as follows:

Exp.   Precipitation Titration   Name
15·2–15·5   of Unknown Chloride

Average of Adsorption Method   ———————

Average of Mohr Method   ———————

If your results are inaccurate, your instructor will want to examine the records pertaining to the analyses. These records should be complete and arranged in an orderly fashion.

POTENTIOMETRIC PRECIPITATION TITRATION

15·5 Preliminary Comments

Potentiometry involves measurements of the *difference* in potential (electromotive force) between the two electrodes of a galvanic cell. A series of potentiometric measurements are obtained at various times in the course of a titration, the repeated small additions of titrant causing changes in the composition of the solution in the cell and coincident changes in the measured potential. By plotting the potential versus ml of titrant, the measurements yield a curved line which reveals information about the composition of the original solution—as well as data useful for calculation of equilibrium constants such as K_{sp}.

pH titration (see Section 12·8) is a special type of potentiometric titration, in which the instrument yields readings directly in units of pH. In ordinary potentiometric titration, the readings are in volts (or millivolts).

As in pH titration, the two electrodes of the galvanic cell are given distinctive names to indicate their respective functions (see Section 12·9). The *indicator electrode* has a potential which is dependent upon the concentration of the substance being titrated. The *reference electrode* maintains a constant potential with respect to the solution, regardless of any change in the concentration of the ions in the solution. Since it is the *difference* in potential between the two electrodes that is measured, any change in potential as a result of titration is evidence of a similar change in the indicator electrode's potential. The galvanic cell as a whole may be represented as

indicator electrode (emf changes during titration)	reference electrode (emf remains constant during titration)

The parallel vertical lines refer to a salt bridge or similar form of electrolytic contact needed between the two electrodes.

Potentiometric methods can be applied to many precipitation titrations, the main limitation being availability of suitable indicator electrodes. In theory, any metal, when bathed by a solution of its ions, should serve this purpose; but in practice, only a few electrodes are satisfactory. (Many metals tend to form insensitive oxide coatings which render the electrode either irreversible or inert.) The silver electrode is one of the most widely used of the indicator electrodes, partly because silver has little tendency to form an oxide coating, and partly because of the large number of insoluble salts formed by silver ion.

The best known precipitation titration with the silver electrode is the determination of the halides, that is, the titration of chloride, bromide, and iodide ions with silver nitrate. The potential of the silver electrode is a logarithmic function of the concentration of the silver ions remaining unprecipitated in solution.

$$E = 0.80 - 0.059 \log \frac{1}{[\text{Ag}^+]} \qquad \text{(at 25°C)}$$

When, during a titration, a rapid change occurs in the logarithm of the Ag^+ concentration, there is also a fast change in the potential measured. Detection of the quick change provides information about the composition of the solution.

Consider the titration of a chloride solution with silver nitrate titrant, as a convenient example.

Early in the titration most of the added silver ions will be precipitated as silver chloride. The high concentration of excess chloride ion keeps the silver ion concentration at a low value according to the K_{sp} relationship for silver chloride:

$$K_{sp} = [\text{Ag}^+][\text{Cl}^-]$$

During the early portion of the titration, only small changes occur in the logarithm of the silver ion concentration, and the potential changes only slightly.

Near the equivalence point the concentration of excess chloride ion rapidly decreases and the concentration of silver ion rapidly increases by several powers of ten. Large changes occur in the logarithm of the concentration, and in the potential measured.

Well beyond the equivalence point the concentration of silver ion is large, because there is now a continual excess of silver ion present in the solution; so again only small changes occur in the logarithm of the concentration. The equivalence point occurs when the concentration of silver ion and chloride ion are equal.

$$K_{sp} = [Ag^+][Cl^-] = 1 \times 10^{-10}$$

At the equivalence point,

$$[Ag^+] = [Cl^-] = \sqrt{K_{sp}}$$
$$= \sqrt{1 \times 10^{-10}}$$
$$= 1 \times 10^{-5} \, F$$

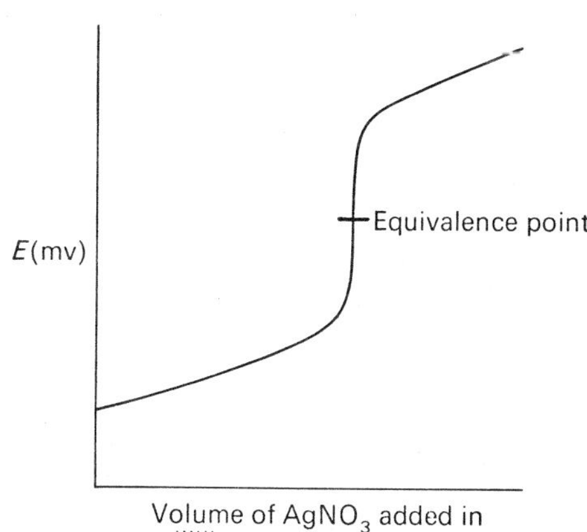

FIGURE 15·2   Potentiometric titration of unknown chloride with standard $AgNO_3$

A typical potentiometric curve is sketched in Figure 15·2. Notice the appreciable change in potential in the vicinity of the equivalence point. It is necessary to obtain numerous potentiometric readings in this region (that is, very small and equal increments of titrant should be added) so as to be able to locate the equivalence point accurately (the midpoint of the steep rise in the curve). The arms of the curve, extending more nearly horizontally, may be determined with only three or four readings each. Therefore, large increments of titrant are sufficient in the regions preceding, and following the vicinity of the equivalence point. In general, the desired

number of data and relative spacing, for which one should strive, are comparable to those for pH titration measurements (Section 12·9).

In Section 15·6 a procedure is given for the titration of an unknown chloride solution with standard silver nitrate solution. A pH meter which permits millivolts to be read from the scale serves as a convenient potentiometer for the titration. The indicator electrode may be either a piece of pure silver wire, a silver "billet", or a platinum gauze plated with silver. A calomel electrode or, more conveniently, a glass electrode may be used as reference electrode, since no significant change in pH occurs during the titration. Use of a side-arm type of calomel electrode requires the presence of a potassium *nitrate* salt bridge, since the usual potassium chloride salt bridge will contaminate the titration solution with extraneous chloride ions. A calomel electrode with a fiber junction is acceptable, since diffusion of chloride ion through the fiber into the titration solution is negligible, but there is likelihood of clogging of the fiber with fine crystals of silver chloride, disrupting the electrolytic contact, and producing unsteady meter readings.

Precipitation titrations require up to five minutes of rapid mechanical stirring before equilibrium is attained, especially near the equivalence point. The approach to equilibrium is indicated when the measured emf does not drift more than 3 or 4 millivolts. The titration must be continued well beyond the equivalence point in order that a graph of the measurements will produce a symmetrical curve.

Many graphical methods have been devised for analyzing data from a potentiometric titration—for example, plotting first or second derivatives of the potential change as the titration proceeds. With manual measurements, none of these methods has any particular advantage over a simple graph of potential (mv) versus volume of titrant added (ml). The methods are useful primarily in automated detection systems in which, for example, the derivatives of the potential change are determined electronically and the results plotted by servo recording devices. Although routine work automation permits more efficient and rapid analyses with less use of manpower, it does so at the expense of accuracy.

15·6 Procedures

The instructor will assign a pH meter to the student and give instructions for its use, and other proper procedures in connection with the assigned experiment. Operate the instrument on the 1400 millivolt setting for emf measurement. As indicator electrode, use a simple silver wire (attached to the instrument by a special connector provided—see Figure 15·3) or a commercial silver metal electrode; plug the electrode into the *reference* terminal at the rear of the instrument. As reference electrode use a glass electrode, but

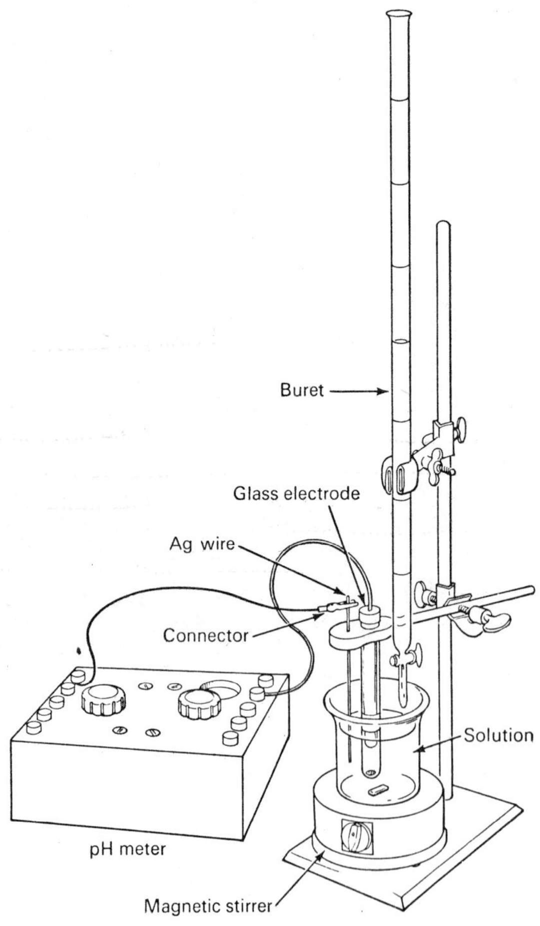

FIGURE 15·3 Apparatus for potentiometric titration. (Ring stands and clamps are required to support the buret and electrodes)

plug the electrode into the large *indicator* terminal.* *Caution:* Keep the glass electrode immersed in pure water whenever it is not in use; avoid striking the glass bulb against any solid object (beaker, stirring bar, etc.). Figure 15·3 shows the arrangement of equipment for the titration.

The titrant is 0.1 F AgNO$_3$ solution, accurately standardized against pure, dry NaCl. See the procedures given in Sections 15·2 and 15·3, but make up only 500 ml of solution containing 8.5 g of analytical grade AgNO$_3$.

Into three 600 ml beakers weigh, to the nearest 0.1 mg, triplicate samples (properly identified) of about 0.5 g each of the unknown chloride, previously dried at 110°C. Dissolve each sample by adding approximately 300 ml of distilled water and stirring thoroughly. Analyze each solution potentiometrically as follows.

Immerse the electrodes into the solution and begin vigorous stirring with magnetic stir bar and motor. Activate the meter (1400 mv setting) and adjust the needle to read 0.2 volts prior to titration.†

Titrate the solution with the standard AgNO$_3$, taking potential readings after each addition of titrant. Start with increments of 4 ml, but decrease to about 0.2 ml near the equivalence point, so that each change in potential never exceeds 20 to 25 millivolts. *Allow time for equilibration.* After the equivalence point has been passed, the increments may gradually be increased again.

Rinse the electrodes thoroughly after each complete titration. Immerse the rinsed glass electrode in fresh pure water at the conclusion of the experiment.

For each titration plot millivolts (emf) versus milliliters of silver nitrate titrant, using rectangular coordinate graph paper. The midpoint of

---

* According to these directions the electrodes are attached to the pH meter in reverse manner, but this is completely satisfactory for the purposes of the experiment, because only *differences* between the two electrode potentials are to be measured. With electrodes assembled this way, the meter reading should increase as titration proceeds.

† The value chosen for this first reading is not important but it should be no greater than 0.6 volts, so that the needle will remain on scale throughout the titration. In the analysis, great interest is placed only on how the *difference* between the two electrode potentials *changes as titration proceeds*, not on the exact value of the difference.

the sharp break in the curve indicates the equivalence point of the titration (see Figure 15·2). From the average of the three titrations, calculate the percentage of chloride in the unknown sample.

$$\frac{\text{ml of AgNO}_3 \times F \text{ of AgNO}_3 \times 0.03545 \times 100}{\text{wt of sample in g}} = \%\ Cl$$

Label the graphs completely and submit them with the averaged result.

GRAVIMETRIC METHODS

15·7 Preliminary Comments

The word *gravimetric* has reference to a weight measurement; consequently, a gravimetric method is one in which an analysis is performed by means of weighing operations. More completely, a gravimetric analysis involves the determination of the weight of a pure substance, which is produced from a given weight of a sample. In the analytical procedure process, the sample is dissolved, and the substance sought is isolated from other constituents by the formation of an insoluble precipitate. Additional operations include the filtration and the ignition (or drying) of the precipitate.

Not all precipitates are suitable for gravimetric procedures. Consideration of the *properties* which render a precipitate adaptable to gravimetric work will be reserved for theoretical discussion elsewhere; as, for example, solubility, purity, filterability, and stability. Even under optimum conditions, the procedures of gravimetric precipitation are usually tedious and time-consuming. On the other hand, some gravimetric analyses are preferable to comparable determination because of greater accuracy. It is also true that many gravimetric precipitations are necessary because no other satisfactory methods are available.

The gravimetric analyses of chlorine in a soluble chloride, and that of iron in ferrous ammonium sulfate, have been selected as typical examples of gravimetric precipitations.

15·8 Gravimetric Determination of Chlorine in a Soluble Chloride

The chlorine content of a soluble chloride is precipitated from a slightly acid (HNO_3) solution as silver chloride:

$$Cl^- + Ag^+ \rightleftarrows AgCl\downarrow$$

When first precipitated, the AgCl particles are in a colloidal state, but they are readily coagulated—when heated—to produce a curdy precipitate. This type of precipitate is easily filtered, and may be readily washed free of impurities. The precipitation is carried out in an acid solution to prevent interference from anions of weak acids, such as carbonate ion, which would coprecipitate in neutral media. A *moderate* excess of silver ion is necessary to reduce the solubility of the precipitate, but a large excess may result in an error through excessive coprecipitation. Other factors are also involved (see Chapter 10).

In the presence of strong light the silver chloride undergoes photodecomposition as follows:

$$AgCl \rightleftarrows Ag\downarrow + Cl_2\downarrow$$

The white precipitate turns violet in color, owing to the presence of finely divided particles of silver. The possible errors resulting from this decomposition are interesting. If the precipitate has been washed, the results will be low because of the loss of chlorine. On the other hand, if the photodecomposition takes place in the precipitating solution, the liberated chlorine molecules react with water and excess silver ions to produce additional AgCl. This will cause the analytical results to be too high. The reaction is as follows:

$$3Cl_2 + 3H_2O + 5Ag^+ \longrightarrow$$
$$5AgCl + ClO_3^- + 6H^+\downarrow$$

The error caused by photodecomposition is far greater in the precipitating medium than in the washed precipitate. Hence, it is necessary to coagulate the precipitate as rapidly as possible, and to protect it from light during the aging process.

Some photodecomposition is unavoidable, but with proper precautions the total error is small. After the washed precipitate is dried, additional decomposition is negligible.

It is customary to filter and wash the AgCl through either a Gooch crucible or a sintered-glass crucible. The latter is more convenient to use, but it is difficult to clean, and should not be heated much above 100°C.

Preparation of Crucibles. Three Gooch crucibles are to be used in the analysis. Before preparing the asbestos mats, examine the crucibles for distinguishing marks. If none are found, mark the unglazed bottoms with identifying letters, or numbers, using a special (heat-insensitive) marking pencil; heat to redness in a colorless flame, and allow to cool in the desiccator.

Prepare asbestos mats in the Gooch crucibles as directed in Section 13·2. It is essential that these mats be made properly. The asbestos soup should be fairly thin and the resulting mats must be uniform in appearance, and of suitable thickness. An unsatisfactory mat may cause the total loss of many hours of work involved in the determination.

Wash the mats thoroughly with distilled water, using a gentle suction, until no loose fibers emerge with the washings. Test by collecting a portion of wash water in a clean test tube. Place each crucible in a small beaker, and dry in the oven at 110°C for one hour. Cool in the desiccator, and weigh. Repeat the drying, cooling, and weighing procedure until each crucible has *reached constant weight* (within 0.0003 g).* Keep a complete record of all weighings. The crucible is now ready for use.

Sintered-glass crucibles of medium or fine porosity may be used in place of the Gooch filters. If crucibles are not new, remove visible dirt with detergent and water. Place in the crucible holder and fill about halfway with concentrated nitric acid; draw through the mat with gentle suction. Repeat once again with

nitric acid, and then wash several times with water. If the crucible mat is not clean, discard the washings from the filter flask, and repeat washing operations using 6 *F* ammonia. Rinse several times with distilled water. Dry crucibles in the oven, and bring to constant weight.

Procedure. Dry the impure chloride at 110°C for at least one hour. Weigh out (to 0.1 mg) three samples between 0.3 and 0.4 g into 400 ml beakers (properly identified). Dissolve in approximately 100 ml of distilled water and add about 1 ml of concentrated nitric acid.† To the cold solution, add slowly, with constant stirring, 80 ml of the 0.1 *F* $AgNO_3$ solution (previously prepared in Section 15·2). Cover with watch glass, and heat nearly to boiling; continue to heat gently until the precipitate is coagulated. Test for complete precipitation by adding a few drops of $AgNO_3$ solution. *Avoid exposing the beaker (containing precipitate and solution) to light any longer than necessary.* Store in your desk overnight, and keep in a darkened area until ready for filtering.

Prepare a wash solution by adding about one ml of concentrated nitric acid to 500 ml of distilled water contained in the wash bottle.

Place a prepared filtering crucible in the filter holder as indicated in Figure 13·1, and apply gentle suction. Carefully decant the liquid above the AgCl precipitate through the filter. Guide the liquid into the crucible by means of a glass rod as shown in Figure 15·1, retaining as much of the precipitate in the beaker as possible.

Add roughly 25 ml of wash solution to the precipitate in the beaker, stir well, and allow to settle. Decant the washing through the filter, still retaining the bulk of precipitate in the beaker. Repeat the washing procedure once more, and then transfer the precipitate to the filter with the aid of a stirring rod and a stream from the wash bottle (Figure 15·1). Particles clinging to the beaker may be scrubbed away by means of a rubber policeman on the end of a stirring rod. Wash such particles into the filter.

Detach the funnel (including the filter and support ring) from the filter flask; insert the

---

* Drying and cooling periods should be approximately the same; otherwise, it will be difficult to attain constant weight. For example, if the crucible is dried for one hour and permitted to cool 20 minutes, the same intervals should be used when the procedure is repeated.

---

† Nitric acid aids in coagulating the precipitate; it also reduces co-precipitation.

funnel into the mouth of a clean test tube. Add 2 or 3 ml of wash water, and allow the water to drip into the tube. Test for complete washing by adding a drop of hydrochloric acid to the filtrate in the test tube. If no cloudiness occurs, the precipitate has been washed sufficiently. In case of cloudiness continue the washing with wash solution until the test produces no cloudiness.

Dry the crucible and contents (in a small beaker) at 110°C until constant weight is attained (within 0.0003 g). Drying and cooling periods should be approximately the same as before: dry for one hour, cool for 20 minutes, and then weigh. Record all data.

Calculations. Compute the weight percentage of chloride by multiplying the gravimetric factor, Cl/AgCl, by the ratio of the dried precipitate over the weight of the sample, as follows:

$$\frac{\text{Cl}}{\text{AgCl}} \times \frac{\text{wt of dried ppt in g}}{\text{wt of sample in g}} \times 100 =$$

$$\% \text{ chlorine}$$

15·9 The Gravimetric Analysis of Iron in Ferrous Ammonium Sulfate

The gravimetric analysis of iron in any type of mineral is a time-consuming process, which may contain a number of pitfalls. On the other hand, the gravimetric determination of iron in an impure salt is relatively simple and straightforward; it gives the student the experience of dealing with a gelatinous precipitate without the pitfalls attendant to the analysis of a mineral.

$FeSO_4 \cdot (NH_4)_2SO_4 \cdot 6H_2O$, ferrous ammonium sulfate, containing known quantities of impurities, is selected for analysis because of three desirable properties: (1) relatively high chemical stability, (2) convenience in handling, and (3) solubility in slightly acid solution.

The iron salt must not be dried in the oven because at an elevated temperature the ferrous ion is slowly converted to the ferric state before weighing, thereby introducing a significant error. It also loses water of hydration.

The ferrous salt is weighed, and then dissolved in distilled water which has been acidified with hydrochloric acid. In contact with air, the iron in ferrous salt solutions is slowly oxidized by dissolved oxygen,

$$4Fe^{++} + O_2 + 2H_2O \rightleftarrows 4Fe^{3+} + 4OH^-$$

and unless the solution contains excess hydronium ions (to remove hydroxide ions) the iron will be partially precipitated in the form of ferric basic sulfate:

$$Fe^{3+} + OH^- + SO_4^{--} \rightleftarrows Fe(OH)SO_4 \downarrow$$

The addition of HCl prevents this precipitation.

The ferrous ion is oxidized to the ferric ion with nitric acid as is indicated by the following equation:

$$3Fe^{++} + NO_3^- + 4H^+ \rightleftarrows$$
$$3Fe^{3+} + NO + 2H_2O$$

Iron in the ferrous state cannot be completely precipitated with ammonia, whereas iron as the ferric ion may be precipitated quantitatively even in a slightly acid solution. For gravimetric precipitation, iron is always oxidized to the +3 oxidation state.

When iron, in the tripositive state, is treated with ammonia, it is quantitatively precipitated as hydrous ferric oxide.

$$2Fe^{3+} + 6NH_3 + xH_2O \rightleftarrows$$
$$Fe_2O_3 \cdot xH_2O \downarrow + 6NH_4^+$$

The precipitate is so gelatinous in nature that it can be regarded as a flocculated colloid. The enormous surface presented by the primary particles is conducive to extensive contamination by adsorption. Since the precipitate cannot be purified appreciably by digestion, it is necessary to resort to double precipitations to remove nonvolatile contaminants.

If the precipitate formed in the ammonia treatment is regarded as $Fe(OH)_3$, it is stabilized by the primary adsorption of hydroxide ions, and the secondary adsorption of any cations (M^+ or M^{++}) which may be present in solution. Consequently, the flocculated particle may be represented as

$$Fe(OH)_3 \cdot OH^- \cdots M^+ \ (or \ M^{++})$$

If a large excess of ammonium ions is present in the precipitating and wash solutions, the counter ion (M^+), during and after the double precipitations, will be largely the ammonium ion. The precipitate can then be indicated as

$$Fe(OH)_3 \cdot OH^- \cdots NH_4^+$$

Since the ammonium is volatilized during the ignition of the washed precipitate, little or no error results from this adsorption.

The precipitate is finally ignited to and weighed as the oxide:

$$Fe_2O_3 \cdot xH_2O \longrightarrow Fe_2O_3 + xH_2O$$

The ignition requires considerable care: the paper must be burned off slowly, with a good circulation of air in the crucible to prevent reduction of the ferric oxide, and the final temperature (about 850°C) must be sufficiently high to remove all water from the oxide.

Preparation of Crucibles. Three porcelain crucibles are to be used as containers for the ignition of the precipitates of hydrous ferric oxide. These crucibles are relatively stable up to 1000°C in the absence of alkalies, alkali carbonates, or fluorides. Above 1000°C, the porcelain glaze melts, and the bodies of the crucibles will begin to soften. An ordinary Tirrill burner (a doubly regulated Bunsen burner) does not give a sufficiently high temperature for ignition. It is customary to use either a Meker burner, or a muffle furnace to attain the desired temperature of about 850°C.

Examine the crucibles for identifying marks, and if none are found, mark the unglazed bottoms with your initials and a sequence of numbers, using a special, heat-resistant marking pencil. Check for hidden cracks in the crucibles.* Heat the bottoms of the dry crucibles to redness in a colorless flame produced by a Tirrill burner, and allow to cool in the desiccator. Iron stains

---

* A sound crucible without lid produces a distinct ringing tone when dropped onto its base from a height of one-half-inch. An infirm crucible produces only a dull thud.

within a crucible from previous determinations do not affect the present analysis.

Bring the crucibles to constant weight (within 0.3 mg) by placing them in a muffle furnace for 20 minutes; allow to cool in open air for 3 minutes (no more), and then cool in the desiccator for 20 minutes. Each of the heating and cooling intervals should be the same; otherwise, it may be difficult to get the crucibles to constant weight. If a muffle furnace is not available, a Meker burner may be used.

Precipitation and Filtration. The unknown samples are *not* to be dried. Weigh out (to 1 mg) three samples between 2.0 and 3.0 g into 400 ml beakers. To each portion add 50 ml of water and 5 ml of concentrated hydrochloric acid; heat until samples are dissolved. Continue to heat until the solutions are boiling, and then add about 30 drops of concentrated nitric acid, drop by drop, by means of a pipet, until the darkened liquid clears to a yellow. Continue to boil for 3 minutes to expel the oxides of nitrogen. Dilute the solution to 200 ml; heat again to boiling, and add slowly with stirring (from graduate) 25 ml of 6 F ammonia. (If the ammonia water from the reagent bottle is not perfectly clear, it should be filtered; otherwise, silica from the glass bottle will cause an appreciable error in the analysis. After the precipitation is complete, a strong odor of ammonia should persist above the solution. If not, add more ammonia. Continue to digest for about 1 minute; then, without delay, decant the clear supernatant liquid through a coarse-grained "ashless" 11 cm filter paper (Whatman No. 41 or equivalent) into a 600 ml beaker, keeping as much as possible of the precipitate in the original beaker. Add to the original beaker 50 ml of ammonium nitrate wash solution (1 g of ammonium nitrate per 100 ml of water), heat, and decant the hot washing through the filter paper. Repeat the washing procedure once more, then remove the 600 ml beaker and discard the clear filtrate.

Place the original 400 ml beaker (containing the bulk of the precipitate) under the filtering funnel. Take care not to get different samples mixed. Now pour 10 ml of 6 F HCl slowly over the filter, so as to dissolve completely any precipitate on the filter. Wash the filter two or three times with hot water. When the last wash water has drained into the beaker, dilute to 200 ml. (If you do not have at least one hour available to complete the double precipitation, stop at this point, and store the beaker covered with watch glasses in your desk). Precipitate once again by adding an excess of ammonia. Repeat the filtration, this time transferring the precipitate quantitatively to the filter paper during the washing operation. With a rubber policeman attached to the end of the stirring rod, rub loose any particles adhering to the beaker. Continue washing the precipitate with hot wash water, containing ammonium nitrate, until a 5 ml portion of the wash water passing through the filter paper gives only a faint cloudiness with silver nitrate. Once started, the filtration cannot be interrupted, because if the hydrous ferric oxide dries, the surface cracks, and washing becomes ineffective. Wipe the beaker with a small piece of dry ashless filter paper to remove the last traces of the precipitate, and add the filter paper to the precipitate in the funnel. Allow the filter and its contents to drain thoroughly. Fold the edge of the filter paper inward to cover the precipitate, then lift it carefully from the funnel, and place upside down in a previously ignited and weighed crucible.

Ignition of Precipitate in Paper. Place the crucible and its contents in a beaker (covered with a watch glass), and dry in the oven for about 20 minutes, or overnight if convenient. Next set the crucible *with cover* in a triangle on a ring stand and heat over a low Tirrill flame. Increase the height of the flame gradually to carbonize the paper. Keep the crucible covered during the carbonization, but you may lift the cover to observe the progress of charring. The paper must not be allowed to burst into flame. When the paper is completely charred, as revealed by its black color, remove the cover, and transfer the uncovered crucible to a muffle furnace. Bring the crucible and contents to constant weight (within 0.3 mg). The time intervals for heating and cooling should be the same as used in getting the crucibles to constant weight: heat 20 minutes, cool in the open air for 3 minutes, and then cool in the desiccator for 20 minutes. When placed in the desiccator, the crucible should be

covered; otherwise, a partial vacuum in the desiccator, produced by the cooling crucible, may cause the contents of the crucible to be sucked out when the top of the desiccator is removed. Always slide the top of the desiccator slowly when opening it.

In the absence of a muffle furnace, a Meker burner may be used in bringing the crucible and its contents to constant weight.

Calculations. The iron content of the original sample is obtained by multiplying the gravimetric factor, $2Fe/Fe_2O_3$, times the weight of ignited precipitate over the weight of the sample, as follows:

$$\frac{2Fe}{Fe_2O_3} \times \frac{\text{wt of ignited ppt in g}}{\text{wt of sample in g}} \times 100 = \% \text{ iron}$$

A COMPLEXATION METHOD (THE DETERMINATION OF HARDNESS IN WATER)

A brief discussion of the uses of ethylenediaminetetraacetic acid in chelometric titrations is given in Sections 8·9 through 8·16. EDTA was first suggested as a reagent for the estimation of hardness in water, and this usage is now a standard procedure in water analysis.

15·10 Preliminary Comments

Surface and well water may contain dissolved materials such as compounds of calcium and magnesium that interfere with the cleansing action of soap. Such water is known as hard water, and the substances responsible for the hardness are removed through a process called water softening. The extent of the hardness determines the magnitude of the softening process, which is frequently a major industrial problem.

The hardness of water is usually measured in parts per million (mg/liter) expressed in terms of $CaCO_3$. Such a calculation is not strictly true since hardness is due to both calcium and magnesium salts. An approximate method for the determination of hardness has been developed in which the sum of these two ions is titrated with EDTA. The indicator most frequently used is Eriochrome Black T, which is sometimes abbreviated to EBT.

The EBT indicator does not give a satisfactory end point for the direct titration of calcium with EDTA in the absence of magnesium. The calcium–EBT complex is weak; consequently, the end point is gradual and appears too soon. On the other hand the magnesium–EBT complex is stronger, and the end point is sharper. When calcium and magnesium ions are present in the same sample, the free calcium and magnesium ions are titrated together. The calcium ions form a more stable complex with EDTA than does magnesium. The calcium ions are first tied up as CaY^{--}, and then the magnesium reacts to give MgY^{--}. Since the strengths of the indicator complexes with calcium and magnesium are opposite to the EDTA complexes, the end point is signaled when all calcium and magnesium ions have combined with the complexing agent, EDTA.

If the disodium salt of EDTA is the titrating agent and EBT is the indicator, the analytical and indicator reactions are as follows:

(analytical reaction)

$$2H_2Y^{--} + Ca^{++} + Mg^{++} \rightleftharpoons$$

$$CaY^{--} + MgY^{--} + 4H^+$$

(indicator reaction)

$$H_2Y^{--} + MgIn^- \rightleftharpoons MgY^{--} + HIn^- + H^+$$
(wine-red)   (blue)

The determination is simplified by adding a small amount of magnesium ions to the EDTA in the preparation of the solution of the complexing agent.

15·11 Reagents

Four reagent solutions are necessary for the analysis.

0.01 F EDTA Solution. Prepare by dissolving 4.0 g of disodium EDTA dihydrate and ap-

proximately 0.1 g $MgCl_2 \cdot 5H_2O$ (triple-beam balance for both weighings) in one liter of distilled water, contained in a clean borosilicate glass bottle. Mix thoroughly by shaking for several minutes.

Standard Calcium Chloride Solution. Accurately weigh to four significant figures (about 1.0 g) one sample of primary standard $CaCO_3$ that has been previously dried at 110°C. Transfer the solid to a 400 ml beaker, and cover with a watch glass; then, cautiously add by means of a pipet approximately 10 ml of 6 F HCl. When the calcium carbonate has dissolved, rinse the watch glass into the beaker. Add 50 ml of distilled water, and transfer the solution quantitatively to a one-liter volumetric flask. Dilute to mark, and mix well.

$$F = \frac{\text{wt of } CaCO_3 \text{ in g}}{1000 \text{ ml} \times 0.1001}$$

The value, 0.1001, is the mfw of $CaCO_3$. The formality should be of a magnitude of 0.0100 F.

Ammonia–ammonium Chloride Buffer Solution. Dilute 8 g of ammonium chloride and 60 ml of concentrated ammonia to 100 ml with distilled water. The pH of the buffer is approximately 10.

Eriochrome Black T Indicator. Dissolve 0.5 g of EBT in 100 ml of ethyl alcohol.

15·12 Standardization of EDTA Solution

Pipet a 25 ml aliquot of the standard calcium solution into a titration flask; add 1 ml of the ammonia–ammonium chloride buffer, and 5 drops of the EBT indicator. Titrate with 0.01 F EDTA until a color change from wine-red to blue signals the end point.

Standard calcium solution		EDTA solution
ml × F	=	ml × F

Repeat the titrations with additional aliquots until the results are within experimental agreement.

15·13 Determination of Total Hardness in Water

Pipet an aliquot portion of a liquid unknown (supplied by the instructor), sufficiently large to require more than 20 ml of the EDTA solution, into a titration flask. Titrate the unknown, following the same procedure used in the standardization of the calcium solution (Section 15·12). Repeat with additional samples until check results are obtained.

$$\frac{\text{ml of EDTA} \times F \text{ of EDTA} \times 0.1001 \times 1000}{\text{vol of sample (ml)}}$$

$$= \text{parts per million of } CaCO_3$$

Appendices

List of Equipment

RETURNABLE ITEMS

2 beakers, 50 ml
2 beakers, 150 ml
2 beakers, 250 ml
2 beakers, 400 ml
2 beakers, 600 ml
1 bottle, 5 pint, plastic top
5 bottles, 1 pint, plastic top
1 bottle, 2 liter, unstoppered
2 bottles, 1 liter, glass-stoppered
1 bottle, $\frac{1}{2}$ liter, glass-stoppered
1 bottle, 250 ml, wide-mouth
3 bottles, weighing
1 brush, test tube
2 burets, 50 ml
2 burners, Tirrill, natural gas
1 casserole, porcelain, 210 ml
1 clamp, Hoffman, screw type
1 clamp, test tube
3 crucibles, porcelain, No. 0
3 crucible covers, porcelain, No. 0
3 crucibles, Gooch, porcelain, No. 3
1 crucible holder, rubber ring for Gooch
1 cylinder, graduated, 25 ml
1 cylinder, graduated, 100 ml
1 desiccator, with porcelain plate
3 flasks, Erlenmeyer, wide-mouth, 500 ml
1 flask, filtering, side-arm, 500 ml
1 flask, Florence, flat-bottom, 500 ml
1 flask, volumetric, 250 ml
1 flask, volumetric, 1000 ml
3 funnels, fluted, long stem, 65 mm
4 glass rods, stirring, 125 × 4 mm
1 lock, combination, with hasp
1 pipet, volumetric, 25 ml
1 rubber stopper, solid, No. 7 or 8
1 rubber stopper, 1-hole, No. 7 or 8
1 rubber stopper, 2-hole, No. 7 or 8
1 rubber stopper, 2-hole, No. 5
2 rubber tubing, ($\frac{1}{4}$ × $\frac{3}{32}$), 3 ft lengths for burners
1 rubber tubing, ($\frac{1}{4}$ × $\frac{3}{32}$), 3 ft length, utility
2 rubber tubing, ($\frac{1}{4}$ × $\frac{3}{16}$), 3 ft lengths, suction, thick wall
1 spatula, micro

1 support, funnel, wood, with clamp
4 test tubes, 150 × 20 mm
1 thermometer, 110° C
1 tongs, crucible, cadmium plated steel
2 triangles, nichrome, wire
2 tripods, iron, threaded legs
2 watch glasses, 50 mm
2 watch glasses, 100 mm
2 watch glasses, 150 mm
2 wire gauzes, 100 × 100 mm
1 wing top for burner

NONRETURNABLE ITEMS

2 corks, No. 3
1 file, triangular
1 labels, box
1 litmus, paper, blue, vial
1 litmus, paper, red, vial
4 matches, boxes
1 pencil, wax
3 rubber policemen
2 rubber shock absorbers for graduates
1 sponge
2 towels
2 tubes, glass, 750 × 6 mm

GENERAL EQUIPMENT

asbestos gloves
balances, analytical
balances, triple beam
buret holders, double, castaloy
colorimeters (with cuvettes)
Filter paper, Quantitative (11 cm, various porosities)
furnace, muffle
oven, drying
pH meters (pH and millivolt scales)
ringstands, with porcelain base
stirrers, magnetic, (with coated bars)
tongs for muffle furnace
wiping tissue

Reagents

ACIDS AND BASES

acetic acid, $17F$
  glacial acetic acid

acetic acid, $6\ F$
  360 ml of glacial acetic acid in a liter of solution

ammonia, $15\ F$
  concentrated NH_3

ammonia, $6\ F$
  400 ml of concentrated NH_3 in a liter of solution

hydrochloric acid, $12\ F$
  concentrated hydrochloric acid

hydrochloric acid, $6\ F$
  500 ml of concentrated HCl in a liter of solution

nitric acid, $15\ F$
  concentrated nitric acid

nitric acid, $6\ F$
  375 ml of concentrated HNO_3 in a liter of solution

50% sodium hydroxide, $19\ F$
  concentrated sodium hydroxide

sodium hydroxide, $1\ F$
  50 ml of concentrated NaOH in a liter of solution

sulfuric acid, $18\ F$
  concentrated sulfuric acid

sulfuric acid, $3\ N$
  160 ml of concentrated H_2SO_4 in a liter of solution

sulfuric acid, $1.5\ N$
  80 ml of concentrated H_2SO_4 in a liter of solution

REAGENT SOLUTIONS

ammonium oxalate, $0.25\ F$
  dissolve 35.5 g of $(NH_4)_2C_2O_4 \cdot H_2O$ in water and
  dilute to 1 liter

asbestos suspension
  4 g of asbestos in 800 ml of distilled water contained
  in a liter bottle

cleaning solution
  20 g of $Na_2Cr_2O_7 \cdot 2H_2O$ in hot, concentrated
  H_2SO_4 (not more than 100° C)

hydroquinone, 1%
  dissolve 10 g in water and dilute to 1 liter

mercuric chloride, saturated
  add 70 g of $HgCl_2$ to 1 liter of water

1,10-phenanthroline, 0.5%
  dissolve 5 g of the monohydrated in water and dilute
  to 1 liter

silver nitrate, $0.1\ F$
  dissolve 17 g of $AgNO_3$ in water and dilute to 1 liter

sodium citrate
  dissolve 250 g of the dihydrate in water and dilute
  to 1 liter

stannous chloride
  dissolve 56 g of $SnCl_2 \cdot 2H_2O$ in 100 ml of conc HCl,
  allow to stand until clear, and then dilute to 1 liter,
  adding a few pieces of metallic tin to the solution

Zimmermann-Reinhardt (preventive solution)
  dissolve 1 lb of $MnSO_4 \cdot H_2O$ in 3 liters of water, and
  combine with a cooled mixture of 600 ml of
  H_2SO_4 and 600 ml of H_3PO_4 in 1800 ml of H_2O

INDICATOR SOLUTIONS

bromcresol green
  dissolve 0.4 g bromcresol green in 6 ml of $0.1\ N$
  NaOH and dilute to 1 liter

dichlorofluorescein
  dissolve 1 g of dichlorofluorescein in 1 liter of ethanol

methyl red
  dissolve 0.4 g of methyl red in 15 ml of $0.1\ N$ NaOH
  and dilute to 1 liter

methyl orange
  dissolve 1 g of methyl orange in 1 liter of water

mixed indicators
  mix 2 parts of methyl red solution with 3 parts of
  bromcresol green solution

phenolphthalein
> dissolve 2g of phenolphthalein in 500 ml of 95% alcohol and dilute to 1 liter

potassium chromate, 0.1 *F*
> dissolve 19.4 g of K_2CrO_4 in water and dilute to 1 liter

starch indicator
> stir 2 g of "soluble" starch with 20 ml of water in a small beaker to form a paste. Pour the paste slowly into 500 ml of boiling water. Add 0.02 g of HgI_2 as preservative

Solid reagents and their uses

Reagent	Use
ammonium nitrate	gravimetric iron
calcium chloride	desiccator
dextrin	volumetric chloride
iodine	antimony
potassium iodide	antimony
potassium permanganate	iron and calcium
potassium sodium tartrate	antimony
potassium thiocyanate	copper
sodium bicarbonate	antimony
sodium thiosulfate	copper
stopcock grease	buret
urea	copper
vaseline	desiccator

Primary standards and their uses

(These reagents should have a purity of 99.90% or better.)

Standard	Use
arsenious oxide	standard iodine
copper, metal	standard copper
ferrous ammonium sulfate	standard iron
potassium acid phthalate	standard base
sodium chloride	standard $AgNO_3$
sodium oxalate	standard $KMnO_4$

UNKNOWN SUBSTANCES FOR ANALYSIS

(These materials may be obtained from Thorn Smith, 847 North Main Street, Royal Oak, Michigan.)

calcium carbonate
copper oxide
ferrous ammonium sulfate
iron ore
potassium acid phthalate
soda ash
soluble antimony
soluble chloride

Ionization Constants of Weak Bases at Room Temperature

Ionization constants of weak bases at room temperature

Base	Formula	Ionization constant
ammonia	NH_3	1.8×10^{-5}
aniline	$C_6H_5NH_2$	3.8×10^{-10}
dimethylamine	$(CH_3)_2NH_2$	5.12×10^{-4}
ethanolamine	$C_2H_5ONH_2$	2.77×10^{-5}
ethylamine	$C_2H_5NH_2$	5.6×10^{-4}
hydrazine	$(NH_2)_2$	3×10^{-6}
methylamine	CH_3NH_2	4.38×10^{-4}
pyridine	C_5H_5N	1.4×10^{-9}
trimethylamine	$(CH_3)_3N$	5.27×10^{-5}

Ionization Constants of Weak Acids at Room Temperature

Ionization Constants of Weak Acids at Room Temperature

Acid	Formula	Ionization constant		
		K_1	K_2	K_3
acetic	$HC_2H_3O_2$	1.75×10^{-5}		
arsenic	H_3AsO_4	5×10^{-3}	8.3×10^{-8}	6×10^{-10}
arsenious	H_3AsO_3	6×10^{-10}		
benzoic	$HC_7H_5O_2$	6.3×10^{-5}		
boric	H_3BO_3	5.8×10^{-10}		
carbonic	H_2CO_3	4.3×10^{-7}	5.6×10^{-11}	
chloroacetic	$HC_2H_2O_2Cl$	1.4×10^{-3}		
chromic	H_2CrO_4	2×10^{-1}	3.2×10^{-7}	
citric	$H_3C_6H_5O_7$	8.7×10^{-4}	1.8×10^{-5}	4×10^{-6}
cyanic	$HCNO$	2×10^{-4}		
dichloroacetic	$HC_2HO_2Cl_2$	5×10^{-2}		
formic	$HCHO_2$	1.77×10^{-4}		
hydrazoic	HN_3	1.7×10^{-5}		
hydrocyanic	HCN	5×10^{-10}		
hydrofluoric	HF	3.53×10^{-4}		
hydrogen sulfide	H_2S	5.7×10^{-8}	1.2×10^{-15}	
hypochlorous	$HOCl$	3.5×10^{-8}		
iodic	HIO_3	1.67×10^{-1}		
lactic	$HC_3H_5O_3$	1.39×10^{-4}		
nitrous	HNO_2	4.6×10^{-4}		
oxalic	$H_2C_2O_4$	5.97×10^{-2}	6.40×10^{-5}	
phenol	HC_6H_5O	1.3×10^{-10}		
phosphoric	H_3PO_4	7.5×10^{-3}	6.2×10^{-8}	4.8×10^{-13}
phosphorous	H_3PO_3	1.6×10^{-2}	2.6×10^{-7}	
phthalic	$H_2C_8H_4O_4$	1.3×10^{-3}	3.9×10^{-6}	
propionic	$HC_3H_5O_2$	1.34×10^{-5}		
succinic	$H_2C_4H_4O_4$	6.4×10^{-5}	2.7×10^{-6}	
sulfuric	H_2SO_4		1.2×10^{-2}	
sulfurous	H_2SO_3	1.72×10^{-2}	1.07×10^{-7}	
tartaric	$H_2C_4H_4O_6$	9.6×10^{-4}	2.9×10^{-5}	

Selected Standard Electrode Potentials

Selected standard electrode potentials

Half-cell reaction	$E°$, volt	Half-cell reaction	$E°$, volt
$Ag^+ + e \rightleftarrows Ag$	0.80	$H_5IO_6 + H^+ + 2e \rightleftarrows IO_3^- + 3H_2O$	1.6
$AgBr + e \rightleftarrows Ag + Br^-$	0.07	$K^+ + e \rightleftarrows K$	-2.92
$AgCl + e \rightleftarrows Ag + Cl^-$	0.22	$Li^+ + e \rightleftarrows Li$	-3.03
$Ag_2CrO_4 + 2e \rightleftarrows 2Ag + CrO_4^{--}$	0.45	$Mg^{++} + 2e \rightleftarrows Mg$	-2.37
$AgI + e \rightleftarrows Ag + I^-$	-0.15	$Mn^{++} + 2e \rightleftarrows Mn$	-1.19
$Ag_2S + 2e \rightleftarrows 2Ag + S^{--}$	-0.71	$MnO_2 + 4H^+ + 2e \rightleftarrows Mn^{++} + 2H_2O$	1.23
$Al^{3+} + 3e \rightleftarrows Al$	-1.66	$MnO_4^- + e \rightleftarrows MnO_4^{--}$	
$Au(CN)_2^- + e \rightleftarrows Au + 2CN^-$	-0.61	(in presence of Ba^{++} ions)	0.56
$Ba^{++} + 2e \rightleftarrows Ba$	-2.90	$MnO_4^- + 4H^+ + 3e \rightleftarrows MnO_2 + 2H_2O$	1.70
$Be^{++} + 2e \rightleftarrows Be$	-1.85	$MnO_4^- + 2H_2O + 3e$	
$Br_2(aq) + 2e \rightleftarrows 2Br^-$	1.09	$MnO_2 + 4OH^-$	0.59
$Ca^{++} + 2e \rightleftarrows Ca$	-2.87	$MnO_4^- + 8H^+ + 5e \rightleftarrows Mn^{++} + 4H_2O$	1.50
$Cd^{++} + 2e \rightleftarrows Cd$	-0.40	$HNO_2 + H^+ + e \rightleftarrows NO + H_2O$	0.99
$Ce^{4+} + e \rightleftarrows Ce^{3+}$ (in 1 F H_2SO_4)	-1.44	$NO_3^- + 3H^+ + 2e \rightleftarrows HNO_2 + H_2O$	0.94
$Cl_2 + 2e \rightleftarrows 2Cl^-$	1.36	$Na^+ + e \rightleftarrows Na$	-2.70
$2HClO + 2H^+ + 2e \rightleftarrows Cl_2 + 2H_2O$	1.63	$Ni^{++} + 2e \rightleftarrows Ni$	-0.23
$Co^{++} + 2e \rightleftarrows Co$	-0.28	$H_2O_2 + 2H^+ + 2e \rightleftarrows 2H_2O$	1.77
$Cr^{3+} + 3e \rightleftarrows Cr$	-0.74	$O_2 + 4H^+ + 4e \rightleftarrows 2H_2O$	1.23
$Cr_2O_7^{--} + 14H^+ + 6e \rightleftarrows 2Cr^{3+} + 7H_2O$	1.33	$O_2 + 2H^+ + 2e \rightleftarrows H_2O_2$	0.69
$Cs^+ + e \rightleftarrows Cs$	-2.95	$Pb^{++} + 2e \rightleftarrows Pb$	-0.13
$Cu^+ + e \rightleftarrows Cu$	0.52	$PbBr_2 + 2e \rightleftarrows Pb + 2Br^-$	-0.28
$Cu^{++} + 2e \rightleftarrows Cu$	0.34	$PbCl_2 + 2e \rightleftarrows Pb + 2Cl^-$	-0.26
$Cu^{++} + I^- + e \rightleftarrows CuI$	0.85	$PbI_2 + 2e \rightleftarrows Pb + 2I^-$	-0.36
$F_2 + 2e \rightleftarrows 2F^-$	2.87	$PbO_2 + 4H^+ + 2e \rightleftarrows Pb^{++} + 2H_2O$	1.47
$Fe^{++} + 2e \rightleftarrows Fe$	-0.44	$PbSO_4 + 2e \rightleftarrows Pb + SO_4$	-0.35
$Fe^{3+} + e \rightleftarrows Fe^{++}$	0.77	$Rb^+ + e \rightleftarrows Rb$	-2.93
$2H^+ + 2e \rightleftarrows H_2$	0.0000	$S + 2H^+ + 2e \rightleftarrows H_2S$	0.14
$Hg_2^{++} + 2e \rightleftarrows 2Hg$	0.79	$S_4O_6^{--} + 2e \rightleftarrows 2S_2O_3^{--}$	0.10
$Hg_2Cl_2 + 2e \rightleftarrows 2Hg + 2Cl^-$	0.27	$Sn^{++} + 2e \rightleftarrows Sn$	-0.14
$Hg_2Br_2 + 2e \rightleftarrows 2Hg + 2Br^-$	0.14	$Sn^{4+} + 2e \rightleftarrows Sn^{++}$ (in 1 F HCl)	0.14
$Hg_2I_2 + 2e \rightleftarrows 2Hg + 2I^-$	-0.04	$Sr^{++} + 2e \rightleftarrows Sr$	-2.89
$Hg^{++} + 2e \rightleftarrows Hg$	0.85	$Tl^+ + e \rightleftarrows Tl$	-0.34
$2Hg^{++} + 2e \rightleftarrows Hg_2^{++}$	0.91	$Tl^{3+} + 2e \rightleftarrows Tl^+$	1.28
$I_2 + 2e \rightleftarrows 2I$	0.54	$V^{++} + 2e \rightleftarrows V$	-1.25
$HIO + H^+ + 2e \rightleftarrows I^- + H_2O$	0.99	$VO^{++} + 2H^+ + e \rightleftarrows V^{3+} + H_2O$	0.34
$2IO_3^- + 12H^+ + 10e \rightleftarrows I_2 + 6H_2O$	1.19	$V(OH)_4^+ + 2H^+ + e \rightleftarrows VO^{++} + 3H_2O$	1.00
		$Zn^{++} + 2e \rightleftarrows Zn$	-0.76

Solubility Product Constants at Room Temperature

Solubility product constants at room temperature

Substance	Formula	Solubility Product
aluminum hydroxide	$Al(OH)_3$	1×10^{-32}
barium carbonate	$BaCO_3$	5.5×10^{-10}
barium chromate	$BaCrO_4$	1.17×10^{-10}
barium fluoride	BaF_2	1.05×10^{-6}
barium iodate	$Ba(IO_3)_2 \cdot 2H_2O$	1.5×10^{-9}
barium oxalate	$BaC_2O_4 \cdot 2H_2O$	1.7×10^{-7}
barium sulfate	$BaSO_4$	1.07×10^{-10}
bismuth hydroxide	$Bi(OH)_3$	4×10^{-31}
bismuth sulfide	Bi_2S_3	1.0×10^{-97}
cadmium carbonate	$CdCO_3$	2.5×10^{-14}
cadmium oxalate	$CdC_2O_4 \cdot 3H_2O$	2.8×10^{-8}
cadmium sulfide	CdS	8.0×10^{-27}
calcium carbonate	$CaCO_3$	4.8×10^{-9}
calcium fluoride	CaF_2	4.0×10^{-11}
calcium iodate	$Ca(IO_3)_2 \cdot 6H_2O$	7.1×10^{-7}
calcium oxalate	$CaC_2O_4 \cdot H_2O$	2.3×10^{-9}
calcium sulfate	$CaSO_4$	1.2×10^{-6}
chromium hydroxide	$Cr(OH)_3$	1×10^{-30}
cobalt sulfide	CoS	1.9×10^{-27}
cupric hydroxide	$Cu(OH)_2$	1.25×10^{-25}
cupric oxalate	CuC_2O_4	2.9×10^{-8}
cupric iodate	$Cu(IO_3)_2$	7.6×10^{-8}
cupric sulfide	CuS	6.3×10^{-36}
cuprous bromide	$CuBr$	3.2×10^{-8}
cuprous chloride	$CuCl$	1.2×10^{-6}
cuprous iodide	CuI	4.0×10^{-12}
cuprous sulfide	Cu_2S	2.5×10^{-48}
ferric hydroxide	$Fe(OH)_3$	2.0×10^{-39}
ferrous hydroxide	$Fe(OH)_2$	8.0×10^{-16}
ferrous oxalate	FeC_2O_4	2.1×10^{-7}
ferrous sulfide	FeS	6.3×10^{-18}
lead carbonate	$PbCO_3$	6.3×10^{-14}
lead chloride	$PbCl_2$	2.0×10^{-5}
lead chromate	$PbCrO_4$	1.8×10^{-14}
lead fluoride	PbF_2	2.7×10^{-8}
lead hydroxide	$Pb(OH)_2$	3.2×10^{-20}
lead iodate	$Pb(IO_3)_2$	3.2×10^{-13}
lead iodide	PbI_2	7.1×10^{-9}

Substance	Formula	Solubility product
lead oxalate	PbC_2O_4	3.4×10^{-11}
lead sulfate	$PbSO_4$	1.6×10^{-8}
lead sulfide	PbS	1.25×10^{-28}
magnesium ammonium phosphate	$MgNH_4PO_4$	2.5×10^{-13}
magnesium carbonate	$MgCO_3$	1×10^{-5}
magnesium fluoride	MgF_2	6.6×10^{-9}
magnesium hydroxide	$Mg(OH)_2$	1.1×10^{-11}
magnesium oxalate	MgC_2O_4	8.6×10^{-5}
manganous carbonate	$MnCO_3$	8.8×10^{-11}
manganous hydroxide	$Mn(OH)_2$	4.5×10^{-14}
manganese sulfide	MnS	1.4×10^{-15}
mercuric sulfide	HgS	1.6×10^{-52}
mercurous bromide	Hg_2Br_2	5.8×10^{-23}
mercurous chloride	Hg_2Cl_2	1.6×10^{-18}
mercurous iodide	Hg_2I_2	4.5×10^{-29}
nickel hydroxide	$Ni(OH)_2$	6.3×10^{-18}
nickel sulfide	NiS	1.1×10^{-27}
silver acetate	$AgC_2H_3O_2$	1.8×10^{-3}
silver bromate	$AgBrO_3$	6×10^{-5}
silver bromide	$AgBr$	5×10^{-13}
silver carbonate	Ag_2CO_3	6.2×10^{-12}
silver chloride	$AgCl$	1.8×10^{-10}
silver chromate	Ag_2CrO_4	1.3×10^{-12}
silver cyanide	$AgCN$	2.2×10^{-16}
silver iodate	$AgIO_3$	3.0×10^{-8}
silver iodide	AgI	4.5×10^{-17}
silver phosphate	Ag_3PO_4	1.25×10^{-20}
silver sulfide	Ag_2S	6.3×10^{-50}
silver thiocyanate	$AgSCN$	1.0×10^{-12}
strontium carbonate	$SrCO_3$	1.1×10^{-10}
strontium fluoride	SrF_2	2.45×10^{-9}
strontium oxalate	SrC_2O_4	5.0×10^{-8}
strontium sulfate	$SrSO_4$	2.9×10^{-7}
zinc hydroxide	$Zn(OH)_2$	7.5×10^{-18}
zinc oxalate	ZnC_2O_4	7.5×10^{-9}
zinc sulfide	ZnS	4.5×10^{-24}

Dissociation Constants of Complex Ions

Dissociation constants of complex ions

Dissociation equilibrium	Dissociation constant
$Cd(NH_3)_4^{++} \rightleftharpoons Cd^{++} + 4NH_3$	1×10^{-7}
$Cd(CN)_4 \rightleftharpoons Cd^{++} + 4CN^-$	8×10^{-18}
$CdI_4^{--} \rightleftharpoons Cd^{++} + 4I^-$	7×10^{-7}
$Co(NH_3)_6^{++} \rightleftharpoons Co^{++} + 6NH_3$	9×10^{-6}
$Co(NH_3)_6^{3+} \rightleftharpoons Co^{3+} + 6NH_3$	2×10^{-34}
$Cr(OH)_4^- \rightleftharpoons Cr(OH)_3(s) + OH^-$	1×10^{-2}
$Cu(NH_3)_4^{++} \rightleftharpoons Cu^{++} + 4NH_3$	4.6×10^{-14}
$Pb(OH)_3^- \rightleftharpoons Pb(OH)_2(s) + OH^-$	50
$HgBr_4^{--} \rightleftharpoons Hg^{++} + 4Br^-$	1×10^{-21}
$HgCl_4^{--} \rightleftharpoons Hg^{++} + 4Cl^-$	1.1×10^{-16}
$HgI_4^{--} \rightleftharpoons Hg^{++} + 4I^-$	5×10^{-31}
$Hg(CN)_4^{--} \rightleftharpoons Hg^{++} + 4CN^-$	4×10^{-42}
$Ni(CN)_4^{--} \rightleftharpoons Ni^{++} + 4CN^-$	1×10^{-22}
$Ni(NH_3)_4^{++} \rightleftharpoons Ni^{++} + 4NH_3$	4.8×10^{-8}
$Ag(NH_3)_2^+ \rightleftharpoons Ag^+ + 2NH_3$	6.8×10^{-8}
$Ag(CN)_2^- \rightleftharpoons Ag^+ + 2CN^-$	1×10^{-20}
$Zn(NH_3)_4^{++} \rightleftharpoons Zn^{++} + 4NH_3$	2.6×10^{-10}
$Zn(CN)_4^{--} \rightleftharpoons Zn^{++} + 4CN^-$	1×10^{-19}
$Zn(OH)_4^{--} \rightleftharpoons Zn(OH)_2(s) + SOH^-$	10

Logarithms

Logarithms

No.	0	1	2	3	4	5	6	7	8	9	1	2	3	4	5	6	7	8	9
10	0000	0043	0086	0128	0170	0212	0253	0294	0334	0374	4	8	12	17	21	25	29	33	37
11	0414	0453	0492	0531	0569	0607	0645	0682	0719	0755	4	8	11	15	19	23	26	30	34
12	0792	0828	0864	0899	0934	0969	1004	1038	1072	1106	3	7	10	14	17	21	24	28	31
13	1139	1173	1206	1239	1271	1303	1335	1367	1399	1430	3	6	10	13	16	19	23	26	29
14	1461	1492	1523	1553	1584	1614	1644	1673	1703	1732	3	6	9	12	15	18	21	24	27
15	1761	1790	1818	1847	1875	1903	1931	1959	1987	2014	3	6	8	11	14	17	20	22	25
16	2041	2068	2095	2122	2148	2175	2201	2227	2253	2279	3	5	8	11	13	16	18	21	24
17	2304	2330	2355	2380	2405	2430	2455	2480	2504	2529	2	5	7	10	12	15	17	20	22
18	2553	2577	2601	2625	2648	2672	2695	2718	2742	2765	2	5	7	9	12	14	16	19	21
19	2788	2810	2833	2856	2878	2900	2923	2945	2967	2989	2	4	7	9	11	13	16	18	20
20	3010	3020	3054	3075	3096	3118	3139	3160	3181	3201	2	4	6	8	11	13	15	17	19
21	3222	3243	3263	3284	3304	3324	3345	3365	3385	3404	2	4	6	8	10	12	14	16	18
22	3424	3444	3464	3483	3502	3522	3541	3560	3579	3598	2	4	6	8	10	12	14	15	17
23	3617	3636	3655	3674	3692	3711	3729	3747	3766	3784	2	4	6	7	9	11	13	15	17
24	3802	3820	3838	3856	3874	3892	3909	3927	3945	3962	2	4	5	7	9	11	12	14	16
25	3979	3997	4014	4031	4048	4065	4082	4099	4116	4133	2	3	5	7	9	10	12	14	15
26	4150	4166	4183	4200	4216	4232	4249	4265	4281	4298	2	3	5	7	8	10	11	13	15
27	4314	4330	4346	4362	4378	4393	4409	4425	4440	4456	2	3	5	6	8	9	11	13	14
28	4472	4487	4502	4518	4533	4548	4564	4579	4594	4609	2	3	5	6	8	9	11	12	14
29	4624	4639	4654	4669	4683	4698	4713	4728	4742	4757	1	3	4	6	7	9	10	12	13
30	4771	4786	4800	4814	4829	4843	4857	4871	4886	4900	1	3	4	6	7	9	10	11	13
31	4914	4928	4942	4955	4969	4983	4997	5011	5024	5038	1	3	4	6	7	8	10	11	12
32	5051	5065	5079	5092	5105	5119	5132	5145	5159	5172	1	3	4	5	7	8	9	11	12
33	5185	5198	5211	5224	5237	5250	5263	5276	5289	5302	1	3	4	5	6	8	9	10	12
34	5315	5328	5340	5353	5366	5378	5391	5403	5416	5428	1	3	4	5	6	8	9	10	11
35	5441	5453	5465	5478	5490	5502	5514	5527	5539	5551	1	2	4	5	6	7	9	10	11
36	5563	5575	5587	5599	5611	5623	5635	5647	5658	5670	1	2	4	5	6	7	8	10	11
37	5682	5694	5705	5717	5729	5740	5752	5763	5775	5786	1	2	3	5	6	7	8	9	10
38	5798	5809	5821	5832	5843	5855	5866	5877	5888	5899	1	2	3	5	6	7	8	9	10
39	5911	5922	5933	5944	5955	5966	5977	5988	5999	6010	1	2	3	4	5	7	8	9	10
40	6021	6031	6042	6053	6064	6075	6085	6096	6107	6117	1	2	3	4	5	6	8	9	10
41	6128	6138	6149	6160	6170	6180	6191	6201	6212	6222	1	2	3	4	5	6	7	8	9
42	6232	6243	6253	6263	6274	6284	6294	6304	6314	6325	1	2	3	4	5	6	7	8	9
43	6335	6345	6355	6365	6375	6386	6395	6405	6415	6425	1	2	3	4	5	6	7	8	9
44	6435	6444	6454	6464	6474	6484	6493	6503	6513	6522	1	2	3	4	5	6	7	8	9

Logarithms (Continued)

No.	0	1	2	3	4	5	6	7	8	9	1	2	3	4	5	6	7	8	9
45	6532	6542	6551	6561	6571	6580	6590	6599	6609	6618	1	2	3	4	5	6	7	8	9
46	6628	6637	6646	6656	6665	6675	6684	6693	6702	6712	1	2	3	4	5	6	7	7	8
47	6712	6730	6739	6749	6758	6767	6776	6785	6794	6803	1	2	3	4	5	5	6	7	8
48	6812	6821	6830	6839	6848	6857	6866	6875	6884	6893	1	2	3	4	4	5	6	7	8
49	6902	6911	6920	6928	6937	6946	6955	6964	6972	6981	1	2	3	4	4	5	6	7	8
50	6990	6998	7007	7016	7024	7033	7042	7050	7059	7067	1	2	3	3	4	5	6	7	8
51	7076	7084	7093	7101	7110	7118	7126	7135	7143	7152	1	2	3	3	4	5	6	7	8
52	7160	7168	7177	7185	7193	7202	7210	7218	7226	7235	1	2	2	3	4	5	6	7	7
53	7243	7251	7259	7267	7275	7284	7292	7300	7308	7316	1	2	2	3	4	5	6	6	7
54	7324	7332	7340	7348	7356	7364	7372	7380	7388	7396	1	2	2	3	4	5	6	6	7
55	7404	7412	7419	7427	7435	7443	7451	7459	7466	7474	1	2	2	3	4	5	5	6	7
56	7482	7490	7497	7505	7513	7520	7528	7536	7543	7551	1	2	2	3	4	5	5	6	7
57	7559	7566	7574	7582	7589	7597	7604	7612	7619	7627	1	2	2	3	4	5	5	6	7
58	7634	7642	7649	7657	7664	7672	7679	7686	7694	7701	1	1	2	3	4	4	5	6	7
59	7709	7716	7723	7731	7738	7745	7752	7760	7767	7774	1	1	2	3	4	4	5	6	7
60	7782	7789	7796	7803	7810	7818	7825	7832	7839	7846	1	1	2	3	4	4	5	6	6
61	7853	7860	7868	7875	7882	7889	7896	7903	7910	7917	1	1	2	3	4	4	5	6	6
62	7924	7931	7938	7945	7952	7959	7966	7973	7980	7987	1	1	2	3	3	4	5	6	6
63	7992	8000	8007	8014	8021	8028	8035	8041	8048	8055	1	1	2	3	3	4	5	5	6
64	8062	8069	8075	8082	8089	8096	8102	8109	8116	8122	1	1	2	3	3	4	5	5	6
65	8219	8136	8142	8149	8156	8162	8169	8176	8182	8189	1	1	2	3	3	4	5	5	6
66	8195	8202	8209	8215	8222	8228	8235	8241	8248	8254	1	1	2	3	3	4	5	5	6
67	8261	8267	8274	8280	8287	8293	8299	8306	8312	8319	1	1	2	3	3	4	5	5	6
68	8325	8331	8338	8344	8351	8357	8363	8370	8376	8382	1	1	2	3	3	4	4	5	6
69	8388	8395	8401	8407	8414	8420	8426	8432	8439	8445	1	1	2	2	3	4	4	5	6
70	8451	8457	8463	8470	8476	8482	8488	8494	8500	8506	1	1	2	2	3	4	4	5	6
71	8513	8519	8525	8531	8537	8543	8549	8555	8561	8567	1	1	2	2	3	4	4	5	5
72	8573	8579	8585	8591	8597	8603	8609	8615	8621	8627	1	1	2	2	3	4	4	5	5
73	8633	8639	8645	8651	8657	8663	8669	8675	8681	8686	1	1	2	2	3	4	4	5	5
74	8692	8698	8704	8710	8716	8722	8727	8733	8739	8745	1	1	2	2	3	4	4	5	5
75	8751	8756	8762	8768	8774	8779	8785	8791	8797	8802	1	1	2	2	3	3	4	5	5
76	8808	8814	8820	8825	8831	8837	8842	8848	8854	8859	1	1	2	2	3	3	4	5	5
77	8865	8871	8876	8882	8887	8893	8899	8904	8910	8915	1	1	2	2	3	3	4	4	5
78	8921	8927	8932	8938	8943	8949	8954	8960	8965	8971	1	1	2	2	3	3	4	4	5
79	8976	8982	8987	8993	8998	9004	9009	9015	9020	9025	1	1	2	2	3	3	4	4	5
80	9031	9036	9042	9047	9053	9058	9063	9069	9074	9079	1	1	2	2	3	3	4	4	5
81	9085	9090	9096	9101	9106	9112	9117	9122	9128	9133	1	1	2	2	3	3	4	4	5
82	9138	9143	9149	9154	9159	9165	9170	9175	9180	9186	1	1	2	2	3	3	4	4	5
83	9191	9196	9201	9206	9212	9217	9222	9227	9232	9238	1	1	2	2	3	3	4	4	5
84	9243	9248	9253	9258	9263	9269	9274	9279	9284	9289	1	1	2	2	3	3	4	4	5
85	9294	9299	9304	9309	9315	9320	9325	9330	9335	9340	1	1	2	2	3	3	4	4	5
86	9345	9350	9355	9360	9365	9370	9375	9380	9385	9390	1	1	2	2	3	3	4	4	5
87	9395	9400	9405	9410	9415	9420	9425	9430	9435	9440	0	1	1	2	2	3	3	4	4
88	9445	9450	9455	9460	9465	9469	9474	9479	9484	9489	0	1	1	2	2	3	3	4	4
89	9494	9499	9504	9509	9513	9518	9523	9528	9533	9538	0	1	1	2	2	3	3	4	4

Logarithms (Continued)

No.	0	1	2	3	4	5	6	7	8	9	1	2	3	4	5	6	7	8	9
90	9542	9547	9552	9557	9562	9566	9571	9576	9581	9586	0	1	1	2	2	3	3	4	4
91	9590	9595	9600	9605	9609	9614	9619	9624	9628	9633	0	1	1	2	2	3	3	4	4
92	9638	9643	9647	9652	9657	9661	9666	9671	9675	9680	0	1	1	2	2	3	3	4	4
93	9685	9689	9594	9699	9703	9708	9713	9717	9722	9727	0	1	1	2	2	3	3	4	4
94	9731	9736	9741	9745	9750	9754	9759	9763	9768	9773	0	1	1	2	2	3	3	4	4
95	9777	9782	9786	9791	9795	9800	9805	9809	9814	9818	0	1	1	2	2	3	3	4	4
96	9823	9827	9832	9836	9841	9845	9850	9854	9859	9863	0	1	1	2	2	3	3	4	4
97	9868	9872	9877	9881	9886	9890	9894	9899	9903	9908	0	1	1	2	2	3	3	4	4
98	9912	9917	9921	9926	9930	9934	9939	9943	9948	9952	0	1	1	2	2	3	3	4	4
99	9956	9961	9965	9969	9974	9978	9983	9987	9991	9996	0	1	1	2	2	3	3	3	4

ANSWERS TO SELECTED PROBLEMS

CHAPTER 1

2. (a) 2.5
   (b) 0.13
3. ± 0.16
4. No result may be rejected
6. Reject 45.99
8. (d) 3.8×10^1

CHAPTER 2

1. (a) $\frac{1}{3}$
   (b) $\frac{1}{2}$
   (c) 1
   (d) $\frac{1}{2}$
   (e) $\frac{1}{6}$
   (f) $\frac{1}{2}$
   (g) $\frac{1}{2}$
   (h) $\frac{1}{2}$
   (i) $\frac{1}{2}$
   (j) 1
   (k) $\frac{1}{2}$
   (l) $\frac{1}{2}$
   (m) $\frac{1}{4}$
3. (d) 3.98 g
4. (a) 0.530
   (e) 2.000
5. (a) 12.5 ml
   (b) 0.05 mfw
   (c) 2.5 ml
   (g) 30
6. (d) 11.93
12. 0.3533

CHAPTER 3

1. (c) 4.12
   (d) -0.18
   (f) 0.7
   (k) 13.7
2. (b) 3.5×10^{-8}
   (e) 4×10^{-9}
3. (d) 12.3
4. (e) 11.38
5. (a) 7.67
6. (a) 4.28
7. (a) 8.92
8. (a) 9.0
9. (d) 3.0
   (f) 8.9
   (h) 9.0

10. 0.22 F
11. 0.65 g
13. 12.4 g
14. 9.14
15. 5.1×10^{-5}
16. 0.014 F
17. 50 ml
18. 2.85 ml
19. 2.29 g

CHAPTER 4

1. (a) Phenol red
3. 4.76
5. 6.8
7. 5.5%
8. 4,000 ml
10. 4.46
11. 10.89
14. 1.75×10^{-5}
15. 11.48

CHAPTER 5

1. (a) $+5$
   (b) $+6$
   (i) $+3$
   (k) $+6$
   (n) $+3$
2. (a) $PbO_2 + Pb + 2SO_4^{--} + 4H^+$
   $\rightleftharpoons 2PbSO_4 + 2H_2O$
   (c) $5Zn + 2NO_3^- + 12H^+$
   $\rightleftharpoons 5Zn^{++} + N_2 + 6H_2O$
3. (b) $NO_2^- + 2Al + OH^-$
   $\rightleftharpoons NH_3 + 2Al(OH)_4^-$
   (c) $2Cu^{++} + 7CN^- + 2OH^-$
   $\rightleftharpoons 2Cu(CN)_3^{--} + CNO + H_2O$
   (e) $2CrO_4^{--} + 3HSnO_2^- + H_2O$
   $\rightleftharpoons 2CrO_2^- + 3HSnO_3^- + 2OH^-$
4. (a) 2
   (b) 2
   (c) 6
   (d) 2
   (e) 12

5. (a) 50 ml
   (c) 4.3 ml
   (d) 0.1978
   (e) 7.8 ml
   (f) 125 ml
6. (a) 5
   (b) 3
   (d) 1
7. (a) 0.1500
   (b) 0.0500
   (c) 0.1000

CHAPTER 6

1. (a) 3.17
   (b) 0.82
   (c) -0.32
   (d) 0.73
2. (a) 0.46 $Zn + Cd^{++} \rightleftarrows Zn^{++} + Cd$
   (b) -0.60 $Sn^{4+} + 2Fe^{++} \rightleftarrows Sn^{++} + 2Fe^{3+}$
   (c) 1.88 $Cl_2 + Cd \rightleftarrows Cd^{++} + 2Cl^-$
3. (a) $K_e = 3 \times 10^9$
   (b) $K_e = 6.3 \times 10^{-7}$
   (e) $K_e = 3 \times 10^{24}$
4. $K_{sp} = 10^{-51}$
   $K_{sp} = 10^{-53}$
6. $K_{sp} = 10^{-20}$
9. $K_d = 10^{-41}$
11. $K_d = 10^{-28}$

CHAPTER 8

2. 0.00642
4. 0.3896

6. 0.7 ml
10. 0.3642 g
15. 0.0724 F
20. 58.68%

CHAPTER 9

1. (a) 4.6×10^{-17}
   (b) 1.3×10^{-5}
   (c) 1.4×10^{-18}
3. (a) $2.3 \times 10^{-6} F$
4. (a) 0.01 F
5. (c) $9 \times 10^{-7} F$
6. (a) 1.2×10^{-5}
8. $6.4 \times 10^{-5} F$
9. 1.56×10^{-20}
12. 0.84 F
14. 5.5×10^{-7} mfw
16. Pb^{++} ppts first
18. $1 \times 10^{-3} F$

CHAPTER 10

1. (a) $3U/U_3O_8$
   (b) $Na_2B_4O_7/2B_2O_3$
   (c) $3Mn_2O_3/2Mn_3O_4$
   (d) $Mn_3O_4/3MnO_2$
   (e) $2MgO/Mg_2P_2O_7$
   (f) $3CaSO_4/Ca_3(PO_4)_2$
   (g) $Pb_3O_4/3PbSO_4$
   (h) $2CuO/Cu_2S$
   (i) $H_2C_2O_4 \cdot 2H_2O/CaO$
2. 32.18%
3. 43.29%
4. 17.49%